AI IN AGRICULTURE
Sensors Application

AI IN AGRICULTURE
Sensors Application

C. S. Narayanamurthy
Abhijeet Kumar

KNOWLEDGE BAKERS

ISBN: 978-93-49964-90-7

Published by : **Knowledge Bakers**
Pune, India
info@knowledgebakers.com

PREFACE

The convergence of artificial intelligence (AI) and sensor technology is transforming agriculture into a data-driven and precision-focused industry. As challenges such as climate variability, resource limitations, and the need for sustainable practices intensify, smart farming solutions are increasingly being adopted to optimize productivity, reduce waste, and ensure environmental stewardship. AI, when integrated with advanced sensors, enables real-time monitoring, predictive analytics, and automated decision-making across various stages of agricultural production.

This handbook, titled AI in Agriculture: Sensors Application, explores the growing role of intelligent systems and sensor-based technologies in revolutionizing modern farming practices. Designed for researchers, practitioners, and students, the volume offers a focused introduction to how AI-powered sensors are applied in crop monitoring, soil analysis, irrigation control, disease detection, yield estimation, and beyond.

Through a blend of theoretical foundations, technical descriptions, and real-world case studies, AI in Agriculture: Sensors Application presents key concepts in sensor integration, data acquisition, machine learning algorithms, and deployment strategies tailored to agricultural environments. The book also considers the challenges related to data accuracy, connectivity, cost-effectiveness, and the adoption of AI in both small- and large-scale operations.

As agriculture becomes increasingly reliant on digital intelligence, this work aims to highlight not only the potential of sensor-based AI applications but also the considerations needed to ensure they are accessible, scalable, and sustainable. It is our hope that this handbook will serve as a valuable guide for stakeholders seeking to leverage technology to meet the evolving demands of global food production.

C. S. Narayanamurthy
Abhijeet Kumar

CONTENTS

Preface v

1. **Latest Advances in Sensor Applications in Agriculture** 1
   1. Introduction: An Overview 1
   2. Summary of the Special Issue 2
   3. Conclusions 10
2. **A Novel Cosmic-Ray Neutron Sensor for Soil Moisture Estimation over Large Areas** 13
   1. Introduction: An Overview 13
   2. Materials and Methods 15
   3. Experimental Sites 20
   4. Result 22
   5. Discussion and Conclusions 25
3. **Calibration and Validation of a Low-Cost Capacitive Moisture Sensor to Integrate the Automated Soil Moisture Monitoring System** 31
   1. Introduction: An Overview 31
   2. Materials and Methods 32
   3. Results and Discussion 36
   4. Conclusions 40
4. **Delineation of Agricultural Drainage Pipe Patterns Using Ground Penetrating Radar Integrated with a Real-Time Kinematic Global Navigation Satellite System** 45
   1. Introduction: An Overview 45
   2. Materials and Methods 47
   3. Results 54

4. Discussion 57
5. Summary and Conclusions 59

5. Assessing Topsoil Movement in Rotary Harrowing Process by RFID (Radio-Frequency Identification) Technique 61

1. Introduction: An Overview 61
2. Materials and Methods 62
3. Results and Discussion 65
4. Conclusions 69

6. Development and Field Evaluation of a Spray Drift Risk Assessment Tool for Vineyard Spraying Application 73

1. Introduction: An Overview 73
2. Materials and Methods 76
3. Results and Discussion 83
4. Conclusions 92

7. Development of an Autonomous Electric Robot Implement for Intra-Row Weeding in Vineyards 101

1. Introduction: An Overview 101
2. Materials and Methods 103
3. Results and Discussion 109
4. Conclusions 113

8. Infrared Thermography Applied to Tree Health Assessment: A Review..117

1. Introduction: An Overview 117
2. Materials and Methods 118
3. Tree Relevance and Risks 119
4. Methods and Techniques of Inspection 120
5. Infrared Thermography Applied to Trees 123
6. Conclusions 131
7. Recommendations 132

9. Contribution to Trees Health Assessment Using Infrared Thermography 139

1. Introduction: An Overview 139
2. Materials and Methods 142
3. Methodology 147
4. Results and Discussion 148

5. Conclusions *154*

6. Recommendations *155*

10. Early Detection of *Zymoseptoria tritici* in Winter Wheat by Infrared Thermography 159

1. Introduction: An Overview *159*

2. Material and Methods *161*

3. Results *164*

4. Discussion *169*

5. Conclusions *170*

11. Utilisation of Ground and Airborne Optical Sensors for Nitrogen Level Identification and Yield Prediction in Wheat........175

1. Introduction: An Overview *175*

3. Material and Methods *177*

4. Results *182*

5. Discussion *185*

6. Conclusions *188*

12. High-Resolution Multisensor Remote Sensing to Support Date Palm Farm Management 193

1. Introduction: An Overview *193*

2. Materials and Methods *195*

3. Results *202*

4. Discussion *211*

5. Conclusions *216*

13. Determination of Cultivated Area, Field Boundary and Overlapping for A Plowing Operation Using ISO 11783 Communication and D-GNSS Position Data........223

1. Introduction: An Overview *223*

2. Materials and Methods *226*

3. Results and Discussion *236*

4. Conclusions *242*

14. Food System Digitalization as a Means to Promote Food and Nutrition Security in the Barents Region 247

1. Introduction: An Overview *247*

2. Impacts of Human Activities and Climate Change in the Barents Region *250*

3. Developing the Infrastructure for Food System Digitalization in the Barents Region 252
4. Transforming the Food System through Digitalization to Promote Food and Nutrition Security 255
5. Sustainable Food System and ICT 260
6. Future Implications for the Barents Region 263
7. Concluding Remarks 266

15. Acquisition of Sorption and Drying Data with Embedded Devices: Improving Standard Models for High Oleic Sunflower Seeds by Continuous Measurements in Dynamic Systems 275

1. Introduction: An Overview 275
2. Materials and Methods 277
3. Results and Discussion 282
4. Conclusions 288

Index 295

Chapter 1

LATEST ADVANCES IN SENSOR APPLICATIONS IN AGRICULTURE

1. INTRODUCTION: AN OVERVIEW

Technologies are playing an important role in the development of crop and livestock farming and have the potential to be the key drivers of sustainable intensification of agricultural systems. In particular, new sensors are now available with reduced dimensions, reduced costs and increased performance, which can be implemented and integrated in production systems, allowing an increase of data and eventually an increase of information. This is of great importance to support digital transformation, precision agriculture and smart farming, and to eventually allow a revolution in the way food is produced. In order to exploit these results, authoritative studies from the research world are still needed to support development and implementation of new solutions and best practices.

Many sensor applications have significant impact in all agricultural practices. For instance, soil moisture sensors support farmers' decisions for irrigation practices which resulted in preventing plants from drought stress and over application of irrigation. Currently, many applications from remotely sensed data are used to assess crop health, drought and yield considering the improved spatial, temporal and spectral resolution and availability of such sensors. Furthermore, the revolution in sensors, information and communications technology resulted in substantial archived data for whole-farm practices and impacted in the whole system management. Farm management is an issue that goes beyond normal everyday agricultural practice. Digital technologies are helping farmers to take more wise decisions by providing a better overview of their farm.

2. SUMMARY OF THE SPECIAL ISSUE

After the review process, 14 out of 21 papers that were submitted to the special issue were accepted for publication. Published articles include ten research articles, two review articles and two technical notes. The topics of published articles discussed different topics related to sensors applications in soil, plant health assessment, farm management and post-harvest process. This editorial classified four sections as follows: soil-related work, plant protection in vineyards, plant health assessment, farm management using digital technologies and, finally, post-harvest application on sunflower seeds.

2.1 Soil-Related Work

2.1.1 Soil Moisture Sensing

Maintaining readily available soil moisture is an essential requirement for optimum plant growing conditions which depends on soil physical properties and surrounding environmental conditions. Several techniques were developed to determine the soil moisture content: among them are the cosmic-ray neutron based sensors which have been proposed in agriculture only in the last years and allow monitoring of wide areas, and capacitance sensors which are relatively cheap and could provide real-time soil moisture content measurements but require precise calibration. The paper by Stevanato et al. [1] titled “A Novel Cosmic-Ray Neutron Sensor for Soil Moisture Estimation over Large Areas”. It introduces the development of an innovative instrument which allows estimation of soil moisture from environmental epithermal neutron counts, thanks to implementation of a composite neutron detector. A second article was authored by Nagahage et al. [2] titled “Calibration and Validation of a Low-Cost Capacitive Moisture Sensor to Integrate the Automated Soil Moisture Monitoring System”. In this technical note, the capacitive soil moisture sensor model: SKU:SEN0193, DFRobot, Shanghai, China, were calibrated under laboratory conditions. The objectives for this study were to examine this sensor under laboratory conditions and integrate it with a data acquisition system. The sensor data were compared with the corresponding measurements from the traditional gravimetric method and with other calibrated sensor model: SM-200, Delta-T Devices Ltd., Cambridge, UK. Results showed that the SKU:SEN0193 sensor can help in maintaining the readily available soil moisture in indoor systems which minimize the risk of both over water application and soil moisture stress.

2.1.2 Drainage Pipes Detection

For over 30 years, a very sizeable amount of agricultural drainage pipe has been installed in the United States and nowadays some farmers need to repair or replace parts of these drain lines that are not functioning properly. For such purposes, a map of the pre-existing lines is needed and in most cases this map is no longer available. The article by Allred et al. [3] is titled "Delineation of Agricultural Drainage Pipe Patterns Using Ground Penetrating Radar Integrated with a Real-Time Kinematic Global Navigation Satellite System". In this study, the authors used a ground-penetrating radar (GPR) integrated with a real-time kinematic global navigation satellite system (RTK-GNSS) to scan and map drain pipe lines within agricultural fields. The idea behind GPR involves directing an electromagnetic radio pulse into the subsurface that reflects partially off a buried feature and by measuring the elapsed time taken, the GPR can detect the depth of targeted objects. A GPR system manufactured by Sensors and Software Inc. (Mississagua, ON, Canada) was used at a central frequency of 250 MHz as recommended by previous studies [4,5] for detecting buried drainage pipes. In order to collect RTK GNSS coordinates, a system consisting of a rover receiver and base station receiver manufactured by Topcon Corporation (Itabashi, Japan) was utilized. Authors tested this system in three different sites where two sites located in Maryland and the third site located in Ohio, USA. GPR settings were adjusted to 5cm distance between signal trace measurements while the depth of investigation was 2 m at sites located in Maryland and 1.5 m at the Ohio site. Results revealed that GPR-RTK/GNSS system could successfully determine drainage pipe lines in all test sites. The detected drainage patterns were in different shapes such as: rectangular, herringbone and random lines in the tested sites. Authors recommended this system as it is faster and causes no pipe damage compared to traditional excavation methods.

2.1.3 Seedbed Preparation

After the application of primary tillage operation, soil surface still needs a secondary tillage operation such as rotary harrow to smooth soil surface and prepare suitable seedbed. Harrowing contributes to soil erosion and many techniques were used to monitor soil movement during tillage operations such as: plastic beads, granite rocks and aluminium cubes. An article by Kayad et al. [6] titled "Assessing Topsoil Movement in Rotary Harrowing Process by RFID (Radio-Frequency Identification) Technique". In this technical note, authors suggested to use RFID systems for assessing topsoil movement after rotary harrowing field trials. The RFID system consists of small RFID tags to be attached to targeted objects and RFID reader that identifies tags. The authors inserted RFID tags inside cork stoppers which have durable material and mostly simulate crop residues such as dry corn stems or corncob. The RFID tags were distributed regularly in soil and after the harrowing operation, the displacement of each tag was measured. Field trials include different working conditions of the rotary harrow such as: tillage depth, forward speed and levelling bar. Authors reported that using RFID system was a proper method for such evaluation and might have further promising application due to its robustness in simulating different materials.

2.2 Plant Rotection in Vineyeards

2.2.1 Spray Drift Evaluation

A research article is by Bourodimos et al. [7] is titled "Development and Field Evaluation of a Spray Drift Risk Assessment Tool for Vineyard Spraying Application". This article starts by describing the negative points of spray drift that are caused by sprayers during the application of plant protection chemicals. Such spray drift is an important cause of environmental pollution and may lead to health risk for farm workers and animals. The objective of this contribution was to evaluate spray drift in vineyards using a drift risk assessment model developed in the framework the framework of the TOPPS-Prowadis project [8]. This model assesses the spray drift by sprayer under specific meteorological conditions such as air relative humidity, temperature, wind speed and direction. Field trials in the vineyard of Agricultural University of Athens were used to evaluate the reliability of this model by assessing ground and airborne spray drift under certain meteorological conditions. Results proved that there are significant differences in both ground and airborne spray drift among different field treatments.

This study highlights that fine-tuning of spraying condition limits support farmers to spray their fields with limited spray drift, improving spray efficiency and reducing the environmental impact.

2.2.2 Weeding Robot

Weeds could be the cause of up to 40% of yield reduction [9,10] and chemical herbicides have a harmful environmental impact which proves the importance of mechanical weeding as a promising alternative. Weeding robots can improve work quality, resources management, labor efficiency and reduce energy consumption [11]. The article by Reiser et al. [12] is titled "Development of an Autonomous Electric Robot Implement for Intra-Row Weeding in Vineyards". This article reports about developing and testing the performance of a rotating electrical tiller weeder to be used for intra-row weeding automatically in vineyards. The developed robot consists of an electric tiller head rotary weeder cultivator designed and manufactured by the University of Hohenheim, Germany and mounted to an autonomous robot called "phoenix" by Caterpillar. The robot is equipped with a 2D laser scanner to follow the tree and vine rows in the front of the vehicle and four security switches for emergency cases. For controlling and recording robot data an open-source robot operating system called ROS Indigo-middleware [13] was used. The developed robot was tested in both indoor at the soil bin laboratory and outdoor at the vineyard of Hohenheim university. Performance for trunk detection were evaluated using two different methods of feeler and sonar where both of them performed well and did not harm any trunks. Moreover, the laser scanner enabled the machine to follow the rows accurately and the overall evaluation proved the ability of the developed robot for intra-row weeding which could save the energy and time of workers.

2.3 Plant Health Assessment

2.3.1 Thermography for Tree Health Assessment

Trees have many benefits to human and environment such as; prevention of desertification and global warming, ecosystems balance and human well-being. In contrast, the probability of trees or branches falling is prone to the risk of damage to people and civil infrastructures especially when trees suffer from defects compromising their health. Monitoring trees health status is essential to evaluate their biological viability, associated risks and support decision making about trees. The eleventh article by Vidal and Pitarma [14] titled "Infrared Thermography Applied to Tree Health Assessment: A Review". In this article, authors reviewed previous studies concerning the effectiveness of thermography in tree health inspection. This review was compiled taking the advantage of several research databases such as Google Scholar, ScienceDirect, Scopus and other databases in Portuguese, English and Spanish languages between October 2018 and February 2019. Authors used several keywords and combinations to select relevant articles followed by classification and filtering for duplicated references resulting in 81 selected papers. The article consists of seven sections starting with an introduction and review methodology then discussing the importance of trees and their associated risks and some methods and techniques for tree inspection. In Section 5, authors reviewed the application of infrared thermography to trees as a non-destructive inspection technique followed by conclusion and recommendations. This article highlighted the efficiency of using infrared thermography in early detection of damages in trees compared to other methods in terms of differentiating between functional and dysfunctional tissues and subsequently evaluating the vitality and health status of trees.

Tree health status verification uses invasive and destructive techniques which interfere with tree structure [15]. It is always recommended to start tree inspection with less-invasive techniques if needed to minimize the damage in the tree [16]. Another research article by Pitarma et al. [17] is titled "Contribution to Trees Health Assessment Using Infrared Thermography". This article focuses on infrared thermography application in trees inspection by using thermograms to differentiate between deteriorated and healthy tissues to observe trees as a functional whole body. Such application is well-established in different fields especially for industrial applications while it is still relatively recent in assessing tree health [16]. The main goal of this study is to provide a qualitative analysis of two different arboreal species based on differences of its thermal images. The two arboreal species are *Quercus pyrenaica* Willd and *Olea europaea* L., and thermal images were acquired by a FLIR T1030sc camera while atmospheric temperature and relative humidity were measured by the thermohygrometer FLIR MR 176. The authors recorded the thermograms at different times along the day besides taking photographs to support the visual inspection then correlating thermal patterns with tree health. Results proved the high potential of the thermography technique for tree inspection which allows early diagnosis of damage and subsequently advances tree maintenance.

2.3.2 Winter Wheat

Fungal infection symptoms usually appear on plants after a period of time according to temperature and humidity conditions and early detection and diagnosis allow farmers to protect the crop before widespread [18]. The article by Wang et al. [19] is titled "Early Detection of *Zymoseptoria tritici* in Winter Wheat by Infrared Thermography". This study reports an application for infrared thermography to detect the fungal infection by Z. tritici in winter wheat crop. The idea behind this application is that plant photosynthesis and transpiration are influenced by Z. *tritici* which led to a change in canopy temperature that could be detected through thermography. The objective of this study was to detect this disease early before the visual symptoms appear on the crop. Twenty-five wheat varieties were tested in a field located in Stuttgart, Germany through a split-plot design experiment. The seeds were sown on 6 October 2011 and part of plants presenting all tested varieties were inoculated artificially on 21 May 2012. Thermal images were acquired by an infrared camera (VarioCAM, InfraTec GmbH, Dresden, Germany) starting from five days before inoculation until 38 days after inoculation. Also, visual scoring was undertaken by experienced staff and all collected data were analysed using SPSS software. Results showed that in some varieties the earliest disease symptoms could be detected as early as three days after inoculation through thermography while firest visual symptoms appeared after 23 days from inoculation. This application highlights the usefulness of thermography for high throughput to improve fungal disease monitoring which could help breeders in selecting disease-resistant varieties.

Several studies on spectral data applications in field crops are available from remote and ground sensors using data-mining techniques for nitrogen status and grain yield [20,21]. Most of these studies were based on measurements acquired by one sensor and there is a lack of available information on how different combinations of sensors are informative. The article by Zecha et al. [22] is titled "Utilisation of Ground and Airborne Optical Sensors for Nitrogen Level Identification and Yield Prediction in Wheat". In this study, an investigation on different fields planted with winter wheat using different nitrogen levels were done through different spectral sensors. The final goal was to test different spectral sensors on field trials conditions and to answer three main questions; How do these sensors perform in field scale? Which calculated features are significant in assessing yield, biomass and nitrogen status? How can sensors' data fusion support farmers' decisions? The investigations took place at different fields related to the University of Hohenheim, Germany where different rates of nitrogen applied within fields between 2011 and 2012. Three ground sensors where two were passive spectrometer sensors and one active fluorescence sensor were mounted on a self propelled carrier. Furthermore, a passive spectrometer mounted on a fixed-wing unmanned aerial vehicle (UAV) to acquire aerial images. All sensors data were processed in form of indices and ratios and correlated with field information and biological parameters such as wheat yield, biomass, leaf area index and available nitrogen using the R statistical software. Results revealed that more robust and higher correlations were obtained from models developed through mixed features from different sensors. Authors suggested that advanced algorithms which consider ambient solar radiation, aerial images, soil electrical conductivity and scoring may result in better yield predictions.

2.3.3 Remote Sensing of Date Palm

Many researchers investigated the possibility of using hyperspectral or thermal imagery for date palm health assessment, however, most of these studies are separate studies. An article by Mulley et al. [23] is titled "High-Resolution Multisensor Remote Sensing to Support Date Palm Farm Management". In this study, authors assessed date palm health using several sensors such as: light detection and ranging (LiDAR), visual red-green-blue (RGB), thermal and hyperspectral images. The ultimate goal for this study was to explore the most proper sensor and indicator for stress detection on date palm plants at different spatial levels. The investigations took place in a 168.8 ha date palm farm located in Al-kharj region in Saudi Arabia divided as rectangular shaped blocks of approximately 10 ha each. This farm has continuous maintenance and irrigation practices and is relatively well managed. The farm manager provided authors with archived records of red palm weevil (*Rhynchophorus ferrugineus*) infestations plus visual assessment to investigate the homogeneity of the canopy area which was considered as an indicator for healthy and unhealthy groups. Furthermore, individual tree analysis was performed from different blocks as well. The ground and the different sources of remotely sensed data were analysed using several statistical, imagery and geographical information systems software at both block and tree levels. Results showed that remote-sensing data could aid plantation management of date palm and provide insight for further site-specific management practices. Finally, authors recommended time-series analysis approach to detect changes in vegetation reflectance properties as an indicator for date palm health and suggested other future interesting topics about exploring within-block parcels to be classified as management zones for the adoption of precision agriculture techniques.

2.4 Farm Management Using Digital Technologies

2.4.1 Plowing Operation Archived Data

Embedded sensors on agricultural machinery that are used for proper diagnostics and communication can also reveal a lot of information regarding the operated tasks. There are many different sensor applications for agricultural equipment that generate huge data sets to monitor machine performance, measure and count agricultural inputs and yield. Currently, these data are combined with GNSS sensors for site specific management practices, and also due to their reduced price this allows archiving data and operations with position references. The article by Heiß et al. [24] is titled "Determination of Cultivated Area, Field Boundary and Overlapping for A Plowing Operation Using ISO 11783 Communication and D-GNSS Position Data". In this study, authors developed an algorithm that deals with georeferenced data recorded during plowing operation in order to calculate different area-related parameters in an automated way. Data were recorded by data logger GL2000 CAN-Bus (Vector Informatik GmbH, Stuttgart, Germany) connected to the diagnostic interface of the tractor used. Recorded data were the wheel-based machine speed and the differential GNSS (D-GNSS) coordinates as well as their timestamps. The MATLAB R2016b (The MathWorks Inc., Natick, MA, USA) software was used to analyse this data through different filtering equations to identify passes and subsequently determine field cultivated area, boundaries and the overlaps between the cultivated tracks. The developed algorithm could detect 58 passes which matched with the number of lifting and lowering points indicating the algorithm s functionality. Furthermore, different common indicators were calculated for the overlapping, cultivated area was quantified and field boundary was detected proving the plausibility of the results. Authors recommended this algorithm for applications such as documenting and invoicing of agricultural tasks and using overlapping analysis as an indicator of efficiency.

2.4.2 Food System Digitalization

Food security is a key factor to develop overall human well-being and human security [25]. The review article by Raheem et al. [26] is titled "Food System Digitalization as a Means to Promote Food and Nutrition Security in the Barents Region". In the Barents region, traditional food includes potatoes, meat, fish, berries and a wide range of dairy products where mostly the processing of these foods is undertaken by small and medium enterprises for better preservation and distribution. Digitalization can improve the added value to traditional food in terms of; improving harvesting process, increasing production, reducing waste and enhancing storage and distribution process. For instance, sensors and data processing applications in food system digitalization are expected to improve prediction accuracy for food value chains in the Barents region. The main objectives for this review article were to identify challenges, improve the sustainability and support for food system digitalization. The article consists of seven sections starting with an introduction then describing the current situation in the Barents region regarding climate change, human activities and food system digitalization until section number 3. From Section 4, authors discussed the role and impact of different digital technologies in food-system components and sustainability followed by future implications for the Barents region and conclusion. This review could help in conceptualizing a framework for food system digitalization and better inform both policy makers and stakeholders in the study region to support food security.

2.5 Post-Harvest Application on Sunflower Seeds

A sunflower drying process is a typical requirement for safe storage where excessive moisture levels may lead to dry matter losses and increase the activity of microorganisms. The article by Munder et al. [27] is titled "Acquisition of Sorption and Drying Data with Embedded Devices: Improving Standard Models for High Oleic Sunflower Seeds by Continuous Measurements in Dynamic Systems". In this study, innovative methods to determine sorption and drying data were used at common temperature during the handling process of agricultural products. The main goal was to develop a robust drying model for high oleic sunflower seeds based on data from sorption and drying experiments. Laboratory experiments were performed to determine a broad set of equilibrium moisture content data through gravimetric analyzer and to collect single-layer drying kinetic data at different drying conditions. The collected data were used for the development of a generalized single-layer drying model according to air conditions. The embedded systems used for this study allowed a large amount of experimental data to be recorded which were used to fit semi-empirical and analytical sorption and drying models. Results showed that the equilibrium moisture content increased at high values of water activity through sorption experiments. This study reported an appropriate model for high oleic sunflower seeds that describes the drying process for a wide range of humidity and temperatures conditions.

3. CONCLUSIONS

This special issue covers a broad range of sensor applications in agriculture and presents some of the recent research results in this topic. Author contributions include applications in soil and plant sensing, farm management and post-harvest application. The articles published in this special issue are considered an addition to the scientific community, and the editors believe that it may stimulate further ideas and new applications for sensors in agriculture.

References

1. Stevanato, L.; Baroni, G.; Cohen, Y.; Cristiano Lino, F.; Gatto, S.; Lunardon, M.; Marinello, F.; Moretto, S.; Morselli, L. A Novel Cosmic-Ray Neutron Sensor for Soil Moisture Estimation over Large Areas. *Agriculture* **2019**, *9*, 202. [CrossRef]
2. Nagahage, E.A.A.D.; Nagahage, I.S.P.; Fujino, T. Calibration and validation of a low-cost capacitive moisture sensor to integrate the automated soil moisture monitoring system. *Agriculture* **2019**, *9*, 141. [CrossRef]
3. Allred, B.; Wishart, D.; Martinez, L.; Schomberg, H.; Mirsky, S.; Meyers, G.; Elliott, J.; Charyton, C. Delineation of agricultural drainage pipe patterns using ground penetrating radar integrated with a real-time kinematic global navigation satellite system. *Agriculture* **2018**, *8*, 167. [CrossRef]
4. Allred, B.J. A GPR Agricultural Drainage Pipe Detection Case Study: Effects of Antenna Orientation Relative to Drainage Pipe Directional Trend. *J. Environ. Eng. Geophys.* **2013**, *18*, 55–69. [CrossRef]
5. Allred, B.J.; Fausey, N.R.; Peters, L.J.r.; Chen, C.; Daniels, J.J.; Youn, H. Detection of Buried Agricultural Drainage Pipe With Geophysical Methods. *Appl. Eng. Agric.* **2004**, *20*, 307–318. [CrossRef]
6. Kayad, A.; Rainato, R.; Picco, L.; Sartori, L.; Marinello, F. Assessing Topsoil Movement in Rotary Harrowing Process by RFID (Radio-Frequency Identification) Technique. *Agriculture* **2019**, *9*, 184. [CrossRef]
7. Bourodimos, G.; Koutsiaras, M.; Psiroukis, V.; Balafoutis, A.; Fountas, S. Development and field evaluation of a spray drift risk assessment tool for vineyard spraying application. *Agriculture* **2019**, *9*, 181. [CrossRef]
8. TOPPS-Prowadis Project. Best Management Practices to Reduce Spray Drift. 2014. Available online: http://www.topps-life.org/ (accessed on 24 July 2020).
9. Peteinatos, G.G.; Weis, M.; Andújar, D.; Rueda Ayala, V.; Gerhards, R. Potential use of ground-based sensor technologies for weed detection. *Pest Manag. Sci.* **2014**, *70*, 190–199. [CrossRef]

10. Andújar, D.; Dorado, J.; Fernández-Quintanilla, C.; Ribeiro, A. An approach to the use of depth cameras for weed volume estimation. *Sensors* **2016**, *16*, 972. [CrossRef] [PubMed]

11. Ge, Z.; Wu, W.; Yu, Y.; Zhang, R. Design of mechanical arm for laser weeding robot. In Proceedings of the 2nd International Conference on Computer Science and Electronics Engineering (ICCSEE 2013), Hangzhou, China, 22–23 March 2013; pp. 2340–2343.

12. Reiser, D.; Sehsah, E.S.; Bumann, O.; Morhard, J.; Griepentrog, H.W. Development of an autonomous electric robot implement for intra-row weeding in vineyards. *Agriculture* **2019**, *9*, 18. [CrossRef]

13. Quigley, M.; Conley, K.; Gerkey, B.; Faust, J.; Foote, T.; Leibs, J.; Berger, E.; Wheeler, R.; Mg, A. ROS: An open-source Robot Operating System. In Proceedings of the IEEE International Conference on Robotics and Automation (ICRA 2009), Kobe, Japan, 12–17 May 2009; Volume 3, pp. 5–11.

14. Vidal, D.; Pitarma, R. Infrared thermography applied to tree health assessment: A review. *Agriculture* 2019, *9*, 156. [CrossRef]

15. Goh, C.L.; Abdul Rahim, R.; Fazalul Rahiman, M.H.; Mohamad Talib, M.T.; Tee, Z.C. Sensing wood decay in standing trees: A review. *Sens. Actuators A Phys.* **2018**, *269*, 276–282. [CrossRef]

16. Bellett-Travers, M.; Morris, S. The relationship between surface temperature and radial wood thickness of twelve trees harvested in nottinghamshire. *Arboric. J.* **2010**, *33*, 15–26. [CrossRef]

17. Pitarma, R.; Crisóstomo, J.; Ferreira, M.E. Contribution to trees health assessment using infrared thermography. *Agriculture* **2019**, *9*, 171. [CrossRef]

18. Lucas, R.; Úbeda, A.; Payá, M.; Alves, M.; Del Olmo, E.; López, J.L.; San Feliciano, A. Synthesis and enzyme inhibitory activities of a series of lipidic diamine and aminoalcohol derivatives on cytosolic and secretory phospholipases A2. *Bioorganic Med. Chem. Lett.* **2000**, *10*, 285–288. [CrossRef]

19. Wang, Y.; Zia-Khan, S.; Owusu-Adu, S.; Miedaner, T.; Müller, J. Early detection of zymoseptoria tritici in winter wheat by infrared thermography. *Agriculture* **2019**, *9*, 139. [CrossRef]

20. Thorp, K.R.; Wang, G.; Bronson, K.F.; Badaruddin, M.; Mon, J. Hyperspectral data mining to identify relevant canopy spectral features for estimating durum wheat growth, nitrogen status, and grain yield. *Comput. Electron. Agric.* **2017**, *136*, 1–12. [CrossRef]

21. Kayad, A.; Sozzi, M.; Gatto, S.; Marinello, F.; Pirotti, F. Monitoring Within-Field Variability of Corn Yield using Sentinel-2 and Machine Learning Techniques. *Remote Sens.* **2019**, *11*, 2873. [CrossRef]

22. Zecha, C.; Peteinatos, G.; Link, J.; Claupein, W. Utilisation of Ground and Airborne Optical Sensors for Nitrogen Level Identification and Yield Prediction in Wheat. *Agriculture* **2018**, *8*, 79. [CrossRef]

23. Mulley, M.; Kooistra, L.; Bierens, L. High-resolution multisensor remote sensing to support date palm farm management. *Agriculture* 2019, *9*, 26. [CrossRef]

24. Heiß, A.; Paraforos, D.S.; Griepentrog, H.W. Determination of cultivated area, field boundary and overlapping for a plowing operation using ISO 11783 communication and D-GNSS position data. *Agriculture* **2019**, *9*, 38. [CrossRef]

25. Hossain, K.; Raheem, D.; Cormier, S. *Food Security Governance in the Arctic-Barents Region*; Springer: Cham, Switzerland, 2018; ISBN 9783319757568.

26. Raheem, D.; Shishaev, M.; Dikovitsky, V. Food system digitalization as a means to promote food and nutrition security in the barents region. *Agriculture* **2019**, *9*, 168. [CrossRef]

27. Munder, S.; Argyropoulos, D.; Müller, J. Acquisition of sorption and drying data with embedded devices: Improving standard models for high oleic sunflower seeds by continuous measurements in dynamic systems. *Agriculture* **2019**, 9, 1. [CrossRef]

Chapter 2

A NOVEL COSMIC-RAY NEUTRON SENSOR FOR SOIL MOISTURE ESTIMATION OVER LARGE AREAS

1. INTRODUCTION: AN OVERVIEW

Water scarcity and drought problems in several parts of the world highlight the necessity for new solutions for better management of water resources. The Food and Agriculture Organization of the United Nations (FAO) calculates that 70% of employed water resources are dedicated to agriculture on the global scale [1]. Sustaining agricultural productivity requires an efficient management of agricultural water resources that involves a clear understanding of the temporal dynamics and spatial variability of soil moisture. Such dynamics are essential to optimize irrigation, preserve water for drought periods, and to optimize other production inputs, such as fertilizer application and water pumping power. The prerequisite is reliable soil moisture data, measured over large-scales and in real-time.

Due to this crucial role, many devices have been developed to measure soil moisture at different spatial and temporal scales [2–4]. Available technologies range from point-scale invasive approaches, like time domain reflectometry (TDR) probes [5], to satellite remote sensing approaches [6]. The former technology can achieve accurate measurements (RMSE, root-mean-square error <2%) at high temporal resolution (minutes) and at different soil depths [5]. However, difficulties arise for the coverage of large areas (e.g., >500 m2), since soil heterogeneity can produce significant misinterpretation of hydrological conditions [4,7]. To monitor a large area, several point-scale probes must be installed, but this might generate technical problems (such as the definition of statistically relevant positions, electrical power supply, data transmission) and high costs, due to multiple instruments and to low accessibility of specific locations (especially in the case of extensive crops) [8–10]. Furthermore, their set-up in agricultural fields is limited by tillage and other land management operations and farming practices. In addition, the detectors are invasive (buried in the soil) and they require high maintenance. For these reasons, they are not suitable for covering heterogeneous and inaccessible sites (mountain sides and cropped fields) and they are expensive for long-term monitoring observatories.

A completely different alternative is represented by remote sensing approaches typically based on microwaves (1 mm^{-1} m). Compared to point-scale methods, satellite remote sensing provides soil moisture observations at a large scale (>km^2) and covers global areas, so it is more suitable for hydrological applications [11]. However, the signal is sensitive only to the very first centimeters of soil interface [12] and the temporal resolution (e.g., weekly measurements) is not always suitable for many applications. Large-scale satellite remote sensing methods have other limitations [13], including limited capability to penetrate vegetation, inability to measure soil ice, and sensitivity to surface roughness [14]. For these reasons, despite the progress over the last decades, accuracy of remote sensing estimation is still too high for several applications (e.g., RMSE >4%) [15] and many studies are focusing on possible improvements [6].

In the last decade, to overcome the aforementioned operational challenges, a proximal geophysical method has been developed in order to fill the gap between point-scale and remote sensing approaches: cosmic-ray neutron sensing (CRNS) [16–18]. CRNS is a valid and robust alternative, offering many advantages: it is contactless, allows quantification of soil moisture averaged over large areas with only one probe, and is not invasive for field agricultural operations. The major advantages of CRNS are its large horizontal footprint (up to tens of hectares) and the penetration depth of tens of centimeters, enough to reach typical roots' depth [19,20].

The technique is based on the natural neutrons detected on the earth's surface, that are mostly generated by cosmic rays, according to various processes. The high energy protons component of cosmic rays, produced by galactic sources, interact with atomic nuclei in the upper atmosphere and produce secondary high-energy particles cascades. Muons and fast neutrons are constituents of such cascades. The secondary neutrons reach the ground level and interact with soil atoms. Another mechanism of neutron production is the so-called spallation effect, due to high-energy secondary muons interacting with ground atoms.

Hydrogen in water molecules becomes the dominant factor for slowing down and absorbing neutrons (also known as neutrons moderation). The fast neutrons, produced in air and soil, travel in all directions within the air–soil–vegetation continuum and therefore an equilibrium concentration of neutrons is established. The equilibrium is shifted in response to changes in the water presence above and below the land surface. For example, a drier soil, having a lower moderation capacity, reflects a greater number of neutrons compared to a more humid soil. In the latter, neutrons are more easily moderated, thus slowed down and partially absorbed; the net effect is an increase of the slow population, in respect to the fast one. The resultant neutron intensity above the land surface is inversely proportional to soil water content.

The portion of the neutrons energy spectrum that is most sensitive to soil moisture is the epithermal/fast region from energies of 0.25 eV to 100 keV [20]. Evaporation neutrons from 100 keV to 10 MeV give additional information especially for snow measurements [21]. Finally, high-energy neutrons over 10 MeV are not dependent from local conditions and are directly proportional to the primary incoming flux. Many studies relied on the performance of a set of CRNS probes for monitoring [22,23], modeling [24,25], or remote-sensing validation purposes [26,27].

It is important to underline that the measured intensity of environmental neutrons depends not only on the water in the soil but also on the incoming cosmic-ray neutrons flux [28]. This component changes with changing atmospheric conditions and also with variation of the incoming flux of galactic cosmic rays [17]. For this reason, on one hand CRNS is typically equipped with sensors for air pressure p, air temperature T, and relative humidity h_{rel}. On the other hand, it is worth noting that normally the correction by the incoming-flux is not directly measured in situ, but it is extrapolated offine using the Neutron Monitor Database (NMDB) [29], which provides data from several stations around the world.

The aim of the present study was to present and assess a new sensor for detecting soil moisture based on the CRNS approach. For this reason, several tests have been conducted in different agro-environmental conditions in comparison to current commercial CRNS probes and point-scale soil moisture measurements. The discussion focuses particularly on the applicability of the new sensor for agricultural applications.

2. MATERIALS AND METHODS

2.1 Instrumentation

For many years neutrons have been detected using ^{3}He proportional counter tubes (for the thermal component) and liquid/plastic scintillators (for the fast component); these two well-established technologies show some important limitations in practical applications. ^{3}He is a nuclide produced almost entirely in artificial contexts, as the product of the tritium decay. The current storage is depleting, and the price is high and rising, since it comes mainly from the production or dismantling of nuclear weapons of past decades [30]. Liquid/plastic scintillators are often made of toxic or hazardous materials, safely used in research contexts, but not suitable for agricultural, civil, or industrial applications. The interest in neutron detection for homeland security applications has triggered, in the last decade, the development of new detectors made up of liquid or plastic scintillation materials with low toxicity, that are safe and easy to use.

We studied a new solution [31], namely Finapp, based on a composite detector made of commercial detectors: EJ-299-33A and EJ-420(6), both manufactured by Eljen Technology (Sweetwater, TX, USA). EJ-420 and EJ-426 are inorganic scintillators, that have proven to have a good response to thermal neutrons [32]. EJ-299-33A was the first plastic detector to become commercially available for gamma/fast-neutron discrimination [33]. The discrimination capability of neutrons from gamma-rays is a fundamental prerequisite, in order to discriminate neutrons from the naturally occurring gamma-ray environmental background, that is not correlated to soil moisture in the same way as neutrons.

We assembled the composite detector by wrapping the plastic scintillator EJ-299-33A (a cylinder of 3×× × 3××) with thermal neutron detectors. In the plane face we used an EJ420 detector with a diameter of 3××. On the lateral face we used EJ426, which has a lower efficiency but is flexible. The optical-grade rubber EJ-560 ensures proper optical contact between the different detectors. A single photomultiplier (PMT) Mod. H6553 (Hamamatsu Photonics, Hamamatsu, Japan) was used to collect light emissions of the scintillators. Figure 1 shows an exploded view of the assembly.

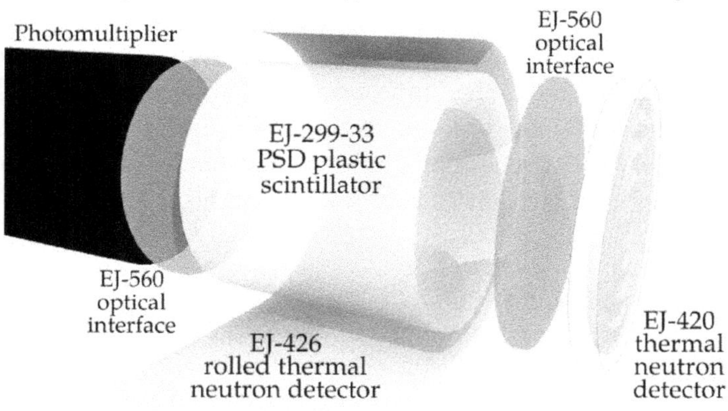

Figure 1. Assembly of the detector, Finapp, to employ in cosmic-ray neutron sensing (CRNS) technique. Abbreviation: PSD, pulse shape discrimination.

Normally EJ420(6) and 3He tubes are able to detect only thermal neutrons with energy below 0.025 eV but, according to Kohli et al. [20], the most sensitive part to soil moisture of the neutron spectra is the epithermal region between 0.025 eV and 100 keV. In order to collect these neutrons, detectors are normally equipped with a few cm of polyethylene, a material enriched with hydrogen that acts as a moderator slowing down neutron energy from the epithermal/fast region to the thermal region.

2.2 Data Acquisition

The data acquisition system (DAQ) was composed of an electronic signal digitizer model DT5790 (CAEN Spa, Viareggio, Italy) featuring: two input channels, that generate digital waveforms from the analog signals, with a 12 bit resolution and a sampling rate of 250 MS/s (samples per second); and two channels of high voltage power supply, for the photomultiplier. The digitizer is interfaced with a low-cost, low-power, embedded computer (Beaglebone Black). The software controlling the digitizer is an open-source, distributed data acquisition system, called ABCD [34,35] developed for the H2020 C-BORD project [36,37]. ABCD is employed to provide a continuous stream of data to a custom analysis system, based on the group's experience developed during the FP7 TAWARA_RTM project [38]. The analysis is fully automatic and performed online. The probe was equipped with internal temperature sensors and was configured to gather weather data from a local weather station installed near the probe. The operating negative voltage for the photomultiplier was set at 1600 V. Data transmission was ensured by a cellphone modem and data were stored locally in a Secure Digital (SD) memory. Figure 2 shows a block diagram of the probe functionality.

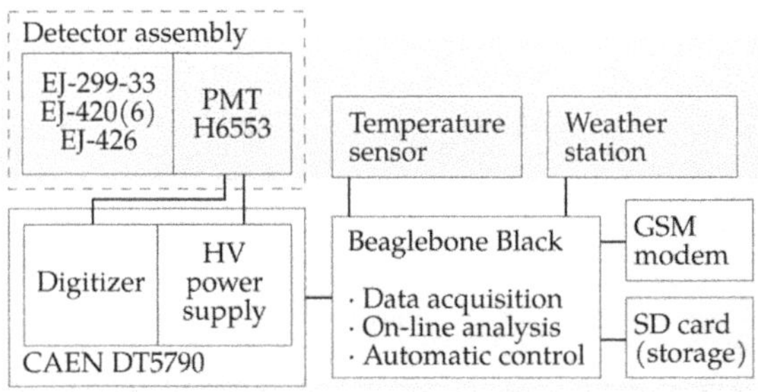

Figure 2. Block diagram of the data acquisition chain of the probe. Abbreviations: PMT, photomultiplier; HV, high voltage; GSM, protocol for data trasmission; SD, Secure Digital.

2.3 Data Processing and Analysis

2.3.1 Particle Discrimination

The significant parameters (e.g., signal integrals), needed for the signals' identification and discrimination, are extracted from the digital waveforms acquired by the digitizer. Together with the auxiliary sensors' data (e.g., temperature, pressure, etc.), the analysis software performs a data merge of the information, in order to correct the parameters and improve the signals identification and discrimination capabilities. Particles are identified and discriminated according to the generated signals, with an algorithm based on the most popular method of pulse shape discrimination (PSD) [39]. This technique exploits the different processes activated by different particles interacting in the scintillator; in particular, the produced light has usually two or more components, characterized by their decay time τ_i. The various light components are excited with different yields, depending on the interacting particle. These processes lead to different waveform shapes, that can be distinguished calculating the so-called PSD parameter, defined as:

PSD = (Long Integral - Short Integral)/Long Integral (1)

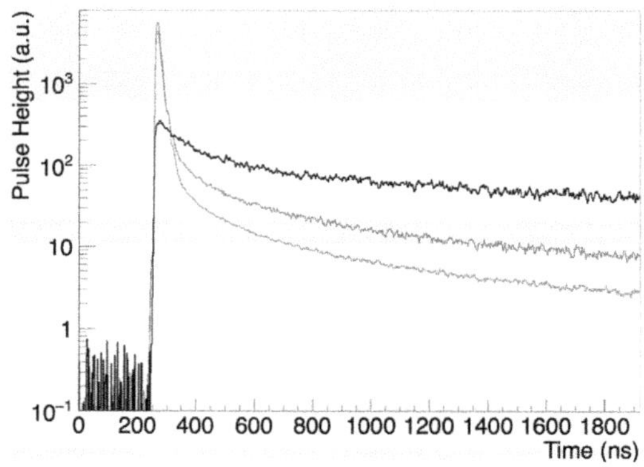

(a)

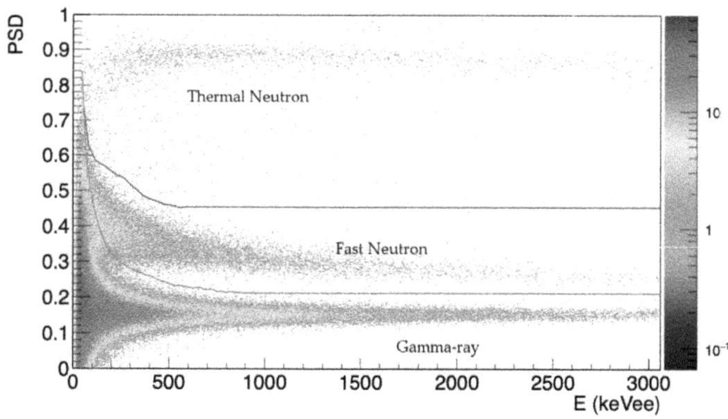

(b)

Figure 3 shows the typical sampled signals from gamma-rays, fast neutrons, and thermal neutrons (Figure 3a) and PSD parameter versus energy (Figure 3b). Discrimination between particles can be made simply by selecting the significant regions, on the PSD plot, in order to separate the three contributes (Figure 3b).

where "Long Integral" is the PMT signal charge-integrated over a long integration gate, while "Short Integral" is calculated on a short integration gate. The "Long

installation site are N 52.410087, E 12.978808. There are no cultivated fields nearby; the probes have been installed on an uncultivated grassy lawn.

3.2 Lagosanto, Italy

The second experimental site was located in Lagosanto, Italy, in an experimental field in the Porto Felloni agricultural company. This company is renowned as being one of the most technologically advanced in Italy, with a continuous and effective implementation of precision farming practices. The area is located in Emilia Romagna, a few km from the sea and 50 km from Ferrara. The climate is warm and temperate with an average temperature of 14 °C and an annual rainfall of 600 mm, with a dry period during the summer months. The probe was installed in a recent orchard with walnut trees, characterized by a small trunk. During summer months water is provided daily by drip irrigation. The probe was positioned as shown in Figure 5a at WGS84 coordinates N 44.752756, E 12.134761. The probe's footprint is an area with a radius of about 150 m, as highlighted by the red circle. Figure 5b shows a picture of the installation; the probe was installed 1.8 m from the ground. At about 90 m from the probe, five classical Sentek sensors for soil moisture were installed at 10, 20, 30, 40, and 50 cm depth, while a weather station was installed less than 1 km from the probe. Data were collected from August 8, 2018 to November 11, 2018.

In the test area, soil is in general homogenous, and characterized by a sandy loamy texture, with a 2% organic matter, as reported in Table 1.

Table 1. Texture of the soil where the Sentek probes were installed.

Loam	Sand	Clay	Organic Matter	ρbulk
42.1%	38.5%	19.4%	2.1%	1.4 g/cm3

(a) (b)

Figure 5. (a) Field site, red point is where the Finapp probe was installed, the red circle is the footprint, the orange point represents where the five Sentek probes are installed at different depths, and the light blue line is the drainage canal. (b) Picture of the installation.df

The drainage of the soil favors infiltration, calling for the need of frequent irrigation. The plant is vulnerable to dehydration and the soil must therefore always be kept damp, but waterlogging must be avoided, as this can cause asphyxia and blossom-end rot of the fruit.

A particularity of this site is the presence of a very shallow saturated zone. The territory is a reclamation area, kept dry thanks to water pumps that operate 24 h a day. This creates a very shallow saturated zone that reaches 50 cm deep. This is clearly visible from Figure 6, where the soil moisture sensors from the Sentek probes reaches 50% of the water equivalent at a depth of 50 cm. Furthermore, the owner of the property has underlined how the nearby reclamation canal (identified in Figure 5a with the light blue line) creates infiltrations in the field and this creates problems of asphyxiation of the plants due to too much water in the soil.

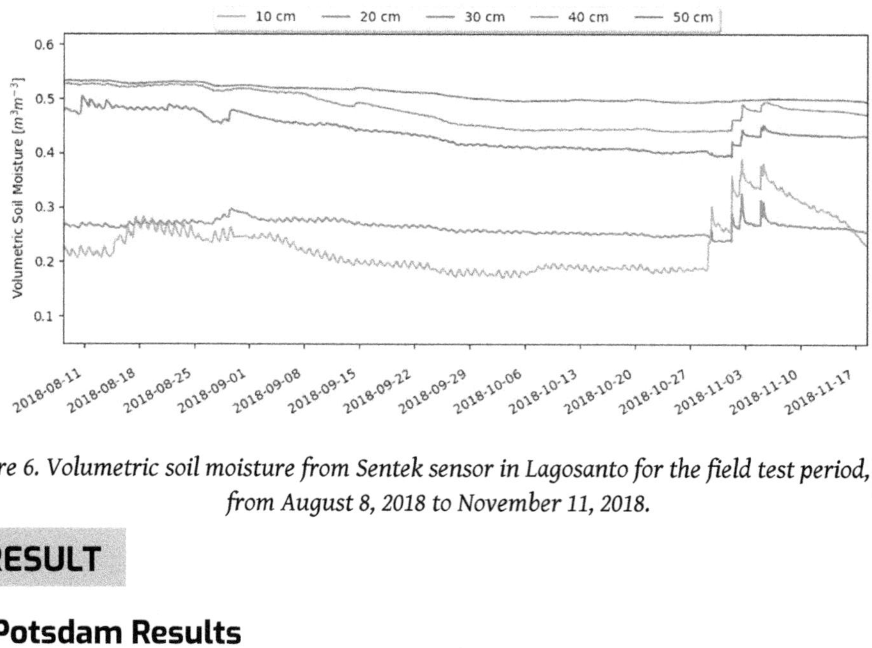

Figure 6. Volumetric soil moisture from Sentek sensor in Lagosanto for the field test period, from from August 8, 2018 to November 11, 2018.

4. RESULT

4.1 Potsdam Results

The Finapp probe was compared with two commercial CRNS probes, one from Hydroinnova (CRS-1000) and another from CANBERRA during the two months of outdoor testing at Potsdam University. Figure 7 shows the time series of epithermal neutron counts corrected for air pressure, for the whole study period. The neutron counts are averaged over 12 h intervals. The y-axis reports the variation in respect to the average counts of the single probes in the whole period.

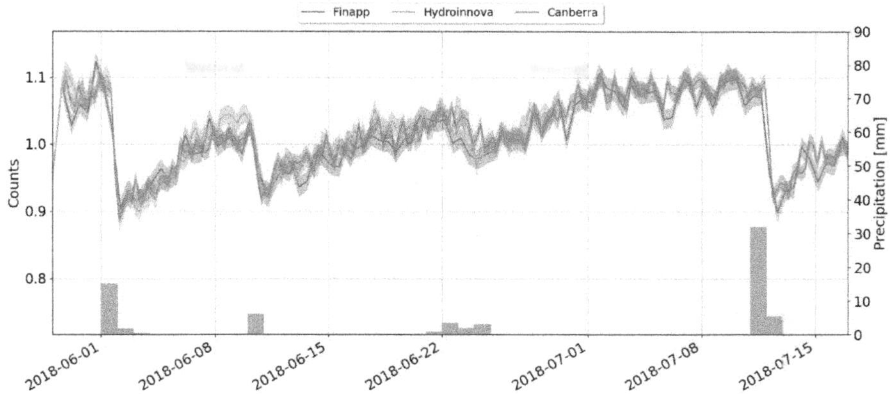

Figure 7. Epithermal counts: Finapp (blue), CRS-1000 from Hydroinnova (orange), and He-3 tube from CANBERRA (green). Data is corrected for atmospheric pressure variations. The colored zone indicates statistic uncertainty. Daily rain is reported in the bottom part of the plot.

Results show very good agreement between epithermal counts. The epithermal neutron counts drop during rain periods when the soil moisture increases. The efficiency of Finapp in respect to CRS-1000 from Hydroinnova, quantified in terms of epithermal counts, was 55%. This aspect means that the Finapp probe needs about twice the time, needed by the Hydroinnova probe, to have the same accuracy in measurements.

Figure 8 shows soil moisture from neutrons count by Equation (3) using data from the Finapp probe. The same figure reports the measurements of classical TDR probes at different depths. It is possible to note that the soil moisture in Potsdam is more uniform at different depths, compared to the Lagosanto data (Figure 6) where the soil moisture increases sensibly with depth. At Potsdam, the soil is much drier and the groundwater system is much deeper, thus it does not affect the conditions in the first 50 cm of the soil.

There is very good agreement between the three rain events (stormy) on June 1, June 11, and July 12. It is worth pointing out the rain event between June 21–26, where light rain increased the soil moisture according to CRNS, but it was not seen from 5TE probes. The behavior could be explained by the small amount of rain which did not infiltrate into the soil but remained on the land surface. For this reason, the 5TE did not record the event while the Finapp detected the water canopy interception or ponding water in the surrounding urban areas. Similar behaviors have been identified in previous studies [40,42].

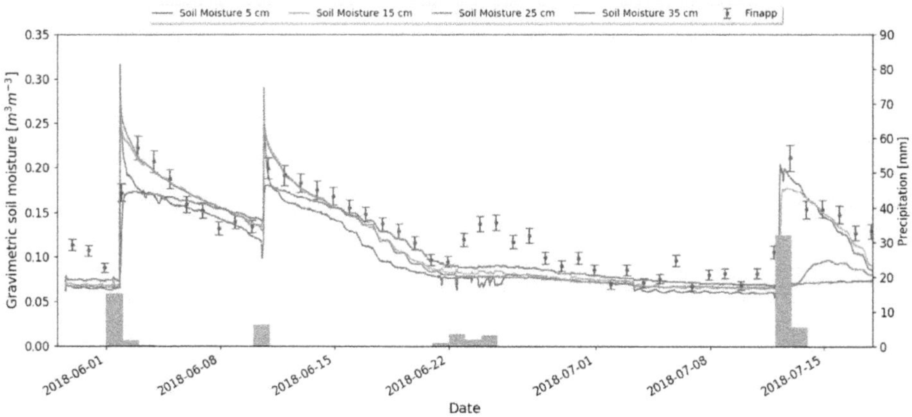

Figure 8. Gravimetric soil moisture from classical time domain reflectometry (TDR) probes (blue, orange, green, red) at different depths. The magenta points are CRNS gravimetric soil moisture using Desilet formula [19] and epithermal counts from Finapp. Daily rain is reported in the bottom part of the plot.

4.2 Lagosanto Results

Raw thermal counts were corrected hourly by absolute humidity and pressure from the near weather station. The reference values were set to h_{ref} = 12 g/m^3 and p_{ref} = 1016.4 hPa. The incoming variation was corrected by the muons measured by Finapp. The flux is mostly stable during the whole period and fluctuation is of the order of 2%. The reference value was set to I_{ref} = 4667.

Furthermore, to improve the comparison with point-scale probes, we computed a weighted average of the measurements (Figure 6) at different depths as described in [43]. CRNS measures the water content in the first 20–70 cm of the soil depending on soil moisture; the higher the water content, less the depth sensitivity, because wet soil has a larger moderation power for neutrons. Finapp acquired data during the second part of summer and first part of autumn.

The end of the vegetative season for the orchard is in September and the irrigation drip was stopped on September 1. September and October were relatively dry months with a rainfall of 60 mm. From October 27 a rainy period significantly increased the soil moisture. Figure 9 shows the averaged soil moisture from the Sentek probe and the soil moisture from Finapp; the lower part of the figure shows the cumulated precipitation with a 24 h interval. Focusing attention on the Finapp data, it is possible to notice a very good agreement between precipitation and soil moisture. Every rain event corresponds to an increase in volumetric soil moisture. Only the events on October 13 report an increase of soil moisture without any precipitation.

This increase is probably due to infiltration from the near reclamation canal, that increased soil moisture in the footprint of our probe, as suggested by the property owner.

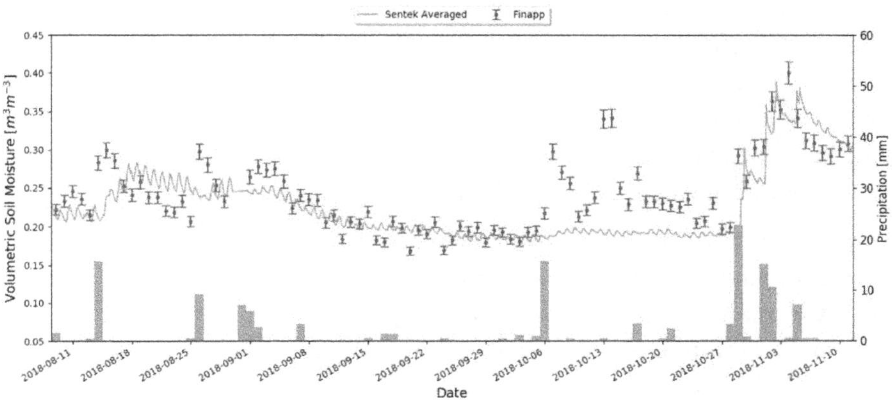

Figure 9. Volumetric soil moisture from Finapp (magenta) and average soil moisture from Sentek (orange); for details see the text. Daily rain is reported in the bottom part of the plot.

On the contrary, there are some differences if we compare Finapp with point-scale probes. First of all, during the irrigation period the point-scale measurements are less sensitive to precipitation because the drip irrigation biases the point measurement due to a heterogeneous water distribution. In September there is good agreement between the Finapp and Sentek probes; the soil moisture decreases in both cases due to the dry period. In October there is again some discrepancy, probably due to a malfunction of the Sentek probe. The property owner notes that in October the probes are no longer checked daily because the data is no longer used, and periods of malfunctioning occur due to various technical problems (power, batteries, data transmission, etc.). In November the agreement is good again and both the Sentek and CRNS probes respond to precipitation in the same way.

5. DISCUSSION AND CONCLUSIONS

The Finapp probe was tested as a novel sensor for soil moisture estimation over a large area. The sensor makes use of the CRNS technique to reconstruct soil moisture from environmental epithermal neutron counts. The results show that Finapp measures the neutrons in the epithermal region like the well-tested 3He tubes. Efficiency as an epithermal neutrons counter is 55%, in respect to the commercial CRS-1000 from Hydroinnova. Efficiency could be increased, in order to reduce statistic uncertainty, by growing the size of the detector and/or using new materials for thermal neutron detection. In particular, good results were achieved by our group [44] and more research is currently underway. We reconstructed soil moisture in Figures 8 and

9 for the Potsdam and Lagosanto field trials, respectively. Data is well correlated with precipitations, and it shows the limit of classical point-like measurements, that can be influenced by the heterogeneity of the soil and/or the irrigation distributions. Averaging soil moisture over a large area overcomes the problem of these heterogeneities, making the measurement better suitable for supporting agricultural management conducted at field scale (e.g., irrigation). The soil moisture in Figures 8 and 9 have a time-interval of 24 h between data points. We use such integration time in order to reduce statistic uncertainty on the final measurements. Typical integration values for CRNS commercial probes are 6 or 12 h.

In respect to the currently-employed technology with 3He tubes, Finapp offers more measured parameters, like gamma-rays and fast neutrons. Future development foresees the analysis of this information. In particular, gamma-rays were found to be correlated to soil moisture as well [45,46], but with a smaller footprint. The fast neutrons and high energy events detected can be used to normalize the incoming galactic fluctuations directly in-situ with the data from the sensor.

Overall, Finapp uses non-toxic and non-carcinogenic plastic materials, is easily recyclable, and its technology can be easily scaled to market size. Scalability will be the key point for future development, in order to reduce the price to the same level as professional point-scale soil moisture probes currently on the market.

References

1. Food and Agriculture Organization (FAO). Introduction. In *Water for Sustainable Food and Agriculture*; FAO: Rome, Italy, 2017; p. 1. ISBN 978-92-5-109977-3.
2. Corradini, C. Soil moisture in the development of hydrological processes and its determination at different spatial scales. *J. Hydrol.* **2014**, 516, 1–5. [CrossRef]
3. Ochsner, T.E.; Cosh, M.H.; Cuenca, R.H.; Dorigo, W.A.; Draper, C.S.; Hagimoto, Y.; Kerr, Y.H.; Njoku, E.G.; Small, E.E.; Zreda, M. State of the Art in Large-Scale Soil Moisture Monitoring. *Soil Sci. Soc. Am. J. 2013*, 77, 1888–1919. [CrossRef]
4. Robinson, D.A.; Campbell, C.S.; Hopmans, J.W.; Hornbuckle, B.K.; Jones, S.B.; Knight, R.; Ogden, F.; Selker, J.; Wendroth, O. Soil Moisture Measurement for Ecological and Hydrological Watershed-Scale Observatories: A Review. *Vadose Zone J. 2008*, 7, 358–389. [CrossRef]
5. Romano, N. Soil moisture at local scale: Measurements and simulations. *J. Hydrol.* **2014**, 516, 6–20. [CrossRef]
6. Bauer-Marschallinger, B.; Freeman, V.; Cao, S.; Paulik, C.; Schaufler, S.; Stachl, T.; Modanesi, S.; Massari, C.; Ciabatta, L.; Brocca, L.; et al. Toward Global Soil Moisture Monitoring with Sentinel-1: Harnessing Assets and Overcoming Obstacles. *IEEE Trans. Geosci. Remote Sens.* **2018**, 57, 520–539. [CrossRef]

7. Susha Lekshmi, S.U.; Singh, D.N.; Baghini, M.S. A critical review of soil moisture measurement. *Measurement* **2014**, 54, 92–105. [CrossRef]

8. Famiglietti, J.S.; Deveraux, J.A.; Laymon, C.A.; Tsegaye, T.; Houser, P.R.; Jackson, T.J.; Graham, S.T.; Rodell, M.; van Oevelen, P.J. Ground-based investigation of soil moisture variability within remote sensing footprints During the Southern Great Plains 1997 (SGP97) Hydrology Experiment. *Water Resour. Res.* **1998**, 35, 1839–1852. [CrossRef]

9. Famiglietti, J.S.; Ryu, D.; Berg, A.A.; Rodell, M.; Jackson, T.J. Field observations of soil moisture variability across scales. *Water Resour. Res.* **2008**, 44, W01423. [CrossRef]

10. Western, A.W.; Grayson, R.B.; Bioshl, G. Scaling of Soil Moisture: A Hydrologic Perspective. *Annu. Rev. Earth Planet. Sci.* **2002**, 30, 149–180. [CrossRef]

11. Zhuo, L.; Han, D. The Relevance of Soil Moisture by Remote Sensing and Hydrological Modelling. *Procedia Eng.* **2016,** 154, 1368–1375. [CrossRef]

12. Petropoulos, G.P.; Ireland, G.; Barrett, B. Surface soil moisture retrievals from remote sensing: Current status, products & future trends. Phys. Chem. Earth **2015**, *83*, 35–56. [CrossRef]

13. Entekhabi, D.; Njoku, E.G.; Houser, P.; Spencer, M.; Doiron, T.; Kim, Y.; Smith, J.; Girard, R.; Belair, S.; Crow, W.; et al. The Hydrosphere State (Hydros) Satellite Mission: An Earth System Pathfinder for Global Mapping of Soil Moisture and Land Freeze/Thaw. *IEEE Trans. Geosci. Remote Sens.* **2004**, 42, 2184–2195. [CrossRef]

14. Wanders, N.; Karssenberg, D.; Bierkens, M.; Parinussa, R.; de Jeu, R.; van Dam, J.; de Jong, S. Observation uncertainty of satellite soil moisture products determined with physically-based modeling. *Remote Sens. Environ.* **2012**, *127*, 341–356. [CrossRef]

15. Fang, B.; Lakshmi, V. Soil moisture at watershed scale: Remote sensing techniques. *J. Hydrol.* **2014**, 516, 258–272. [CrossRef]

16. Zreda, M.; Desilets, D.; Ferré, T.; Scott, R. Measuring soil moisture content non-invasively at intermediate spatial scale using cosmic-ray neutrons. *Geophys. Res. Lett.* **2008**, 35, L21402. [CrossRef]

17. Zreda, M.; Shuttleworth, W.; Zeng, X.; Zweck, C.; Desilets, D.; Franz, T.; Rosolem, R. COSMOS: The COsmic-ray Soil Moisture Observing. *Hydrol. Earth Syst. Sci.* **2012**, 16, 4079–4099. [CrossRef]

18. Franz, T.E.; Zreda, M.; Rosolem, R.; Ferre, T.P.A. Field Validation of a Cosmic-Ray Neutron Sensor Using a Distributed Sensor Network. Vadose Zone J. 2012, 11. [CrossRef]

19. Desilets, D.; Zreda, M. Footprint diameter for a cosmic-ray soil moisture probe: Theory and Monte Carlo simulations. *Water Resour.* Res. **2013**, *49*, 3566–3575. [CrossRef]

20. Kohli, M.; Schron, M.; Zreda, M.; Schmidt, U.; Dietrich, P.; Zacharias, S. Footprint characteristics revised for field-scale soil moisture monitoring with cosmic-ray neutrons. *Water Resour. Res.* **2015**, *51*, 5772–5790. [CrossRef]

21. Desilets, D.; Zreda, M.; Ferré Ty, P.A. Nature's neutron probe: Land surface hydrology at an elusive scale with cosmic rays. *Water Resour. Res.* **2010**, *46*, W11505. [CrossRef]

22. Villarreyes, R.C.; Baroni, G.; Oswald, S. Integral quantification of seasonal soil moisture changes in farmland by cosmic-ray neutrons. *Hydrol. Earth Syst. Sci.* **2011**, *15*, 3843–3859. [CrossRef]

23. Evans, J.G.; Ward, H.C.; Blake, J.R.; Hewitt, E.J.; Morrison, R.; Fry, M.; Ball, L.A.; Doughty, L.C.; Libre, J.W.; Hitt, O.E.; et al. Soil water content in southern England derived from a cosmic-ray soil moisture observing system—COSMOS-UK. *Hydrol. Process.* **2016**, *30*, 4987–4999. [CrossRef]

24. Baatz, R.; Bogena, H.R.; Franssen, H.-J.H.; Huisman, J.A.; Montzka, C.; Vereecken, H. An empirical vegetation correction for soil water content quantification using cosmic ray probes. *Water Resour. Res.* **2015**, *51*, 2030–2046. [CrossRef]

25. Andreasen, M.; Jensen, H.K.; Zreda, M.; Desilets, D.; Bogena, H.; Looms, C.M. Modeling cosmic ray neutron field measurements. *Water Resour. Res.* **2016**, 52, 6451–6471. [CrossRef]

26. Holgate, C.M.; De Jeu, R.A.M.; van Dijk, A.I.J.M.; Liu, Y.Y.; Renzullo, L.J.; Kumar, V.; Dharssi, I.; Parinussa, R.M.; Van Der Schalie, R.; Gevaert, A.; et al. Comparison of remotely sensed and modelled soil moisture data sets across Australia. *Remote Sens. Environ.* **2016**, *186*, 497–500. [CrossRef]

27. Montzka, C.; Bogena, H.; Zreda, M.; Monerris, A.; Morrison, R.; Muddu, S.; Vereecken, H. Validation of Spaceborne and Modelled Surface Soil Moisture Products with Cosmic-Ray Neutron Probes. *Remote Sens.* **2017**, *9*, 103. [CrossRef]

28. Schrön, M.; Rosolem, R.; Köhli, M.; Piussi, L.; Schröter, I.; Iwema, J.; Kögler, S.; Oswald, S.E.; Wollschläger, U.; Samaniego, L.; et al. Cosmic-ray Neutron Rover Surveys of Field Soil Moisture and the Influence of Roads. *Water Resour. Res.* 2018, *54*, 6441–6459. [CrossRef]

29. Mavromichalaki, H.; Papaioannou, A.; Plainaki, C.; Sarlanis, C.; Souvatzoglou, G.; Gerontidou, M.; Papailiou, M.; Eroshenko, E.; Belov, A.; Yanke, V.; et al.

Applications and usage of the real-time Neutron Monitor Database. *Adv. Space Res.* **2011**, *47*, 2210–2222. [CrossRef]

30. Kouzes, R. The ^{3}*He Supply Problem*; PNNL-18388; Pacific Northwest National Laboratory: Richland, WA, USA, 2009.
31. Cester, D.; Lunardon, M.; Moretto, S.; Nebbia, G.; Pino, F.; Sajo-Bohus, L.; Stevanato, L.; Bonesso, I.; Turato, F. A novel detector assembly for detecting thermal neutrons, fast neutrons and gamma rays. *Nucl. Instrum. Methods A* **2016**, *830*, 191–196. [CrossRef]
32. Pino, F.; Stevanato, L.; Cester, D.; Nebbia, G.; Sajo-Bohus, L.; Viesti, G. Study of the thermal neutron detector ZnS(Ag)/LiF response using digital pulse processing. *J. Instrum.* **2015**, *10*, T08005. [CrossRef]
33. Cester, D.; Nebbia, G.; Stevanato, L.; Pino, F.; Viesti, G. Experimental tests of the new plastic scintillator with pulse shape discrimination capabilities EJ-299-33. *Nucl. Instrum. Methods A* **2014**, *735*, 202–206. [CrossRef]
34. Fontana, C.L. A distributed data acquisition system for signal digitizers with on-line analysis capabilities. In Proceedings of the IEEE Nuclear Science Symposium and Medical Imaging Conference, Atlanta, GA, USA, 21–28 October 2017; pp. 1–5. [CrossRef]
35. Fontana, C.L. A distributed data acquisition system for nuclear detectors. *Int. J. Mod. Phys. Conf. Ser.* **2018**, *48*, 1860118. [CrossRef]
36. Sardet, A.; Pérot, B.; Carasco, C.; Sannié, G.; Moretto, S.; Nebbia, G.; Fontana, C.; Moszyn′ ski, M.; Sibczyn′ ski, P.; Grodzicki, K.; et al. Design of the rapidly relocatable tagged neutron inspection system of the C-BORD project. In Proceedings of the IEEE Nuclear Science Symposium, Medical Imaging Conference and Room-Temperature Semiconductor Detector Workshop, Strasbourg, France, 29 October–5 November 2016; pp. 1–5. [CrossRef]
37. Sibczynski, P.; Dziedzic, A.; Grodzicki, K.; Iwanowska-Hanke, J.; Mianowska, Z.; Moszyñski, M.; Swiderski, L.; Syntfeld-Kaz˙ uch, A.; Szawłowski, M.; Wolski, D.; et al. C-BORD—An overview of efficient toolbox for high-volume freight inspection. In Proceedings of the IEEE Nuclear Science Symposium and Medical Imaging Conference (NSS/MIC), Atlanta, GA, USA, 21–28 October 2017; pp. 1–3. [CrossRef]
38. Fontana, C.L. A Distributed Data Acquisition System for the Sensor Network of the TAWARA_RTM Project. *Phys. Procedia* **2017**, *90*, 271–278. [CrossRef]
39. Cester, D.; Lunardon, M.; Nebbia, G.; Stevanato, L.; Viesti, G.; Petrucci, S.; Tintori, C. Pulse shape discrimination with fast digitizers. *Nucl. Instrum. Methods A* **2014**, *748*, 33–38. [CrossRef]

40. Schrön, M.; Zacharias, S.; Womack, G.; Köhli, M.; Desilets, D.; Oswald, S.E.; Bumberger, J.; Mollenhauer, H.; Kögler, S.; Remmler, P.; et al. Intercomparison of cosmic-ray neutron sensors and water balance monitoring in an urban environment. *Geosci. Instrum. Methods Data Syst.* **2018**, 7, 83–99. [CrossRef]

41. Bogena, H.R.; Huisman, J.A.; Baatz, R.; Hendricks Franssen, H.-J.; Vereecken, H. Accuracy of the cosmic-ray soil water content probe in humid forest ecosystems: The worst case scenario. **Water Resour. Res. 2013**, 49, 5778–5791. [CrossRef]

42. Baroni, G.; Oswald, S.E. A scaling approach for the assessment of biomass changes and rainfall interception using cosmic-ray neutron sensing. *J. Hydrol.* **2015**, 525, 264–276. [CrossRef]

43. Franz, T.E.; Zreda, M.; Ferre, T.P.A.; Rosolem, R.; Zweck, C.; Stillman, S.; Zeng, X.; Shuttleworth, W.J. Measurement depth of the cosmic ray soil moisture probe affected by hydrogen from various sources. *Water Resour. Res.* **2012**, 48, W08515. [CrossRef]

44. Carturan, S.M.; Vesco, M.; Bonesso, I.; Quaranta, A.; Maggioni, G.; Stevanato, L.; Zanazzi, E.; Marchi, T.; Fabris, D.; Cinausero, M.; et al. Flexible scintillation sensors for the detection of thermal neutrons based on siloxane 6LiF containing composites: Role of 6LiF crystals size and dispersion. *Nucl. Instrum. Methods A* **2019**, 925, 109–115. [CrossRef]

45. Baldoncini, M.; Albéri, M.; Bottardi, C.; Chiarelli, E.; Raptis, K.G.C.; Strati, V.; Mantovani, F. Investigating the potentialities of Monte Carlo simulation for assessing soil water content via proximal gamma-ray spectroscopy. *J. Environ. Radioact.* **2018**, 192, 105–116. [CrossRef]

46. Baldoncini, M.; Albéri, M.; Bottardi, C.; Chiarelli, E.; Raptis, K.G.C.; Strati, V.; Mantovani, F. Biomass water content effect on soil moisture assessment via proximal gamma-ray spectroscopy. *Geoderma* **2018**, 335, 67–77. [CrossRef].

Chapter 3

CALIBRATION AND VALIDATION OF A LOW-COST CAPACITIVE MOISTURE SENSOR TO INTEGRATE THE AUTOMATED SOIL MOISTURE MONITORING SYSTEM

1. INTRODUCTION: AN OVERVIEW

Readily available soil moisture is a key requirement for the growth and development of plants and depends on the physical properties of the soil and the meteorological conditions of the surrounding environment. The upper and lower limits of the readily available soil moisture are known as the field capacity and the permanent wilting percentage, respectively. At the field capacity, sufficient water and air are retained in the soil, resulting in optimum plant growth. The effect of the meteorological conditions on soil moisture is minimal in indoor systems. However, the readily available soil moisture content varies throughout the soil, owing to the differences in transpiration and moisture loss from the soil, even in controlled environments [1]. Thus, continuous monitoring of the soil moisture content at different locations is required in indoor systems. These practices are costly, time-consuming, and labor-dependent. There are several techniques to determine the soil moisture content, including the destructive gravimetric method as well as nuclear, electromagnetic, tensiometric, hygrometric, and remote sensing processes [2]. Among them, sensors employing an electromagnetic technique are widely utilized to measure soil moisture levels. Time-domain reflectometry (TDR), time-domain transmission (TDT), and capacitance sensors are the most commonly used sensors based on an electromagnetic technique [3,4]. TDR and TDT sensors are accurate but have limited large-scale applicability owing to the cost of investment. In contrast, capacitance sensors are less expensive but require precise calibration. The capacitance of a sensor is determined by the dielectric constant [5,6] and the volume fraction of each phase (bulk water, water vapor, air, solid minerals, etc.) [7,8]. It provides real-time soil moisture data according to the changes in the moisture content of the soil.

Many researchers have focused on the calibration of low-cost sensors which are used in different sensing techniques, such as capacitance-based sensors [9,10], resistivity-based granular matrix sensors [9,11], and sensors based on a tensiometer technique [9], to measure the soil water content in fields [2,12–14] and to develop low-cost automated irrigation systems [15]. Some studies have considered the effects of the physical and chemical properties of soil on the performances of soil moisture sensors [4,13,14,16–18]. These studies provide valuable insight into the necessity of soil-specific calibration.

A variety of capacitance sensors have become increasingly popular because they are less expensive than TDR and TDT high-frequency (GHz range) sensors and sufficiently reliable. Considering the cost of investment, a low-cost sensor even with a relatively weak accuracy is preferred in agriculture [19]. The SKU:SEN0193 sensor is a commercially available, low-cost capacitive soil moisture sensor which operates in low-power consumption. However, this soil moisture sensor has not been properly investigated for its accuracy and repeatability under laboratory conditions. Hence, the present study was performed (i) to investigate the accuracy and reliability of the SKU:SEN0193 low-cost capacitive soil moisture sensor under laboratory conditions, (ii) to develop a calibration function, and (iii) to integrate this sensor with a data acquisition system.

2. MATERIALS AND METHODS

The SKU:SEN0193 capacitive soil moisture sensor, (DFRobot, Shanghai, China) with dimensions of 9.8 × 2.3 cm (L × W) was used for this study. The SKU:SEN0193 sensor can be powered from a voltage source in the range of 3.3 to 5.5 V. Thus, it can be interfaced with low-power microcontrollers. In addition, the sensor is made of corrosion-resistant material which increases durability. In parallel, SM-200 (Delta-T Devices Ltd., Cambridge, UK) a well-established commercially available soil moisture sensor was used for comparison to evaluate the accuracy of the SKU:SEN0193 sensor. The SM-200 sensor operates at 100 MHz, which measures a material (soil, water, air) response to polarization in an electromagnetic field (i.e., permittivity). The permittivity of the soil can be detected as a voltage output corresponding to the soil moisture content. The measuring accuracy is ±0.3% for volumetric water content, θ from 0 to 0.50 m3 m−3. The SM-200 sensor has been used in studies as a reliable soil moisture sensor to determine the soil moisture content in the plant root zone [20] and as a reference sensor to compare the accuracy of other capacitive-type soil moisture sensors [21]. In addition, the SHT30 temperature and humidity sensor was used to obtain the temperature and humidity of the air.

Commercially available organic-rich gardening soil was obtained from a local producer (Gardening. Pro, Maruki, Japan), hereafter referred to as organic-rich soil. This soil was originally obtained from the surface of the kanto loam layer (Black soil, Kanuma city, Tochigi Prefrecture, Japan). The measured organic matter content and mineral content were 24.8 and 75.2%, respectively [22]. The air-dried soil was sieved through a 2 mm mesh to remove aggregated soil clumps, and the <2 mm fraction of the soil was used in this study. In addition, a laboratory soil mixture was prepared by mixing the organic-rich soil with vermiculite in a 1:1 ratio [23,24]. This soil mixture was used as a growing medium for laboratory plants.

2.1 Data Acquisition and Analysis via an Internet-Based Platform

The data acquisition system used in this study consisted of two main components called the microcontroller unit and the Wi-Fi module. The main microcontroller STC89C52RC was operated at a speed of 11.0592 MHz. A software-implemented I^2C bus [25] was used to interface a 16 × 2 LCD (1602 character-type liquid crystal display) module, an ADS1115 16-bit analog-to-digital converter (ADC), and temperature and humidity sensor with the microcontroller. An ESP8266-12E low-cost serial-to-WiFi module was interfaced through STC89C52RC inbuilt UART. The analog data output pin of the SKU:SEN0193 sensor was connected to the ADS1115 with a full-scale range of ±4.096 V. The ADS1115 converted the voltage of SKU:SEN0193 sensor to raw counts (raw). ThingSpeak API, an open IoT (Internet of Things) platform, was used to collect and analyze data with MATLAB@ analytics. The assembly program for the microcontroller (Supplementary Material: Microcontroller program code) was written by using Keil µVision 5 IDE, and AT commands were used to control the WiFi module (Figures 1 and 2). The total cost of the developed prototype was around $45.7 (Table 1).

Table 1. The total cost of the developed soil moisture monitoring system (US$ in 2019).

Component	Units	Unit Cost ($)	Subtotal ($)	Total ($)
STC89C52RC	1	1.12	1.12	
ADS1115	1	2.73	2.73	
ESP8266-12E	1	1.79	1.79	
SKU:SEN0193	4	7.24	28.96	
SHT30	1	3.98	3.98	
LCD1602	1	2.12	2.12	
Other components			5.00	
				45.7

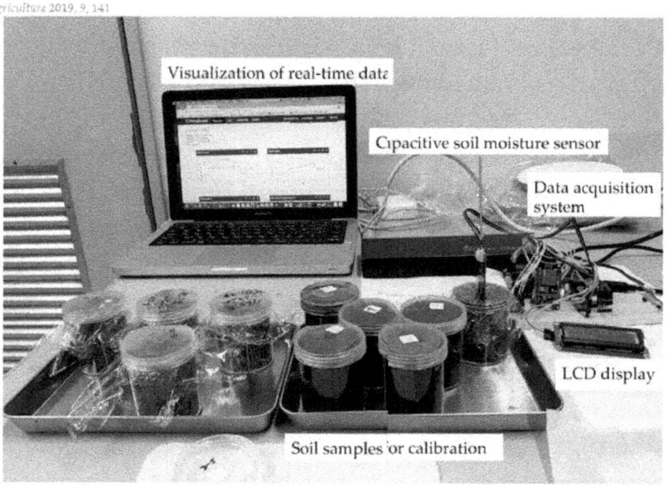

Figure 1. Acquisition and visualization of real-time data during the soil sample calibration process.

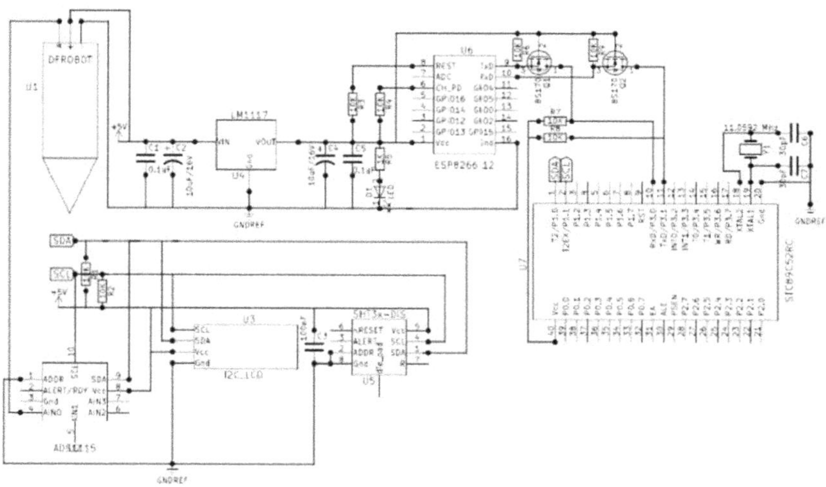

Figure 2. Circuit diagram of the data acquisition system.

2.2 Soil Sample Preparation

Soil samples with different soil moisture contents were prepared by adjusting the soil moisture content gravimetrically. The moisture-adjusted samples were packed into polycarbonate containers (average volume 110 cm^3 (Area: 18.08 cm^2 × Height: 6.1 cm)) with an average dry bulk density, ρ_d, of 0.6 g cm^{-3} for organic-rich gardening soil and 0.3 g cm^{-3} for the laboratory soil mixture. Then the containers were covered with plastic wrappings to avoid moisture loss during the calibration, and the samples were stored after sealing with a lid. The sensor was inserted to the center of the soil core at a depth of 5–6 cm from the upper soil surface. Then, the samples were let achieve equilibrium for 15 to 20 min, before performing the measurements. The noise of the data, i.e., data fluctuation, was very limited during the measurements. All experiments were performed at room temperature (25°C). The raw counts of the samples for each gravimetric water content were stored using the Thingspeak platform.

2.3 Sensor-to-Sensor Variability Study

Prior to the calibration, the sensor-to-sensor reading variability (raw counts) of the SKU:SEN0193 sensors for predetermined soil moisture contents was evaluated (Table 2). The sensor-to-sensor variability study was performed only for the organic-rich soil. Samples with two different soil moisture contents (40 and 80%, g g^{-1} of organic-rich soil), in duplicate, and water were used to test sensor-to-sensor variability. Repeated measurements (twenty replications of sensor measurements) were obtained for every duplicated soil sample with four SKU:SEN0193 sensors (S_1, S_2, S_3, and S_4). The sensor-to-sensor variability was analyzed using a one-way Analysis of Variance (ANOVA) statistical test [26]. Further, to estimate the measurement noise of the SKU:SEN0193 sensors, the coefficient of variance (CV) of raw counts was determined. The CV values thus obtained were CV_{S1} = 0.05%, CV_{S2} = 0.10%, CV_{S3} = 0.06%, CV_{S4} = 0.09% for 80%, g g-1 sample, i.e., an organic-rich soil sample in a field capacity range.

2.4 Calibration of the Sensor

Considering the sensor-to-sensor variability, the samples were measured repeatedly using two selected SKU:SEN0193 sensors to obtain a calibration function. The soil-specific calibration was performed using the organic-rich gardening soil and the laboratory soil mixture. The moisture contents of the tested soil and soil mixture were adjusted gravimetrically using distilled water. Moisture-adjusted samples were prepared with gravimetric water content of 0 (oven dried), 10, 20, 30, 40, 50, 60, 70, 80, 90, and 100 g g^{-1}. The moisture-adjusted samples were kept in closed plastic bags for 24 h for equilibration, and the gravimetric water content was verified by evaluating the moisture content of each sample by oven-drying at 105 ∘C [27]. The raw counts for the samples of the two different materials with each gravimetric water content were stored using the Thingspeak platform.

2.5 Validation of the Developed Calibration Function

The developed calibration function was validated by measuring organic-rich soil samples with different moisture contents: oven-dried samples, air-dried samples, samples with a moisture content of up to 100%, and saturated samples (>100%, g g^{-1}). The samples with different moisture contents were packed into polycarbonate containers. The soil water content of the samples was measured by the SKU:SEN0193 and SM-200 sensors. These values were confirmed gravimetrically by oven-drying. Similarly, data validation was performed in the selected values of the field capacity using samples with soil water contents of 60, 65, 70, 75, and 80% g g^{-1}. A soil-specific calibration was performed for the SM-200 sensor in accordance with its user manual [28]. The calculated coefficients a0 and a1, which conveniently parameterize the dielectric properties of soils, were 1.3 and 6.1, respectively, for the organic-rich soil (Equations (1)–(3)).

$$\sqrt{\varepsilon} = a_0 + a_1\theta \tag{1}$$

$$a_0 = \sqrt{\varepsilon_{dry\_soil}} \tag{2}$$

$$\sqrt{\varepsilon} = 1.0 + 16.103V - 38.725V^2 + 60.881V^3 - 46.032V^4 + 13.536V^5 \tag{3}$$

where ϵ is the dielectric permittivity of the soil, θ (cm^3 cm^{-3}) is the volumetric water content, and V is the SM-200 reading in Volts of the corresponding soil moisture content. A polynomial conversion (Equation (4)) was performed to calculate the volumetric water content θ (cm^3 cm^{-3}) after soil-specific calibration for the organic-rich soil:

$$\theta = \frac{\left[1.0 + 16.103V - 38.725V^2 + 60.881V^3 - 46.032V^4 + 13.536V^5\right] - a_0}{a_1} \tag{4}$$

3. RESULTS AND DISCUSSION

The data measured by the low-cost SKU:SEN0193 sensor and the temperature and humidity sensor were stored and visualized via an internet-based platform. The soil water contents predicted using the linear equation (which we developed by considering the raw counts of the sensor for bulk water and air with average values of 10,653 and 20,240, respectively) were inaccurate for both the organic-rich soil and the laboratory soil mixture. Hence, new calibration functions were developed by plotting the volumetric water content θ of the tested materials as a function of the raw count.

The volumetric water content θ (cm^3 cm^{-3}) of the samples can be obtained by the measured gravimetric water content w (g/g, %) of the samples, the dry bulk density of the material ρb (g cm^{-3}), and the density of water ρw (g cm^{-3}) as follows (Equation (5)) [29]:

$$\theta = w\frac{\rho_b}{\rho_w} \tag{5}$$

3.1 Results of the Sensor-to-Sensor Variability Study

The results of the ANOVA test at a 5% significance level for 20 replications of sensor response measurements (for each duplicated soil sample) of 4 SKU:SEN0193 sensors indicated that the mean sensor response was significantly different for the 4 SKU:SEN0193 sensors (see Table 2), demonstrating significant sensor-to-sensor variability.

Table 2. Results of the ANOVA test for 20 replications of sensor response measurements of 4 SKU:SEN0193 sensors for 3 different conditions.

Condition	Consideration	Df *	Sum of Squares (Raw counts)	Mean Square (Raw counts)	*F* Value
40%, g g^{-1}	Sensor-to-sensor variability	3	14888295	4962765	30091
	Noise	76	12535	165	
80%, g g^{-1}	Sensor-to-sensor variability	3	33893919	11297973	9813
	Noise	76	87499	1151	
Water	Sensor-to-sensor variability	3	7610824	2536941	1121
	Noise	76	171977	2263	

* df: degrees of freedom.

3.2 Calibration of the SKU:SEN0193 Sensor for Organic-Rich Soil and Laboratory Soil Mixture

The sensor-to-sensor variability study suggested that the responses of the tested four sensors were different, and therefore, their accuracy should be improved by using a sensor-specific calibration model for each sensor [29]. Considering the sensor-to-sensor variability, the samples were measured repeatedly using two selected SKU:SEN0193 sensors to obtain a calibration function.

The average raw count of the two SKU:SEN0193 sensors was obtained for each moisture-adjusted sample. Soil moisture, temperature, and humidity data were recorded at intervals of 1 min. After the equilibrium, the data were recorded over 15–20 min for each sample (n = 15–20). The averaged raw count as a function of volumetric water content, θ, was plotted in Figure 3. Here, we show the results for three moisture categories for organic-rich soil: dry to moderately wet samples with volumetric water content around 0.0–0.26 cm^3 cm^{-3}, field capacity samples with volumetric water content around 0.34–0.50 cm^3 cm^{-3}, and saturated samples with volumetric water content around 0.62–0.74 cm^3 cm^{-3}. In addition, these data indicate that the sensor could distinguish the three different soil moisture levels of dry to moderately wet, field capacity, and saturated. Hence, the minimum and maximum values of the readily available soil moisture (permanent wilting percentage and field capacity) can be maintained using the SKU:SEN0193 sensor in organic-rich soil. Interestingly, the calibration curve obtained for the laboratory soil mixture did not show a clear difference among the three moisture categories. The laboratory soil mixture was prepared by mixing organic-rich soil and vermiculite. It has been reported that also high-cost sensors such as TDR do not accurately predict the soil moisture content of soil substitutes based on mineral media (vermiculite, tuff, perlite) [30]. Furthermore, it is known that less expensive capacitance moisture sensors operate in low frequencies and are thereby more sensitive to effects of soil textural variances and salinity [4,31]. Thus, the low sensitivity of the SKU:SEN0193 sensor for the soil moisture content of the laboratory soil mixture could be due to the high mineral content of the laboratory soil mixture.

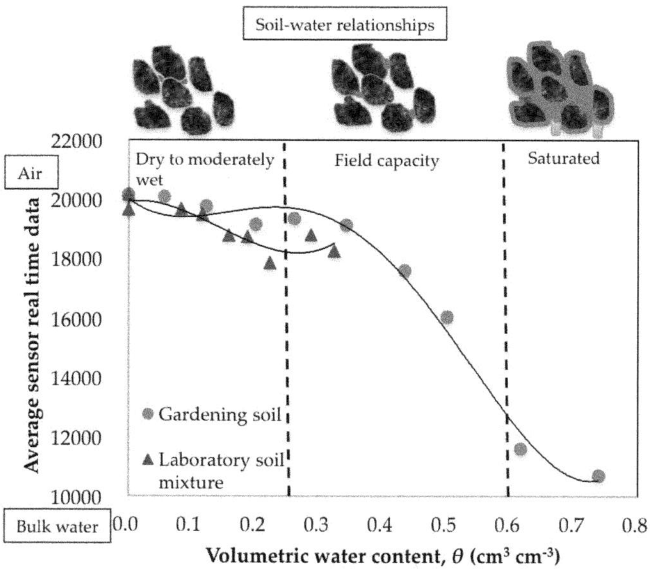

Figure 3. Average raw count as a function of the volumetric water content Θ of the tested porous materials (n = 20). The raw count varies with the dielectric constants of the bulk water and air and ranges between 10,000 and 20,000 according to the sensor output.

3.3 Soil-Specific Calibration Function

The derived soil-specific calibration function for the organic-rich soil accurately predicted the soil moisture values of known samples, while that for the laboratory soil mixture did not. Therefore, further studies were carried out using organic-rich soils.

The derived calibration function (polynomial function) of the SKU:SEN0193 sensor for the organic-rich soil is (R2 = 0.98):

$$\theta = 13.248 - 2.576 \times 10^{-3}\, raw + 1.726 \times 10^{-7}\, raw^2 - 3.839 \times 10^{-12}\, raw^3 \tag{6}$$

3.4 Data Validation with the Commercially Available SM-200 Sensor and the Low-Cost SKU:SEN0193 Capacitive Sensor

Soil samples were analyzed for their soil water content using the commercial available SM-200 sensor and the low-cost SKU:SEN0193 capacitive sensor. The volumetric water contents derived from the polynomial function of the SKU:SEN0193 sensor, the polynomial function of the SM-200 sensor for organic soil (manufacturer's function), and the soil-specific calibration (the function obtained from the measured coefficients) are plotted in Figure 4 as a function of the measured volumetric water content. The root-mean-square error (RMSE) was calculated for the soil water contents derived by the different calibration (polynomial) functions. The RMSE values of the SKU:SEN0193 sensor function, the SM-200 manufacturer's function, and the SM-200 soil-specific calibration function were 0.09, 0.07, and 0.06 $cm^3\ cm^{-3}$, respectively, for the samples with dry to saturated levels of soil water content. The soil water content at 0.45 $cm^3\ cm^{-3}$ (74% gravimetric water content) was predicted precisely by the derived polynomial functions of both sensors (Figure 4a). Hence, data prediction at the field capacity (a gravimetric water content of 60–80%) was performed by measuring the moisture-adjusted samples (Figure 4b). The RMSE was calculated as before. The RMSE values of the SKU:SEN0193 sensor function, the SM-200 manufacturer's function, and the SM-200 soil-specific calibration function were 0.05, 0.08, and 0.03 $cm^3\ cm^{-3}$, respectively, in the field capacity range. The data prediction ability of the sensor appeared similar to that of the SM-200 sensor (with soil-specific calibration) at a lower investment cost.

This study highlights the potential use of the low-cost SKU:SEN0193 capacitive moisture sensor for predicting the water content of soil.

Prior to deploying the system in the field, it is necessary to evaluate the performance of the derived function for continuous operation. Thus, soil moisture loss from a sample in the field capacity range was evaluated using the derived polynomial function (gravimetric basis) (Figure 5). The experiment was performed for 12 h, from day to nighttime, under laboratory conditions. The room temperature and the relative humidity of the room ranged from 25 to 26°C and 72 to 75%, respectively, during the measurements. The 3D scatter-plot in Figure 5 illustrates the soil moisture loss with time. It shows a gradual decrease in soil moisture during 12 h of measurements.

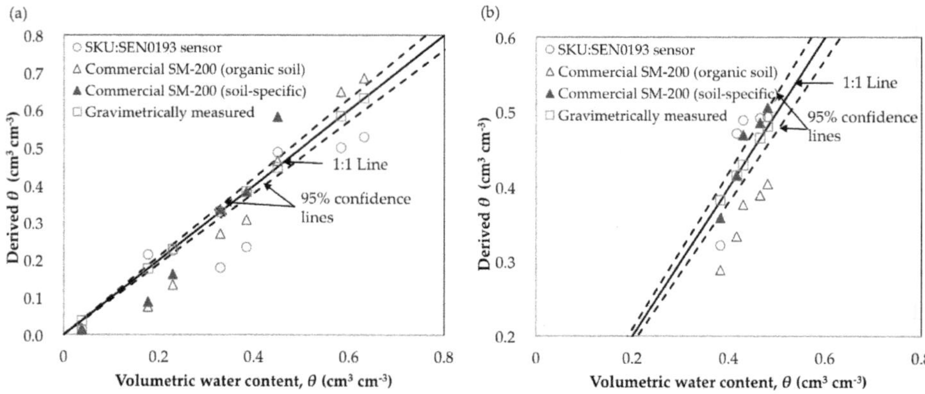

Figure 4. The samples' soil water contents θ derived from the polynomial function of the SKU:SEN0193 sensor, the polynomial function of the SM-200 sensor for organic soil (manufacturer's function), and the soil-specific calibration function as a function of volumetric water content: (a) in the range of soil water contents from dry to saturated, (b) in the range of the field capacity.

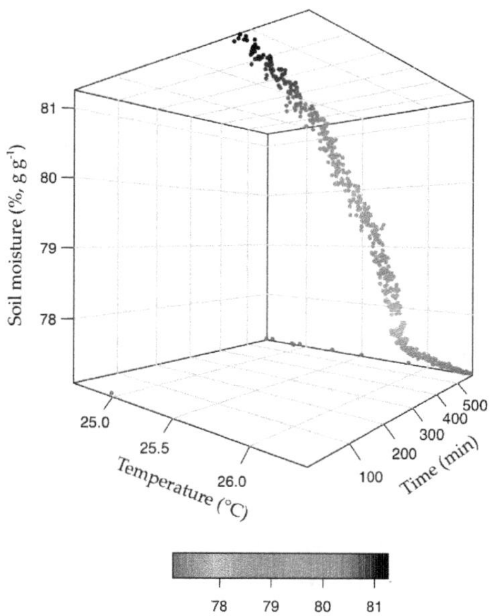

Figure 5. Soil moisture loss with time. The developed calibration function (gravimetric basis) was used to derive the soil moisture loss every minute (Equation (6)).

4. CONCLUSIONS

We evaluated the accuracy and reliability of the SKU:SEN0193 low-cost capacitive soil moisture sensor under laboratory conditions. The developed soil-specific calibration function for gardening soil performed satisfactorily during the sensor validation procedure for the prediction of soil water content. Furthermore, our data suggest that the soil-specific calibration function of the SKU:SEN0193 sensor can be used to predict the soil water contents in three different ranges of soil moisture. Hence, it can be used to maintain the minimum and maximum values of readily available soil moisture in indoor systems. In contrast, the SKU:SEN0193 capacitive soil moisture sensor did not perform acceptably for the laboratory soil mixture in predicting the soil moisture content. This result suggests that the accuracy of the sensor depends on the soil mixture constituents.

However, it is necessary to investigate the effects of soil temperature, bulk density of the soil profile, and salinity levels on the accuracy of the sensor measurements. Further studies should be undertaken to assess its behavior in a real working scenario under field conditions.

References

1. Kramer, P.J.; Boyer, J.S. Soil and water. In *Water Relations of Plants and Soils*; Academic Press: San Diego, CA, USA, 1995; pp. 84–114. Available online: http://udspace.udel.edu/handle/19716/2830. (accessed on 15 December 2018).
2. Zazueta, F.S.; Xin, J. *Soil Moisture Sensors*; Florida Cooperative Extension Service, Institute of Food and Agricultural Science; University of Florida: Gainesville, FL, USA, 1994.
3. Bogena, H.R.; Huisman, J.A.; Schilling, B.; Weuthen, A.; Vereecken, H. Effective calibration of low-cost water content sensors. *Sensors* **2017**, 17, 208. [CrossRef]
4. Vaz, C.M.P.; Jones, S.; Meding, M.; Tuller, M. Evaluation of Standard Calibration Functions for Eight Electromagnetic Soil Moisture Sensors. *Vadose Zone J.* **2013**, 12, 1–16. [CrossRef]
5. Terzic, E.; Terzic, J.; Nagarajah, R.; Alamgir, M. Capacitive sensing technology. In *Neural Network Approach to Fluid Quantity Measurement in Dynamic Environments*; Springer: London, UK, 2012.
6. Robbins, A.; Miller, W. *Circuit Analysis: Theory and Practice*; Delmar: Albany, NY, USA, 2000.
7. Fen-Chong, T.; Fabbri, A.; Guilbaud, J.; Coussy, O. Determination of liquid water content and dielectric constant in porous media by the capacitive method. *Comptes Rendus Mécanique* **2004**, 332, 639–645. [CrossRef]
8. Kaatze, U. The dielectric properties of water in its different states of interaction. *J. Solution Chem.* **1997**, *26*, 1049–1112. [CrossRef]

9. Ganjegunte, G.K.; Sheng, Z.; Clark, J.A. Evaluating the accuracy of soil water sensors for irrigation scheduling to conserve freshwater. *Appl. Water Sci. 2012*, 2, 119–125. [CrossRef]

10. Parvin, N.; Degré, A. Soil-specific calibration of capacitance sensors considering clay content and bulk density. *Soil Res.* **2016**, 54, 111–119. [CrossRef]

11. Payero, J.O.; Mirzakhani-Nafchi, A.; Khalilian, A.; Qiao, X.; Davis, R. Development of a low-cost Internet-of-Things (IOT) system for monitoring soil water potential using Watermark 200SS sensors. *Adv. Internet Things* **2017**, 7, 71–86. [CrossRef]

12. Archer, N.A.L.; Rawlins, B.R.; Marchant, B.P.; Mackay, J.D.; Meldrum, P.I. Approaches to calibrate in-situ capacitance soil moisture sensors and some of their implications. *SOIL Discuss.* **2016**. [CrossRef]

13. Fares, A.; Awal, R.; Bayabil, H.K. Soil water content sensor response to organic matter content under laboratory conditions. **Sensors 2016**, 16, 1239. [CrossRef]

14. Mittelbach, H.; Lehner, I.; Seneviratne, S.I. Comparison of four soil moisture sensor types under field conditions in Switzerland. **J. Hydrol. 2012**, 430–431, 39–49. [CrossRef]

15. Ferrarezi, R.S.; Dove, S.K.; Van Iersel, M.W. An automated system for monitoring soil moisture and controlling irrigation using low-cost open-source microcontrollers. *HortTechnology* **2015**, 25, 110–118. [CrossRef]

16. Cardenas-Lailhacar, B.; Dukes, M. Effect of temperature and salinity on the precision and accuracy of landscape irrigation soil moisture sensor systems. *J. Irrig. Drain. Eng.* **2015**, 141, 7. [CrossRef]

17. Inoue, M.; Ould Ahmed, B.A.; Saito, T.; Irshad, M.; Uzoma, K.C. Comparison of three dielectric moisture sensors for measurement of water in saline sandy soil. *Soil Use Manag.* **2008**, 24, 156–162. [CrossRef]

18. Nemali, K.S.; Montesano, F.; Dove, S.K.; Van Iersel, M.W. Calibration and performance of moisture sensors in soilless substrates: ECH2O and Theta probes. *Sci. Hortic.* **2007**, 112, 227–234. [CrossRef]

19. Kojima, Y.; Shigeta, R.; Miyamoto, N.; Shirahama, Y.; Nishioka, K.; Mizoguchi, M.; Kawahara, Y. Low-cost soil moisture profile probe using thin-film capacitors and a capacitive touch sensor. *Sensors* **2016**, 16, 1292. [CrossRef] [PubMed]

20. Puértolas, J.; Alcobendas, R.; Alarcón, J.J.; Dodd, I.C. Long-distance abscisic acid signalling under different vertical soil moisture gradients depends on bulk root water potential and average soil water content in the root zone. *Plant Cell Environ.* **2013**, 36, 1465–1475. [CrossRef]

21. Kodešová, R.; Kodeš, V.; MRáz, A. Comparison of two sensors ECH2O EC-5 and SM200 for measuring soil water content. *Soil Water Res.* **2011**, 6, 102–110. [CrossRef]

22. ASTM D 2974. *Standard Test Methods for Moisture, Ash, and Organic Matter of Peat and Organic Soils;* ASTM International: West Conshohocken, PA, USA, 2007.

23. Ding, S.; Zhang, B.; Qin, F. Arabidopsis RZFP34/CHYR1, a ubiquitin E3 ligase, regulates stomatal movement and drought tolerance via SnRK2.6-mediated phosphorylation. *Plant Cell.* **2015**, 27, 3228–3244. [CrossRef] [PubMed]

24. Ma, Q.; Xia, Z.; Cai, Z.; Li, L.; Cheng, Y.; Liu, J.; Nian, H. GmWRKY16 enhances drought and salt tolerance through an ABA-mediated pathway in *Arabidopsis thaliana. Front Plant Sci.* **2019**, 9, 1979. [CrossRef]

25. Quarles, S.D. How to implement I2C serial communication using Intel MCS-51 microcontrollers; 1993. Intel Corporation: P.O. Box 7641, Mt. Prospect, IL 60056-7641. Available online: http://electro8051.free.fr/I2C/ 27231901.pdf. (accessed on 8 March 2018).

26. Rosenbaum, U.; Huisman, J.A.; Weuthen, A.; Vereecken, H.; Bogena, H.R. Sensor-to-sensor variability of the ECH2O, EC-5, TE, and 5TE sensors in dialectic liquids. *Vadose Zone J.* **2010**, 9, 181–186. [CrossRef]

27. ASTM D 2216. *Standard Test Method for Laboratory Determination of Water (Moisture) Content of Soil, Rock, and Soil-Aggregated Mixture;* ASTM International: West Conshohocken, PA, USA, 2010.

28. Delta-T. *Devices Ltd. User Manual for the SM200 Soil Moisture Sensor;* SM200-UM-1; 1 May 2006; Delta-T. Devices Ltd.: Cambridge, UK, 2006.

29. Hillel, D. *Introduction to Environmental Soil Physics*; Academic press, Elsevier Science: San Diego, CA, USA, 2004; pp. 14–15. Available online: https://dewagumay.files.wordpress.com/2011/12/environmental-soil-physics.pdf (accessed on 16 December 2018).

30. Da Silva, F.F.; Wallach, R.; Polak, A.; Chen, Y. Measuring water content of soil substitutes with time-DOMAIN reflectometry (TDR). *J. Am. Soc. Hortic. Sci.* **1998**, 123, 734–737. [CrossRef]

31. Kizito, F.; Campbell, C.S.; Campbell, G.S.; Cobos, D.R.; Teare, B.L.; Carter, B.; Hopmans, J.W. Frequency, electrical conductivity and temperature analysis of a low-cost capacitance soil moisture sensor. *J. Hydrol.* **2008**, 352, 367–378. [CrossRef]

Chapter 4

DELINEATION OF AGRICULTURAL DRAINAGE PIPE PATTERNS USING GROUND PENETRATING RADAR INTEGRATED WITH A REAL-TIME KINEMATIC GLOBAL NAVIGATION SATELLITE SYSTEM

1. INTRODUCTION: AN OVERVIEW

A 1985 economic survey showed that the states comprising the Midwest U.S. (Illinois, Indiana, Iowa, Ohio, Minnesota, Michigan, Missouri, and Wisconsin) by that year had approximately 12.5 million ha that contained subsurface drainage systems, with cropland accounting for the large majority of areas possessing these buried drainage pipe networks [1]. In the more than 30 years that have elapsed since the 1985 survey, a very sizeable amount of additional agricultural drainage pipe has been installed throughout the Midwest U.S. In order to improve crop yields, farmers within this region often need to repair drain lines that are not functioning properly or install new drain lines between the old ones to improve soil water removal efficiency.

Whether for system repairs or efficiency improvements, locations of the preexisting drain lines are required; however, in most cases, a map of the original subsurface drainage system installation is no longer available.

Present methods of locating drain lines involve the use of hand-held tile probes or heavy trenching equipment. Finding drainage pipes with a tile probe (Figure 1a) is both time consuming and extremely tedious. With a tile probe, it is sometimes difficult at a particular spot beneath the ground surface to distinguish a clay tile drainage pipe from the large stones that are commonly present in glacially derived soils of the Midwest U.S. A tile probe also easily penetrates through a corrugated plastic tubing (CPT) drainage pipe, thereby making the presence of the CPT drainage pipes difficult to ascertain, while at the same time potentially causing pipe collapse problems. Locating drainage pipes with heavy trenching equipment (Figure 1b) is effective, but causes costly pipe damage that has to be repaired before the trench is backfilled. Consequently, an effective, efficient, and noninvasive (i.e., nondestructive) method is needed for mapping subsurface drainage systems in agricultural fields.

Figure 1. Present methods for finding drainage pipes: (a) hand-held tile probe; (b) heavy trenching equipment with inset photo showing an example of the drainage pipe damage caused during this excavation operation.

In a preliminary study [2], an unmanned aircraft system (UAS) with a thermal infrared camera exhibited promise in regard to drainage pipe mapping at an agricultural field in central Ohio, USA; however, the technology is now only just beginning to be rigorously evaluated for this particular application. Prior research has demonstrated the feasibility of using ground penetrating radar (GPR) in agricultural settings to locate buried clay tile and CPT drainage pipe over a range of soil types and shallow hydrologic conditions [3–5]. Additional research has shown that GPR, under certain circumstances, can isolate the location of drainage pipe obstructions [6]. The strength of the GPR drainage pipe response for specific soil moisture conditions was found to depend on the orientation of the GPR antennas relative to the drain line directional trend [7].

All prior GPR drainage pipe detection research focused on testing the feasibility of this geophysical technology on small agricultural experiment plots. The practicality of employing GPR for determining drain line patterns in larger farm fields has not been adequately assessed. Ground penetrating radar integrated with a Real-Time Kinematic (RTK) Global Navigation Satellite System (GNSS) may be an effective, efficient, and noninvasive (i.e., nondestructive) means of mapping drain lines within agricultural fields. Consequently, GPR-RTK/GNSS was evaluated at three agricultural fields. The overall process involved mapping the latitude and longitude coordinates of points in the field beneath which drainage pipes were potentially detected and then using this map to provide insight on drain line patterns. The guiding hypothesis for this research can be stated; "GPR integrated with RTK/GNSS is an effective, efficient, and noninvasive tool for delineating drain line patterns in agricultural fields." The results of this study confirmed this hypothesis by successfully demonstrating GPR-RTK/GNSS mapping of complex subsurface drainage systems at three agricultural field sites.

2. MATERIALS AND METHODS

2.1 Site Descriptions

Testing of GPR integrated with an RTK/GNSS for drainage pipe mapping was carried out at three agricultural field sites. Two of the sites are located in Beltsville, MD, USA, and the third site is near Columbus, OH, USA. Aerial images of the three sites, with soil map overlays, are shown in Figure 2. Note: Throughout this article, aerial images obtained via Google Earth (Google LLC., Mountain View, CA, USA). were used as a base map to overlay soils and GPR information. Soil map overlays were obtained via the SoilWeb Earth App (UC Davis, California Soil Resource Lab, Davis, CA, USA). Descriptions for each test site are provided as follows.

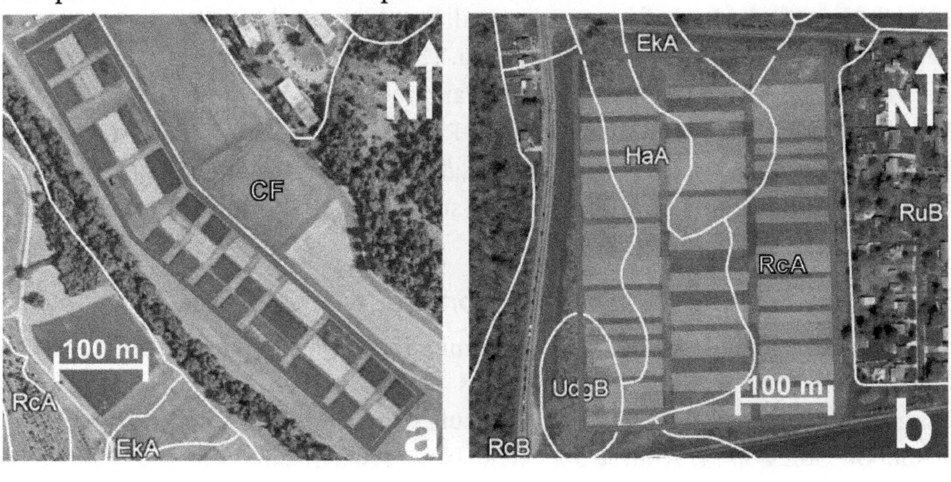

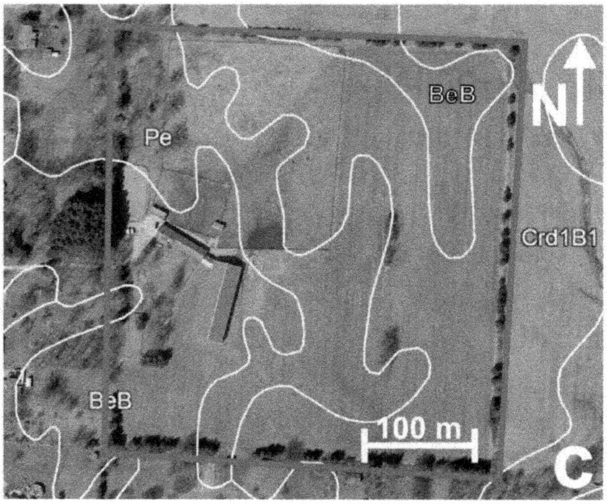

Figure 2. Aerial images with soil map overlays of the three agricultural sites that were investigated with GPR-RTK/GNSS: (a) Site 1; (b) Site 2; (c) Site 3. Purple lines mark site boundaries.

2.1.1. Site 1 (Latitude: 39.012555, Longitude: -76.940204)

Site 1 is located at the U.S. Department of Agriculture (USDA)—Agricultural Research Service (ARS)—Beltsville Agricultural Research Center (BARC). Site 1 is comprised of numerous small experiment plots (Figure 2a). The total area of this agricultural field is 5.6 ha. Cordorus and Hatboro soils, CF, cover Site 1. Both the Cordorus (*mesic Fluvaquentic Dystrudept*) and Hatboro (*mesic Fluvaquentic Endoaquept*) are classified as loams.

2.1.2. Site 2 (Latitude: 39.025426, Longitude: -76.899118)

Site 2 is also located at BARC and is approximately 4 km northeast of Site 1. Site 2, as with site 1, is comprised of numerous small experiment plots (Figure 2b). The total area of the Site 2 agricultural field is 11.3 ha. The soils that cover Site 2 include Elkton silt loam, EkA, (*mesic Typic Endoaquults*), Hammonton loamy sand, HaA, (*mesic Aquic Hapludults*), Russett—Christiana complex, RcA or RcB, (Russett loam—mesic A*quic Hapludults, Christiana—mesic Aquic Hapludults*), and Udorthents, UdgB, (reclaimed gravel pits). Udorthents cover only a very small part of Site 2, just in the southwest corner.

2.1.3. Site 3 (Latitude: 40.029482, Longitude: -82.727897)

Site 3 is a horse boarding, riding, and training facility located near Columbus, Ohio. The northeastern portion of the site has fenced-in areas for horses to exercise and graze (Figure 2c). The building structures contain offices, boarding stalls, and an indoor riding area. A large hay field is present within southern and western portions of the site. The total area of Site 3 is 13.0 ha. The soils in this agricultural setting include Bennington silt loam, BeB, (*mesic Aeric Epiaqualfs*), Cardington silt loam, Crd1B1, (*mesic Aquic Hapludalfs*), and Pewamo silty clay loam, Pe, (*mesic Typic Argiaquolls*).

2.2. Equipment

A Sensors & Software Inc. (Mississagua, ON, Canada) Noggin® GPR system having 250 MHz antennas (Figure 3a,c) was employed to detect buried drainage pipes. The antennas used were shielded and had a frequency range from 125 to 375 MHz, with a center frequency of 250 MHz. Previous research [3–7] indicated that 250 MHz center frequency antennas worked best for finding clay tile and CPT drainage pipes in agricultural settings. Integration of RTK/GNSS with the GPR system allowed accurate latitude and longitude coordinates to be obtained at field locations where potential drainage pipes were detected. In order to obtain RTK/GNSS coordinates at Sites 1 and 2, a Topcon Corporation (Itabashi, Japan) GNSS (Global Positioning System—GPS and Globalnaya Navigazionnaya Sputnikovaya Sistema—GLONASS) FC-200 rover receiver (Figure 3a) and Topcon Corporation GNSS (GPS and GLONASS) HiPer XT base station receiver (Figure 3b) were utilized. At Site 3, a Topcon Corporation GRS-1 dual-frequency, GNSS (GPS and GLONASS) rover receiver and Topcon Corporation PG-S1 hand-held controller were employed together (Figure 3c) with a virtual base station set-up through the Ohio Department of Transportation's (ODOT) network of Continuously Operating Reference Stations (CORS) [8]. Time domain reflectometry (TDR) soil water content values were collected using a Spectrum Technologies, Inc. (East Plainfield, Illinois) Field Scout TDR-300 with 20 cm waveguides (Figure 3d). The TDR water content measurements were utilized to determine soil dielectric constant values [9], that in turn were used to calculate preliminary soil radar velocities [10,11], which were then employed to convert GPR two-way travel times to depth values.

2.3 Field Data Collection

Almost all of the GPR-RTK/GNSS data for Site 1 was collected on 5 April 2017, with just a few additional short GPR transects carried out on 24 January 2018. All GPR-RTK/GNSS data for Site 2 was obtained on 25 January 2018. For Site 3, linear GPR transects were completed on 6 November 2017, followed by the spiral/serpentine GPR transects on 5 February 2018. At each site, between four to six hours total was needed to complete the GPR-RTK/GNSS surveys. The GPR equipment settings included a 5 cm distance between signal trace measurement points along a transect (i.e., station interval = 5 cm). At each measurement point on the GPR transect, 16 signal traces were collected and averaged (i.e., stacking = 16). The GPR two-way travel time for each signal trace was set to provide a depth of investigation of 1.5 m (Site 3) or 2.0 m (Sites 1 and 2). The GPR signal trace two-way travel time needed to achieve 1.5 m or 2.0 m investigations depths were calculated from soil radar velocity values based on TDR measurements.

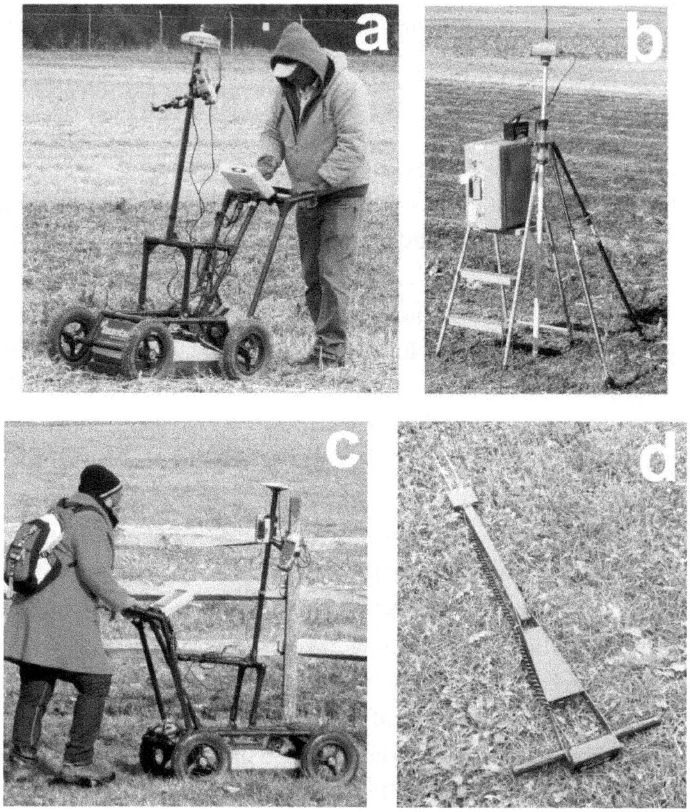

Figure 3. Equipment used in investigation: (a) Noggin GPR system (250 MHz antennas) with GNSS FC-200 rover receiver; (b) GNSS HiPer XT base station receiver; (c) Noggin GPR system (250 MHz antennas) with GNSS GRS-1 rover receiver and PG-S1 hand-held controller; (d) Field Scout TDR-300.

One main component of the Noggin® GPR unit is the Digital Video Logger (DVL), which is used to input equipment settings (station interval, stacking, radar velocity, depth of investigation, etc.) and store GPR-RTK/GPS data. The DVL also has a display screen that provides a real-time GPR cross-section view of the subsurface as data is being collected along a measurement transect (Figure 3a,c). Consequently, the DVL can provide an almost instant indication of a buried drainage pipe. Suspected locations of the buried drainage pipes can then be flagged in the field, with flagged locations that line-up, pointing to the presence and trend of a drain line. Additionally, spiral/serpentine GPR transects or segments of transects can be carried out, on the spot, based on DVL indications of a buried drainage pipe, directly for the purpose of confirming drain line presence and determining its directional trend.

Site maps showing GPR measurement transects (yellow lines—obtained through RTK/GNSS data) are provided in Figure 4. The physical GNSS base station locations at Sites 1 and 2 are marked with orange square symbols (Figure 4a,b). (Again, a virtual base station was employed at Site 3, so there is no orange square symbol depicted in Figure 4c) At Sites 1 and 2, there were essentially two sets of GPR transects, one that coincided or paralleled the longest side boundaries at the sites and then a second set that was oriented perpendicular to the first set. At Site 2, most of the GPR transects contained spiral path segments that provided insight on drain line directional trends (Figure 4b). At Site 3, linear GPR transects around the perimeter of the property and within the fenced-in grazing/exercise areas or hay field were used for initial indications of where drainage pipes might exist. These linear GPR transects at Site 3 were then followed-up with spiral or serpentine GPR transects that were again used to determine the trend and extent of individual drain lines.

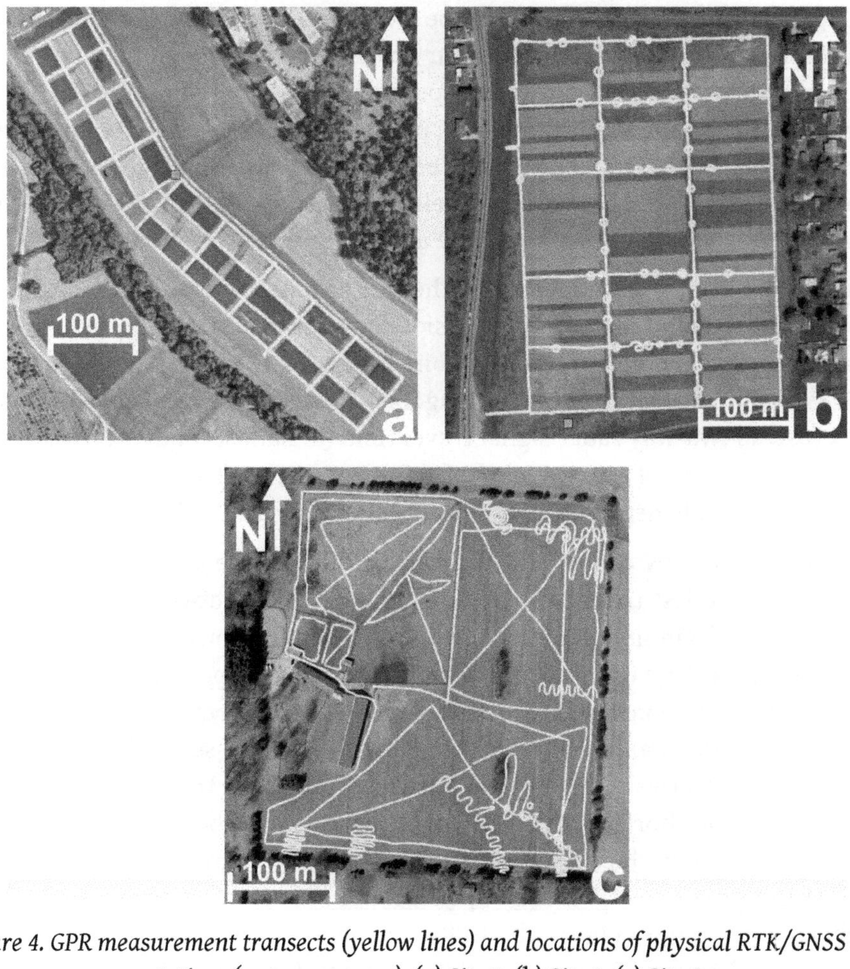

Figure 4. GPR measurement transects (yellow lines) and locations of physical RTK/GNSS base stations (orange squares): (a) Site 1; (b) Site 2; (c) Site 3.

2.4 Data Processing and Interpretation

The principle of GPR operation is conceptually simple and involves directing an electromagnetic radio energy (radar) pulse into the subsurface, followed by measurement of the elapsed time taken by the signal as it is travels downwards from the transmitting antenna, partially reflects off a buried feature, and is eventually returned to the surface, where it is recorded by a receiving antenna. For each point of measurement along a GPR transect, reflections from different depths produce a signal trace, which is a function of the radar wave amplitude (and energy) versus two-way travel time. (Note: As previously indicated, soil water content measurements can be used to convert two-way travel times to depth values.) Antenna frequency, soil moisture conditions, clay content, salinity, and the amount of iron oxide present all have a substantial influence on the distance beneath the surface to which the radar signal penetrates. Differences in the dielectric constant across a buried feature discontinuity govern the amount of radar energy that reflects off the buried feature and then returns to surface to be recorded by the receiving antenna. Previous research [5] shows that the GPR drainage pipe response (i.e., amount of radar energy reflected from a buried pipe) does not depend on the type of pipe (clay tile or CPT) but rather the difference between the dielectric constant of the soil surrounding the pipe versus the dielectric constant of the air and/or water inside the pipe.

A GPR profile (i.e., cross-section) of the subsurface beneath a GPR transect is generated by sequentially combining, side-by-side, one after another, the signal traces obtained at each measurement point along the transect. The horizontal axis on a GPR profile represents distance along the transect (in meters), while the vertical axis represents two-way radar signal travel time (in nanoseconds) and/or depth (in meters). With depths of investigation of 1.5 or 2.0 m, the GPR profiles generated for this study essentially depict GPR responses only within the soil profile.

Figure 5 depicts the two types of GPR profile drainage pipe responses. Where there is a somewhat modest to large angle (i.e., 15° < x° < 90°) between the GPR transect orientation and drain line directional trend, the GPR response is that of an upside-down U-shaped feature (i.e., reflection hyperbola), and the position for the top of the drainage pipe corresponds with the apex of the reflection hyperbola (Figure 5a). The drainage pipe reflection hyperbola is horizontally compressed if the angle is closer to 90°, while alternatively, if the angle is closer to 15°, then the reflection hyperbola becomes spread out horizontally. It is important to note other buried features, such as large stones can produce a reflection hyperbola, but in map view, the location of these features are isolated, while mapped locations of actual drainage pipe reflection hyperbolas form a line. Where the GPR transect is essentially over top and along trend (x° < 15°) of a drain line, the GPR response formed is that a banded linear feature, with the position for the top of the drainage pipe corresponding to the top of the banded linear feature (Figure 5b). Obtaining the banded linear GPR drainage pipe response is fairly uncommon because this response requires, without any prior knowledge of drain line locations and directional trends, that a GPR transect just happens to be oriented over the top and along the trend of a drain line.

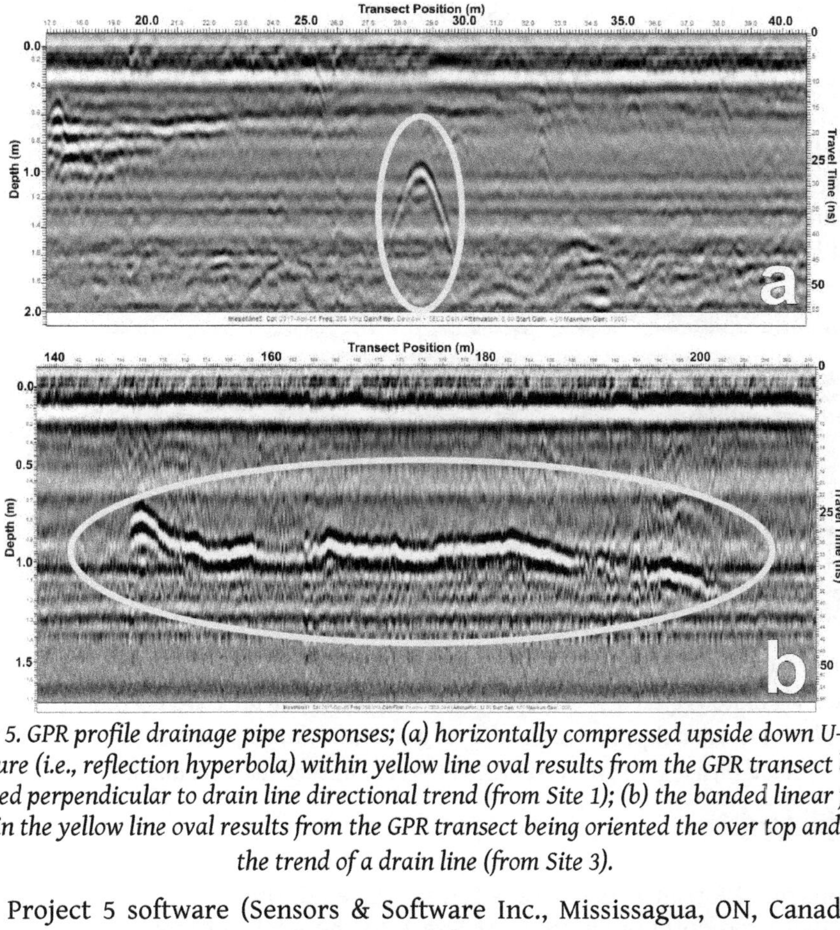

Figure 5. GPR profile drainage pipe responses; (a) horizontally compressed upside down U-shaped feature (i.e., reflection hyperbola) within yellow line oval results from the GPR transect being oriented perpendicular to drain line directional trend (from Site 1); (b) the banded linear feature within the yellow line oval results from the GPR transect being oriented the over top and along the trend of a drain line (from Site 3).

EKKO Project 5 software (Sensors & Software Inc., Mississagua, ON, Canada) was used to process and interpret the GPR-RTK/GNSS data acquired at Sites 1, 2, and 3, starting with the generation of GPR profiles from the GPR measurement transects. EKKO Project 5 reflection hyperbola curve fitting procedures, employed only in cases where the GPR transect was perpendicular to the drain line directional trend, allowed soil radar velocities to be refined at each site (from the original TDR measurements), and in turn, improve the accuracy of GPR profile depth scales. Soil radar velocities, refined via EKKO Project 5, were 0.072 m/ns on 5 April 2017 and 0.063 m/ns on 24 January 2018 at Site 1, 0.067 m/ns on 25 January 2018 at Site 2, and 0.060 m/ns on both 6 November 2017 and 5 February 2018 at Site 3. The only processing steps used to produce GPR profiles via EKKO Project 5 were (1) application of a signal saturation correction filter (i.e., Dewow) to remove slowly decaying low frequency noise and (2) utilization of a spreading and exponential calibrated compensation gain function to amplify potential GPR drainage pipe responses. An interpretation module in EKKO Project 5 allowed potential drainage pipe responses in a GPR profile to be marked (Figure 6). With GPR and RTK/GNSS data collected together, EKKO Project 5 was then employed to save a spreadsheet file of the latitudes, longitudes, and depths corresponding to the marked potential drainage pipe responses. Furthermore, because GPR and RTK/GNSS data were collected concurrently, EKKO Project 5 was then able to generate a KMZ file that stored a site map of the GPR transects along with the marked potential drainage pipe locations. Opening this KMZ file in Google Earth allowed GPR measurement transects and the possible pipe locations to be overlaid on an aerial image.

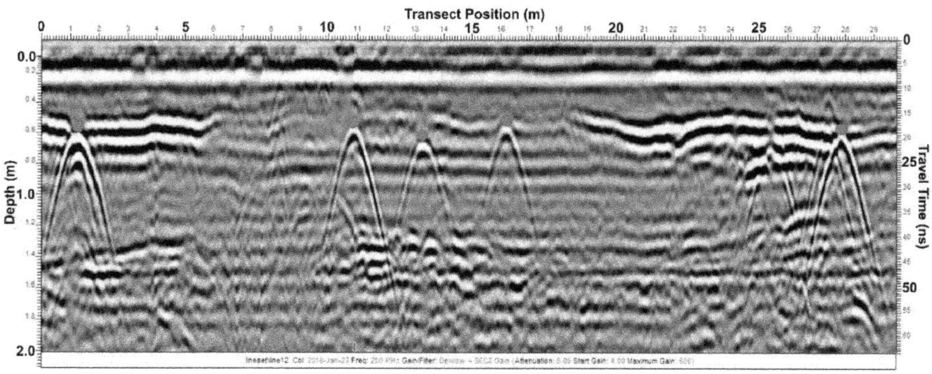

Figure 6. GPR profile from Site 1 with potential reflection hyperbola drainage pipe responses marked via EKKO Project 5. Because GPR and RTK/GNSS data are collected concurrently, the latitude, longitude, and depth corresponding to a potential drainage pipe response mark (red dot) could then all be saved (via EKKO Project 5) into spreadsheet and KMZ files.

3. RESULTS

Locations on GPR transects where potential buried drainage pipes were detected are shown (with red dots) in Figure 7. At Site 1, to the west of the white dashed line in Figure 7a, the overall soil profile had a relatively homogeneous appearance regarding GPR response (as exhibited in GPR profiles). Conversely, to the east of the dashed line, the overall soil profile had a far more heterogeneous appearance regarding GPR response. The overall GPR appearance of the soil profile (i.e., homogeneous vs. heterogeneous) definitely impacted Site 1 drainage pipe detection. In the more homogeneous soils on the west side of Site 1, reflection hyperbola potential drainage pipe responses were fairly easy to isolate within GPR profiles (Figure 8a), while in the more heterogeneous soils on the east side of Site 1, isolating the reflection hyperbola potential drainage pipe responses from the surrounding clutter was much more difficult (Figure 8b).

Figure 7. GPR measurement transects (yellow lines) and potential drainage pipe locations (red dots): Site 1; (b) Site 2; (c) Site 3.

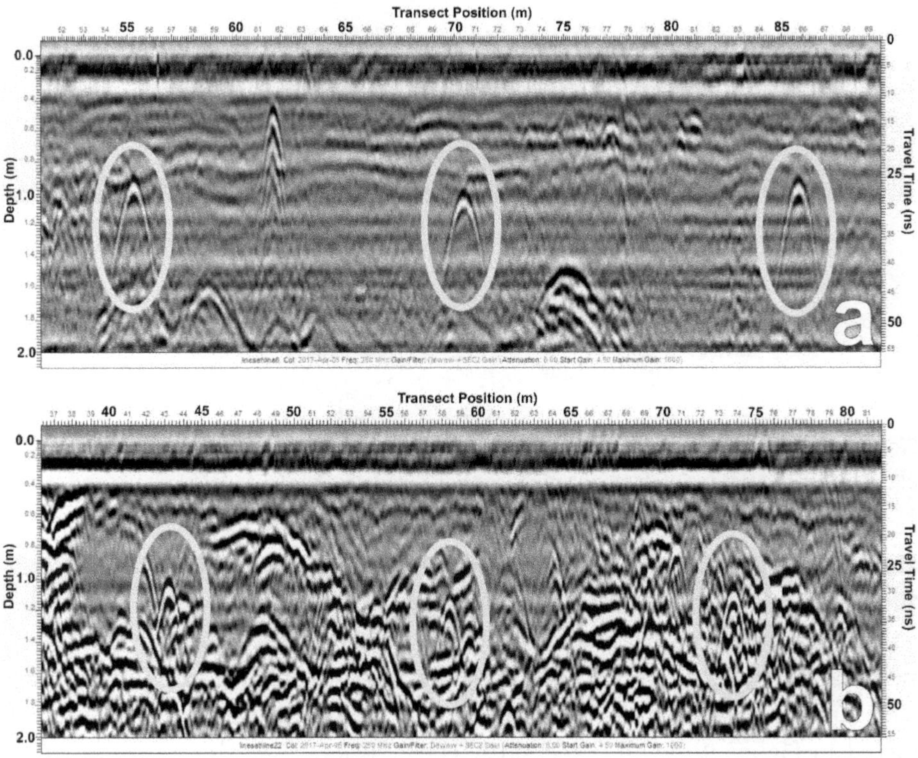

Figure 8. GPR profiles showing reflection hyperbola potential drainage pipe responses highlighted by yellow line ovals; (a) homogeneous soils on west side of Site 1; (b) heterogeneous soils on east side of Site 1. The reflection hyperbola potential drainage pipe responses are far easier to discern in the homogeneous soils as opposed to the heterogeneous soils.

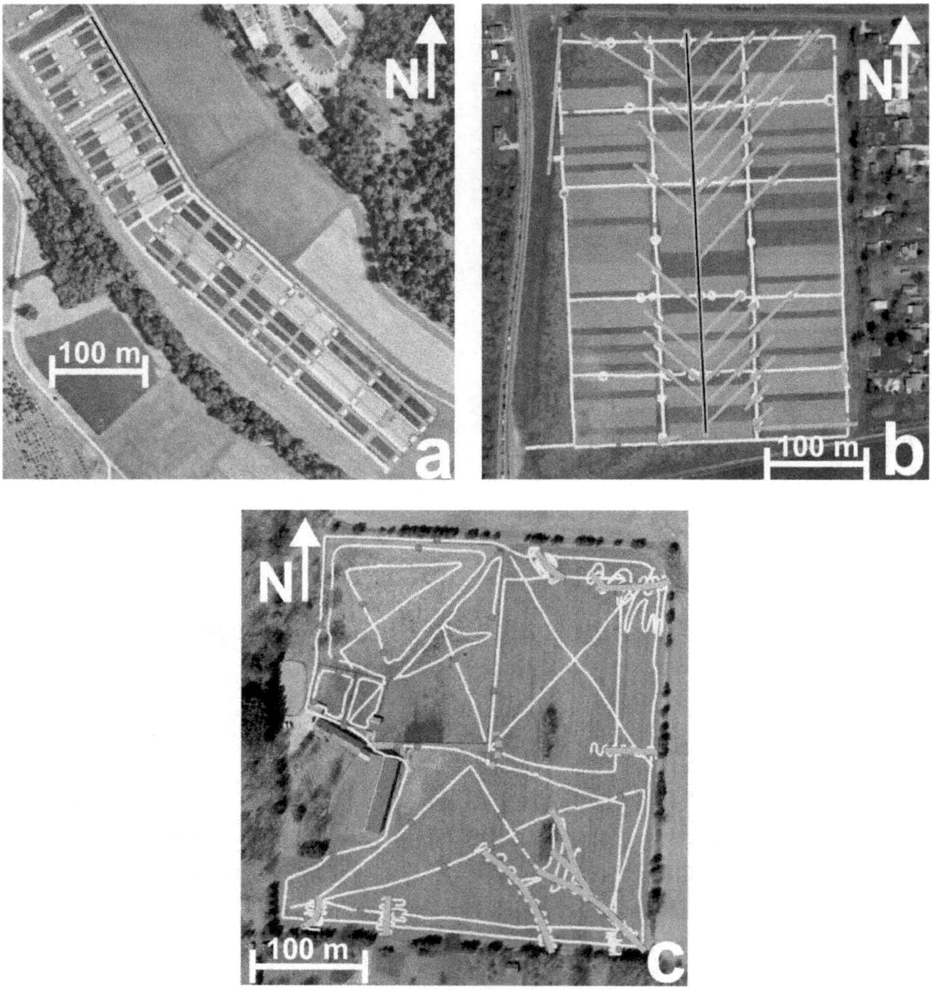

Figure 9. Interpreted drain line patterns (blue lines) overlaid on GPR measurement transects (yellow lines) and potential drainage pipe locations (red dots): (a) Site 1; (b) Site 2; and (c) Site 3. A thin black line within a thicker blue line indicates a main collector pipe.

Drain lines were interpreted to exist in places where mapped locations of potential drainage pipe responses clearly followed a line. Again, a potential drainage pipe location that is isolated likely implies that the corresponding GPR profile reflection hyperbola response is not due to a drainage pipe, but rather some solitary buried object, such as a large stone. The interpreted drain line patterns (blue lines) are depicted in Figure 9 for all three sites. The GPR-RTK/GNSS data collected in this study delineated a complex rectangular drainage pipe system at Site 1 (Figure 9a), with a set of southwest-northeast drain lines present on the west side of the site and a southeast-northwest drain line set present on the east side of the site.

Figure 9b indicates that although not all the drainage pipe present at Site 2 was detected, GPR integrated with RTK/GNSS did find a sufficient number of drain lines to unambiguously determine that a herringbone subsurface drainage system had been installed within this agricultural field. Random drain lines were found at Site 3 (Figure 9c), which are likely to be very old and installed for the purpose of draining wet areas in the hay field.

4. DISCUSSION

At Site 1, GPR profiles (Figure 8) showed soils to the west of the white dashed line in Figure 7a to have a relatively homogeneous GPR response, while soils to the east of the white dashed line exhibited a far more homogeneous GPR response. Prior research [5] has shown that fine-textured silty/clayey soils tend to have a homogeneous appearance in GPR profiles, while coarse-textured sandy soils tend to have a more heterogeneous, or cluttered, appearance in GPR profiles. This finding could indicate for Site 1 that there were generally fine-textured silty/clayey soils west of the white dashed line in Figure 7a and generally coarse-textured sandy soils east of the white dashed line. Textural analysis of soil cores obtained across Site 1 will be required in order to confirm this supposition.

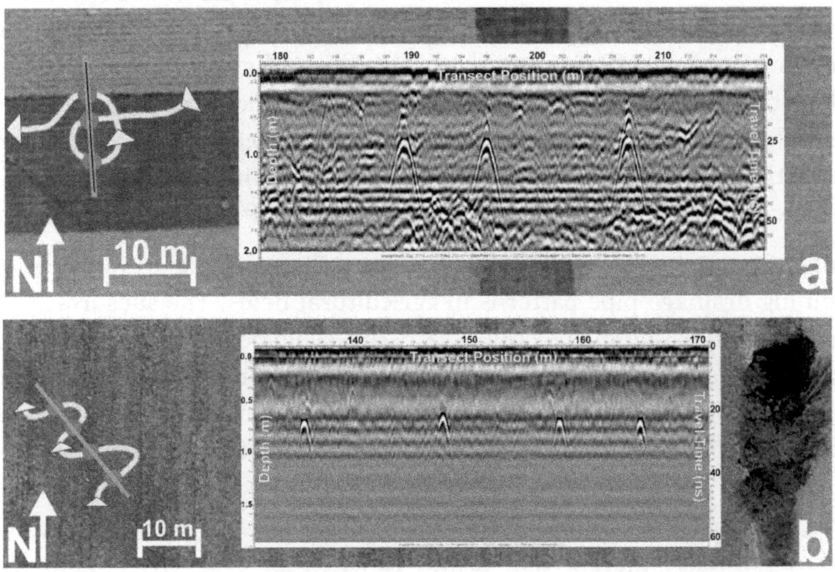

Figure 10. Spiral/serpentine segments of a GPR transect can be employed to determine drain line presence and direction: (a) spiral transect segment example from Site 2 (main collector pipe); serpentine transect segment example from Site 3. Drainage pipe reflection hyperbolas interpreted with EKKO Project 5 (red dots on GPR profile insets) were mapped (red dots) along with transect segments paths (yellow lines with directional arrowheads). The mapped drainage pipe locations are in a line, thereby indicating the presence and direction of a buried agricultural drain line.

When GPR is integrated with RTK/GNSS, spiral or serpentine GPR transects, or spiral/serpentine segments of GPR transects, can be employed where needed to determine drain line directional trends.

Field data collection using GPR integrated with RTK/GNSS to map subsurface drainage was quite efficient, requiring only four to six hours at each of the three sites. This study shows that a dense grid of GPR transects is not needed to determine drainage pipe patterns. Actually, all that is required in a particular agricultural field to find drain lines and determine their directional trends are a few select GPR transects (usually with incorporated spiral/serpentine segments). The efficiency of GPR-RTK/GNSS for subsurface drainage mapping could be improved further over what was obtained in this investigation by using an all-terrain vehicle (ATV) pulling a sled to mount all the GPR and RTK/GNSS equipment instead of mounting the equipment on the push cart shown in Figure 3a,c.

The GPR-RTK/GNSS drainage pipe mapping approach described is applicable for mapping subsurface drainage systems in agricultural fields up to tens of hectares in size. With suitable soil/hydrologic conditions and widespread adoption of this approach over large areas, it might be possible to gain insight on drainage practice intensity at small watershed scales. As indicated, the types of soils present can be important. Soils having very high clay contents oftentimes, but not always, severely limit radar signal penetration into the subsurface, thereby making GPR drainage pipe detection difficult [4,5]. Transient shallow hydrologic conditions also have a substantial impact. Soils that are not too dry or too wet are best for GPR drainage pipe detection, especially good are moist soils surrounding air-filled drainage pipes, which is a common situation occurring a day or more after a large rainfall event [5].

The integration of GPR with RTK/GNSS can therefore be a valuable tool for determining drainage pipe patterns in agricultural fields. The subsurface drainage system information obtained by GPR-RTK/GNSS could be very useful for retrofitting a subsurface drainage system for subirrigation or to improve soil water removal efficiency, with both scenarios requiring installation of new drain lines in between pre-existing drain lines [12–14], which in turn cannot be accomplished without knowing where the preexisting drain lines are located. Knowledge of drain line patterns obtained via GPR-RTK/GNSS could also be used for repairing part of a subsurface drainage system that is not functioning properly. Finding drainage pipes with GPR-RTK/GNSS is certainly an improvement over tile probe and trench excavation methods, because GPR-RTK/GNSS field data collection, compared to these other two methods, is much faster, is better at determining drain line directional trends, and causes no pipe damage. Consequently, in areas where GPR is capable of detecting buried drainage pipes, GPR-RTK/GNSS can supplant tile probe and trench excavation methods for mapping drain line patterns.

5. SUMMARY AND CONCLUSIONS

Ground penetrating radar integrated with a Real-Time Kinematic (RTK) Global Navigation Satellite System (GPR-RTK/GNSS) successfully determined drainage pipe patterns at three agricultural field sites. At one site, a rectangular pattern of drain lines was mapped, at the second site, a herringbone pattern was delineated, and at the third site, random drain lines were found. A novel field data collection approach was developed, taking advantage of GPR-RTK/GNSS, by which spiral or serpentine GPR measurement transects (or spiral/serpentine segments of GPR measurement transects) were employed to establish the presence and direction of drain lines. Using GPR-RTK/GNSS to map drain lines can therefore provide useful information needed for retrofitting or repair of agricultural subsurface drainage systems. Furthermore, in areas where GPR can detect drainage pipes, GPR-RTK/GNSS is a substantial improvement (faster, better able to resolve drain line trends, and causes no pipe damage) over the traditional tile probe and trench excavation methods for finding drainage pipes.

References

1. Pavelis, G.A. Economic survey of farm drainage. In *Farm Drainage in the United States: History, Status, and Prospects, Miscellaneous Publication Number 1455*; Pavelis, G.A., Ed.; U.S. Dept. of Agriculture, Economic Research Service: Washington, DC, USA, 2007; pp. 110–136.
2. Allred, B.; Eash, N.; Freeland, R.; Martinez, L.; Wishart, D. Effective and efficient agricultural drainage pipe mapping with UAS thermal infrared imagery: A case study. *Agric. Water Manag.* **2018**, *197*, 132–137. [CrossRef]
3. Chow, T.L.; Rees, H.W. Identification of subsurface drain locations with ground-penetrating radar. *Can. J. Soil Sci.* **1989**, *69*, 223–234. [CrossRef]
4. Allred, B.J.; Fausey, N.R.; Peters, L., Jr.; Chen, C.; Daniels, J.J.; Youn, H. Detection of buried agricultural drainage pipe with geophysical methods. *Appl. Eng. Agric.* **2004**, 20, 307–318. [CrossRef]
5. Allred, B.J.; Daniels, J.J.; Fausey, N.R.; Chen, C.; Peters, L., Jr.; Youn, H. Important considerations for locating buried agricultural drainage pipe using ground penetrating radar. *Appl. Eng. Agric. 2005*, *21*, 71–87. [CrossRef]
6. Allred, B.J.; Redman, D. Agricultural drainage pipe assessment using ground penetrating radar: Impact of pipe condition, shallow hydrology, and antenna characteristics. *J. Environ. Eng. Geophys.* **2010**, 15, 119–134. [CrossRef]

7. Allred, B.J. A GPR agricultural drainage pipe detection case study: Effects of antenna orientation relative to drainage pipe directional trend. *J. Environ. Eng. Geophys.* **2013**, 18, 55–69. [CrossRef]

8. ODOT's VRS RTK Network. Available online: http://www.dot.state.oh.us/Divisions/Engineering/ CaddMapping/Survey/Pages/VRSRTK-.aspx (accessed on 9 September 2018).

9. Topp, G.C.; Davis, J.L.; Annan, A.P. Electromagnetic determination of soil water content—Measurements in coaxial transmission lines. *Water Resour. Res.* **1980**, 16, 574–582. [CrossRef]

10. Conyers, L.B. *Ground-Penetrating Radar for Archaeology;* AltaMira Press: Walnut Creek, CA, USA, 2004; pp. 23–80.

11. Sharma, P.V. *Environmental and Engineering Geophysics*; Cambridge University Press: Cambridge, UK, 1997; pp. 309–328.

12. Brown, L.A.; Zucker, L.A. *Agricultural Drainage: Water Quality Impacts and Subsurface Drainage Studies in the Midwest*; Extension Bulletin 871; The Ohio State University: Columbus, OH, USA, 1998; pp. 8–9.

13. Allred, B.J.; Brown, L.C.; Fausey, N.R.; Cooper, R.L.; Clevenger, W.B.; Prill, G.L.; La Barge, G.A.; Thornton, C.; Riethman, D.T.; Chester, P.W.; et al. A water table management approach to enhance crop yields in a wetland reservoir subirrigation system. *Appl. Eng. Agric.* **2003**, 19, 407–421. [CrossRef]

14. Allred, B.J.; Gamble, D.L.; Clevenger, W.B.; LaBarge, G.A.; Prill, G.L.; Czartoski, B.J.; Fausey, N.R.; Brown, L.C. Crop yield summary for three wetland reservoir subirrigation systems in northwest Ohio. *Appl. Eng. Agric.* **2014**, *30*, 889–903.

Chapter 5

ASSESSING TOPSOIL MOVEMENT IN ROTARY HARROWING PROCESS BY RFID (RADIO-FREQUENCY IDENTIFICATION) TECHNIQUE

1. INTRODUCTION: AN OVERVIEW

Soil tillage is an agronomic practice that effects on both soil and crop properties. The main objective for tillage operations is to improve the soil environment for seed germination and, subsequently, improve crop yield [1–3]. There are several kinds of tillage methods and tools such as conventional tillage, which is commonly divided into primary and secondary tillage. In primary tillage, moldboard or chisel plows make the major part of the tillage operation. Meanwhile, soil surface after primary tillage still needs further operation in order to smooth and reduce clods size. Therefore, secondary tillage operation by rotary harrow is used for preparing suitable seedbed [4].

Tillage operations have a great impact on topsoil in terms of aggregate size and crop residue cover, which plays an important role in sustainable agriculture. A minimum amount of residue is needed to protect soil erosion, reduce greenhouse gas emissions and maintain soil carbon level [5–8]. The impact of tillage operations on topsoil depend on tillage operational depth and speed, soil characteristics, initial amount of residues, and type of used equipment [9,10].

The contribution of harrowing in soil erosion could be realized from weed dispersal studies [4]. Soil movement by tillage operations was investigated through different instruments [11] and different approaches [12,13]. Researchers used different techniques to monitor soil movement such as plastic beads [14], granite rocks [13], or aluminum cubes [4]. It is in the authors' opinion that Radio-Frequency Identification (RFID) systems can be successfully applied for this scope.

The RFID is a system that transmits the identity of an object using radio waves and consists of RFID tags and RFID readers. RFID tags are small electronic devices attached to objects and RFID reader is an antenna that identifies the tags [15]. The passive tags have no power source and discovered by antenna within 0.5–1.0 m radius while other tags may include batteries to be detected at higher distances [16,17]. In recent years, due to relatively low costs, many applications have been developed based on RFID such as location identification for many shipping and postal services, security purposes in shops and companies, retailers and supply chain for confirming that products on shelf, payment systems for payments at drive-through windows, livestock and agricultural production monitoring [18–21]. With reference to the soil, some recent advancements have been reported, mainly dealing with the assessment of erosion, landslide displacement monitoring, or sediment mobility [22–25], while the authors are not aware of RFID applications in relation to soil tillage practices.

The main objective of this study is to assess the movements of topsoil layer after the application of a tillage operation such as rotary harrowing. In addition, the study highlights how RFID technology can be effectively applied in order to simulate crop residues or even soil clods for tillage experiments as more robust technique especially in open field trials (Supplementary Materials S1).

2. MATERIALS AND METHODS

This study was conducted at a 5 ha field in Agripolis experimental field (University of Padova, Italy). The soil can be defined, according to the United State Department of Agriculture (USDA), as loamy and containing 46% sand, 30% silt, and 24% clay. Primary tillage was applied at the end of the winter season and followed by a harrowing process as secondary tillage on May 2018 in order to prepare the soil for soybean sowing. It is worth noting how most of crop residues are buried after primary tillage, especially in the case of mouldboard ploughing. Such conditions were helpful in order to maximize the understanding of implemented RFID tags dynamics.

A rotary harrow with two horizontal rollers from Alpergo Co., Lonigo, VI, Italy (Model: Rotodent DP) was used to perform the harrowing process. The implement consisted of a series of 20 pairs of tines (indicated by A in Figure 1) which rotate about a vertical axis in order to produce soil disturbance over a working width of 5 m. The working volume is limited on the back by a levelling bar (B in Figure 1): its height relatively to the ground can be adjusted in order to limit the flow of clods through the machine and allow a better control on aggregates dimensions. The levelling bar thus allows the soil to be hold for a certain time (from a few tenths up to a few seconds depending on its relative position) within the shell where the tines operate, allowing aggregates mixing and reduction. Two horizontal cage rollers (C in Figure 1) were positioned on the back end of the rotary harrow to allow soil compaction and levelling. The implement was driven by a 160 hp tractor through the three-point hitch and Power Take Off (PTO) shaft at 1000 rpm.

A RFID package from Oregon RFID Co., Portland, OR, USA was implemented in this work. This package consists of RFID reader and passive tags. Tags are in a cylindrical shape with 3.65 mm diameter and 23 mm length. Being small in dimensions and weight (less than 1 gram) allows their integration into bigger envelopes. In fact, in order to simulate topsoil components such as crop residues and soil clods, a cork stopper was drilled laterally and one RFID tag inserted inside each stopper then closed by glue. Cork stoppers were chosen, due to their wide availability, durability, and high similarity (in terms of shape, size, and density) specially with crop residues such as dry corn stems or corncob. Furthermore, they allow to damp the effects of rotary harrow tools, minimizing the eventuality of damages to the delicate RFID tags. Each tag (and thus each cork stopper) had an identification number which allowed individual monitoring. Additionally, cork trackers were painted with shining fluorescent color and tag identification number in order to facilitate recognition after the harrowing process (Figure 2).

Figure 1. On the left schematic view and on the right a picture of the implemented harrowing machine, with indication of the working depth h of the rotary tines (A) and of the height k of the levelling bar (B) closing the back part of the working volume; Soil is eventually levelled by a couple of cage rollers (C).

Figure 2. On the left a picture if the RFID tag, and on the right a cork stopper with RFID inside, painted and numbered.

The experimental field was divided into two lateral strips (5 m × 80 m) and each strip consisted of four different working conditions, as summarized in Table 1 and depicted in Figure 3. The different applied working conditions were as follows: two forward speeds of 1 and 3 km/h, two working depths (namely h in Figure 1) of 6 and 10 cm, and two levelling bar positions (namely k in Figure 1) of 0 (i.e., at ground level) and 10 cm (i.e., lifted up from the soil). Eight working conditions were investigated by using the RFID trackers. In each treatment, 18 trackers were distributed in six lines and three rows. The distance between rows was 1 m while the distance between lines were 40, 80, 220, 280, 420, and 460 cm starting from one of the harrow sides in order to maintain a symmetric measurement between the two horizontal rollers (Figure 3). The three rows of trackers act as replicates for each treatment.

Table 1. Experimental conditions.

Acronym	Forward Speed (km/h)	Working Depth (cm)	Levelling Bar Position (cm)	Reference
V1D10	1	10	0	Figure 3B a
	1	10	10	Figure 3B e
V3D10	3	10	0	Figure 3B b
	3	10	10	Figure 3B f
V1D6	1	6	0	Figure 3B c
	1	6	10	Figure 3B g
V3D6	3	6	0	Figure 3B d
	3	6	10	Figure 3B h

Trackers' positions were recorded in conjunction with their identification number and tested with the RFID antenna before applying the experimental treatments. After performing each treatment, the position of each tracker was recorded in order to investigate its movement in terms of distance, depth, and direction. At the start of each treatment, two stick markers were fixed to act as a datum for trackers position recognition (Figure 3C).

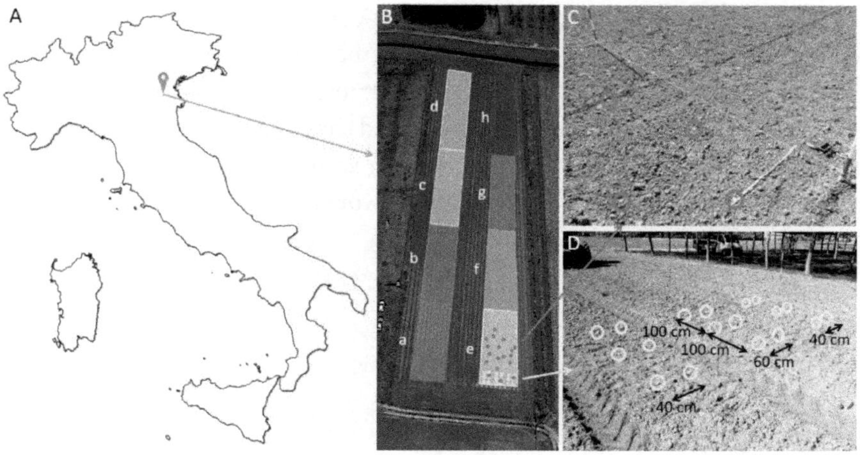

Figure 3. (A)—Geographical localization of the experimental site in Italy. (B)—Experimental area with representation of the experimental design. (C)—Some of the tags after tillage operation. (D)—Trackers distribution on undisturbed soil, before harrowing process.

3. RESULTS AND DISCUSSION

A total number of 144 trackers were tracked after applying eight different treatments of rotary harrowing. Trackers were localized by naked eye in the case of surface positions, conversely buried trackers (more than eighty) were localized taking advantage of the RFID reader. During the experiments, only two of the trackers were lost: in such cases they were not revealed by the RFID reader most probably due to some breakage occurred during harrowing operation. Hence, the overall recovery rate was 98.6%. Results showed that there was a clear movement in the machine direction with an average value of 2.3 m, but exceeding 5 m in different situations. On the other hand, the lateral movement was limited to an average of 0.2 m. The average movements for applied treatments are shown in Figure 4.

The major direction for trackers movement was in the machine direction. The major factor that effected on movement distance was noticed from the levelling bar. The average movement were 2 and 0.8 m at bar level of 0 and 10 cm, respectively. Also, many trackers moved more than 5 m in the machine direction with the bar at ground level, compared to less than 2.5 m in the case of lifted up bar. This is ascribable to the fact that when the levelling bar is down, the amount of soil kept within the chamber between tines and levelling bar itself increases. Concurrently, there is an increase in the amount of time soil remains within such chamber: aggregates exit only after their size has been sufficiently reduced to pass under the levelling bar. Such phenomenon results in a longer distance done by clods or residues (and thus also by the RFID tags) and in a higher chance of burial effects. Moreover, as tines are rotating, there is a probability of hitting the cork trackers which may throw trackers to a forward or backward direction. In case of backward thrown, the levelling bar prevents trackers from moving beyond that barrier. In contrary, there is no limit for the forward thrown where there is no barrier and the rugged material of cork trackers may explain the long forward displacement in some cases. Figure 5 illustrates the forward movement at different rotary harrow working conditions.

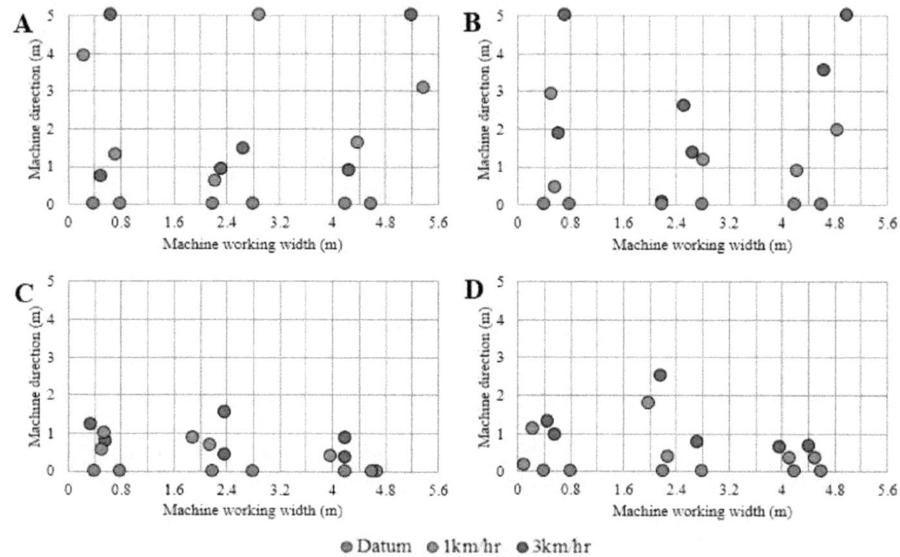

Figure 4. Average trackers movement at different rotary harrow working conditions where: (A) at level bar of zero and depth of 6 cm, (B) at level bar of zero and depth of 10 cm, (C) at level bar of 10 cm and depth of 6 cm, and (D) at level bar of 10 cm and depth of 10 cm.

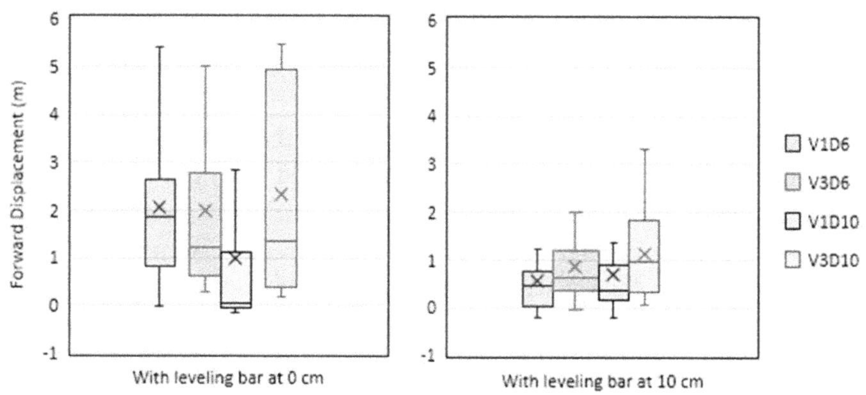

Figure 5. Average trackers forward movement at different working conditions where V1 and V3 refer to forward speed of 1 and 3 km/h. Also, D6 and D10 refer to working depth of 0.06 and 0.1 m.

Shifts in lateral direction were limited to 0.03 m left on average, with a standard deviation as high as 0.40 m. This is in agreement with the rotary harrow symmetric construction, where clockwise and anticlockwise rotating tine pairs are alternated: such working mode allows avoidance of soil piling phenomena, as confirmed by RFID distribution. In this case, the levelling bar position plays only a marginal role statistically irrelevant: 0.12 m right lateral displacement in the case of bar at ground level and 0.20 m left shift in the case of lifted up position. A slightly different situation is highlighted in the case of analyses related to tags depth. In fact, lower position of the levelling bar caused an average increase on burial phenomenon from -0.8 cm to -1.5 cm on average. Such apparently low values (compared to the working depth) are most probably ascribable to some kind of floating phenomenon the cork stoppers underwent due to their density, sensibly lower than soil clods density, but comparable to crop residues one. Figures 6 and 7, illustrates the lateral movement and burial depth at different rotary harrow working conditions.

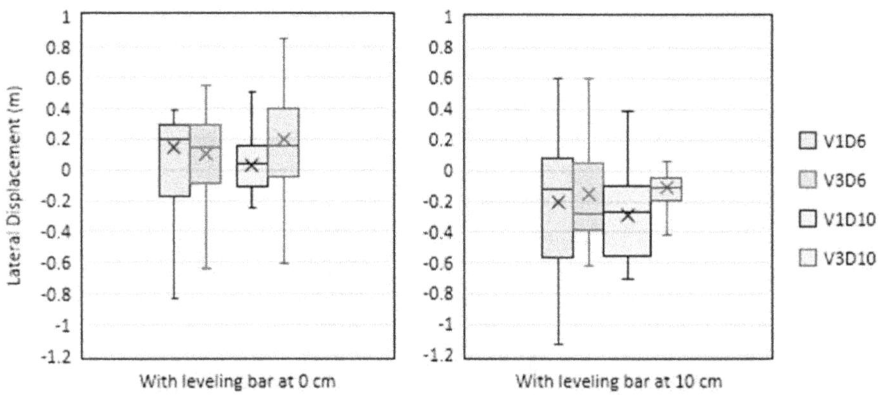

Figure 6. Average trackers lateral movement at different working conditions.

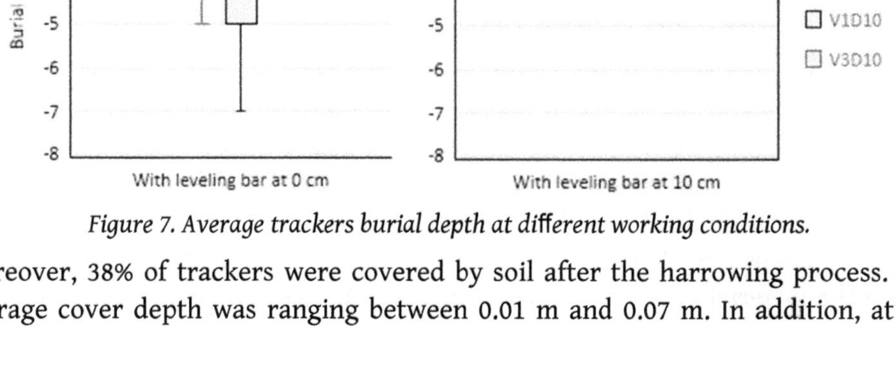

Figure 7. Average trackers burial depth at different working conditions.

Moreover, 38% of trackers were covered by soil after the harrowing process. The average cover depth was ranging between 0.01 m and 0.07 m. In addition, at the working depth of 0.1 m and levelling bar at ground level, more than 60% of trackers were covered by soil, which highlights a higher mixing rate at these conditions. Such higher mixing rate is due to the higher interaction between the soil and machine in terms of higher tines depth and longer contact compared with other studied working conditions. Also, it can be noticed how higher tillage speed resulted in higher trackers displacements, in agreement with Liu et al. [26] in sweep trials at soil bin

The experimental results exhibited a large variability especially when the bar was set at a null height, which was the main reason for no significant differences between variables. Such large variations were due to a number of reasons such as; tracker (rigid material of cork) might be hit by tines and thrown at long distances, or tracker dragging through tillage operation, in agreement with similar results reported by different relevant research works in soil bin trials [26,27].

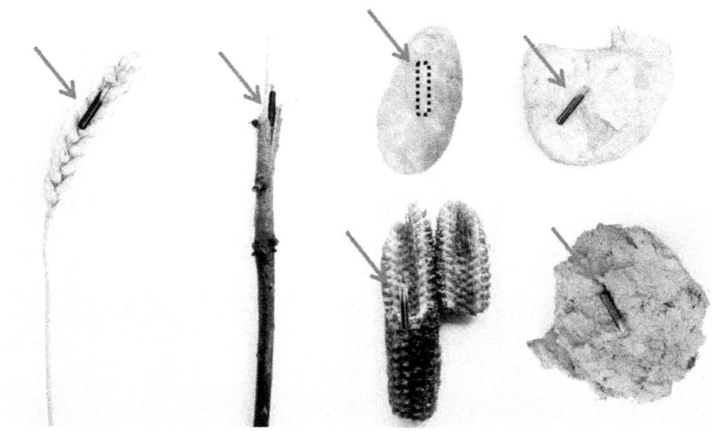

Figure 8. Suggested applications for the use of Radio-Frequency Identification (RFID) to simulate or monitor crop residues, straw, soil clods and potato in agricultural experiments.

Besides characterization of the topsoil movement after the application of rotary harrow tillage operation, the proposed study demonstrates the effectiveness of the RFID technique soil movements monitoring operations. Differently, from metal or polymer trackers, RFID allows minimization of tags dimensions, allowing univocal identification and maximizing localization process both in terms of high detection percentage (close to 100%, and lower in case of harsh operation possibly affecting RFID tags integrity) and low detection time (a few seconds per tag). Additionally, it is also worth noting how the RFID approach can be in principle applied in order to simulate different materials in similar trials. The very tiny shape of RFID tags offers a wide range of simulations such as soil clods, stones, and crop residues of different shapes and densities. The best simulation for any studied object is the object itself where the material properties, such as density, surface roughness, and shape, will simulate and act naturally. Figure 8 illustrates the possibility of integrating the RFID tags in wheat spikes, straw, corn residues, potato, soil clods, or even small rocks for suggested future applications. In addition, it is worth mentioning that the easy recognition of RFID by the antenna and the unique ID for each tag will reduce the source of error and limitations in open field experiments, which will also lead to more reliable and understandable results.

4. CONCLUSIONS

Field trials were performed to investigate the effect of different rotary harrow working conditions of forward speed, levelling bar, and tillage depth on topsoil aggregates displacement using RFID technique. Results showed no significant difference between treatments, while the higher displacement was noticed from the levelling bar factor. The average movement was about 2.3 m in the machine direction and exceeded 5 m in different cases. Lateral movements were limited to 0.03 m on average because of the different rotating directions between each tine pairs. Also, at the levelling bar of 0.1 m, 60% of trackers were buried because of the high mixing rate. Furthermore, using RFID tags and the antenna was a promising application in this field, since it is providing a robust way to simulate different materials.

References

1. Torabian, S.; Farhangi-Abriz, S.; Denton, M.D. Do tillage systems influence nitrogen fixation in legumes? A review. *Soil Tillage Res.* **2019**, *185*, 113–121. [CrossRef]
2. Zeyada, A.M.; Al-Gaadi, K.A.; Tola, E.; Madugundu, R.; Kayad, A.G. Impact of soil firmness and tillage depth on irrigated maize silage performance. *Appl. Eng. Agric.* **2017**, *33*, 491–498. [CrossRef]
3. Pezzuolo, A.; Dumont, B.; Sartori, L.; Marinello, F.; de Antoni Migliorati, M.; Basso, B. Evaluating the impact of soil conservation measures on soil organic carbon at the farm scale. *Comput. Electron. Agric.* **2017**, *135*, 175–182. [CrossRef]
4. Van Muysen, W.; Govers, G. Soil displacement and tillage erosion during secondary tillage operations: The case of rotary harrow and seeding equipment. *Soil Tillage Res.* **2002**, *65*, 185–191. [CrossRef]
5. Cherubini, F.; Ulgiati, S. Crop residues as raw materials for biorefinery systems—A LCA case study. *Appl. Energy* **2010**, *87*, 47–57. [CrossRef]
6. Graham, R.L.; Nelson, R.; Sheehan, J.; Perlack, R.D.; Wright, L.L. Current and potential U.S. corn stover supplies. *Agron. J.* **2007**, 99, 1–11. [CrossRef]

7. Cillis, D.; Maestrini, B.; Pezzuolo, A.; Marinello, F.; Sartori, L. Modeling soil organic carbon and carbon dioxide emissions in different tillage systems supported by precision agriculture technologies under current climatic conditions. *Soil Tillage Res.* **2018**, *183*, 51–59. [CrossRef]

8. Pezzuolo, A.; Basso, B.; Marinello, F.; Sartori, L. Using SALUS model for medium and long term simulations of energy efficiency in different tillage systems. *Appl. Math. Sci.* **2014**, 8, 129–132. [CrossRef]

9. Chen, Y.; Monero, F.V.; Lobb, D.; Tessier, S.; Cavers, C. Effects of six tillage methods on residue incorporation and crop performance in a heavy clay soil. *Trans. ASAE* **2004**, *47*, 1003–1010. [CrossRef]

10. Wagner, L.E.; Nelson, R.G. Mass reduction of standing and flat crop residues by selected tillage implements. *Trans. ASAE* **1995**, *38*, 419–427. [CrossRef]

11. Dubbini, M.; Pezzuolo, A.; de Giglio, M.; Gattelli, M.; Curzio, L.; Covi, D.; Yezekyan, T.; Marinello, F. Last generation instrument for agriculture multispectral data collection. *CIGR* J. **2017**, *19*, 158–163.

12. Lindstrom, M.J.; Nelson, W.W.; Schumacher, T.E.; Lemme, G.D. Soil movement by tillage as affected by slope. *Soil Tillage Res.* **1990**, *17*, 255–264. [CrossRef]

13. Thapa, B.B.; Cassel, D.K.; Garrity, D.P. Assessment of tillage erosion rates on steepland Oxisols in the humid tropics using granite rocks. *Soil Tillage Res.* **1999**, *51*, 233–243. [CrossRef]

14. Marshall, E.J.P.; Brain, P. The horizontal movement of seeds in arable soil by different soil cultivation methods. *J. Appl. Ecol.* **1999**, 36, 443–454. [CrossRef]

15. Kaur, M.; Sandhu, M.; Mohan, N.; Sandhu, P.S. Technology principles, advantages, limitations and its applications. Int. *J. Comput. Electr. Eng.* **2011**, 3, 1793–8163. [CrossRef]

16. Want, R. An introduction to RFID technology. *IEEE Pervasive Comput.* **2006**, 5, 25–33. [CrossRef]

17. Nikitin, P.V.; Rao, K.V.S. Theory and measurement of backscattering from RFID tags. *IEEE Antennas Propag. Mag.* **2006**, 48, 212–218. [CrossRef]

18. Tian, F. An agri-food supply chain traceability system for China based on RFID & blockchain technology. In Proceedings of the 13th International Conference on Service Systems and Service Management (ICSSSM), Kunming, China, 24–26 June 2016.

19. Vanderroost, M.; Ragaert, P.; Devlieghere, F.; de Meulenaer, B. Intelligent food packaging: The next generation. *Trends Food Sci. Technol.* **2014**, 39, 47–62. [CrossRef]

20. Prinsloo, J.; Malekian, R. Accurate vehicle location system using RFID, an internet of things approach. *Sensors* **2016**, 16, 825. [CrossRef]
21. Abdullah, M.F.F.; Ali, M.T.B.; Yusof, F.Z.M. Rfid application development for a livestock monitoring system. In *Bioresources Technology in Sustainable Agriculture*; Apple Academic Press: Palm Bay, FL, USA, 2018; pp. 103–116.
22. Parsons, A.; Cooper, J.; Onda, Y.; Sakai, N. Application of RFID to soil-erosion research. *Appl. Sci.* **2018**, 8, 2511. [CrossRef]
23. Rainato, R.; Mao, L.; Picco, L. Near-bankfull floods in an Alpine stream: Effects on the sediment mobility and bedload magnitude. *Int. J. Sediment. Res. 2018*, 33, 27–34. [CrossRef]
24. Picco, L.; Tonon, A.; Rainato, R.; Lenzi, M.A. Bank erosion and large wood recruitment along a gravel bed river. *J. Agric. Eng.* **2016**, 47, 72. [CrossRef]
25. Le Breton, M.; Baillet, L.; Larose, E.; Rey, E.; Benech, P.; Jongmans, D.; Guyoton, F.; Jaboyedoff, M. Passive radio-frequency identification ranging, a dense and weather-robust technique for landslide displacement monitoring. *Eng. Geol.* **2019**, 250, 1–10. [CrossRef]
26. Liu, J.; Chen, Y.; Kushwaha, R.L. Effect of tillage speed and straw length on soil and straw movement by a sweep. *Soil Tillage Res.* **2010**, 109, 9–17. [CrossRef]
27. Liu, J.; Chen, Y.; Lobb, D.A.; Kushwaha, R.L. Soil-straw-tillage tool interaction: Field and soil bin study. *Can. Biosyst. Eng. Genie Biosyst. Can.* **2007**, 49, 2.

Chapter 6

DEVELOPMENT AND FIELD EVALUATION OF A SPRAY DRIFT RISK ASSESSMENT TOOL FOR VINEYARD SPRAYING APPLICATION

1. INTRODUCTION: AN OVERVIEW

Chemical crop protection is one of the most important factors in agricultural production, as global potential crop yield is diminished by pests up to 40%, a figure that would be twice as large if no plant protection products (PPPs) were used [1]. The benefits from crop protection are also undeniably significant in regards to improved food security and the reduction of labor [2]. However, as most PPPs are applied in the field by spraying, due to low cost and efficient performance [3], serious PPP losses are a side effect of chemical crop protection due to run-off, leaching, evaporation, and spray drift, putting the health of the environment, animals, and humans at risk [4].

Spray drift is the quantity of PPPs that is carried out by air currents from the sprayed area during the spraying application [5]. Spray drift is an important and costly (environmentally and economically) problem that is hard to control and may cause the PPPs to be deposited in off-target areas. The consequences can be serious, such as surface water contamination, air pollution, damage to sensitive nearby crops and other susceptible off-target areas, residues of chemical substances in food and feed commodities, health risks for animals and people (farm workers, bystanders, and passers-by), nearby urban or natural area contamination, and reduced PPP effectiveness due to lower dose than intended on the targeted crops [6–10]. Additionally, the financial burden resulting from spray drift due to increased inputs is very high [11].

There are many factors that contribute to spray drift and several of them are interrelated. Some factors can be controlled by the sprayer operator, while others cannot be controlled. These factors can be grouped into the following categories: (i) Equipment and application techniques (i.e., sprayer type, size and type of nozzles, spray pressure, spray volume rate, air flow rate, driving speed, sprayer's setup, etc.); (ii) weather conditions during application (i.e., wind speed and direction, temperature, relative humidity, and stability of air at the application site); (iii) spray characteristics, (i.e., volatility and viscosity of the PPP formulation; (iv) operator's care, attitude, and skill; and (v) characteristics and geometry of the crop (i.e., foliage, density, dimensions, etc.) [12–17].

Among the technical factors that affect spray drift, the most important is droplet size [18] and more particularly the percentage of fine spray droplets [19–22]. Considering the pressure atomization, droplet size depends on the nozzle design, orifice size, operating pressure, and the physical properties of the PPP formulation and spray additives [23,24]. The smaller a spray droplet, the longer it remains airborne, and the higher the possibility for it to be carried away by crosswind [25]. Droplets with diameter smaller than 100 μm contribute significantly to drift losses [26,27]. Forward speed has also a clear effect on spray drift, where the higher the driving speed the greater the spray drift, both for airborne drift and for ground deposition [28]. Regarding bush and tree sprayers, among the various application operating parameters that affect spray drift, those that concern spray generation (nozzle type and pressure) and droplet transport to the canopy (air fan volume, speed, orientation) are of great importance [29].

Environmental conditions influence spray drift and cannot be controlled by the sprayer operator. These factors need to be taken into consideration and be monitored before and during the PPP application. Among the meteorological factors affecting spray drift, wind velocity has the greatest impact, while wind direction plays also an important role. There is a strong positive correlation between wind speed and spray drift deposition [30]. Higher wind speeds result in more drift at greater distances [31]. Summer [32] pointed out that wind speed should not exceed 16 km h-1 (4.44 m s-1) to continue spraying. According to da Cunha et al. [33], the maximum permissible wind speed for spraying is 12 km h^{-1} (3.33 m s^{-1}), while Maciel et al. [34] also proposed that spraying operation should be performed with wind speeds ranging between 2 and 12.8 km h–1 (0.55 to 3.55 m s^{-1}). It should be noted that different wind speeds and directions have also been reported to contribute to uneven spray distribution between the left and right side of the sprayer [35]. In addition, large fluctuations in wind direction increase the unpredictability of droplet travel direction and the amount of dilution due to atmospheric turbulence [36].

Air temperature and relative humidity during the spraying operation play also key role in spray drift. Low relative humidity and/or high air temperature can reinforce evaporation by decreasing the droplet size, especially small droplets, having, as a result, decreased sedimentation velocity and droplets more prone to drift [37]. Low relative humidity combined with high temperatures contributes to the evaporation of the spray liquid with an impact both on the environment and on the economic viability of the farm [34]. ISO 22866 standard [5] considers acceptable conditions for field measurement of spray drift at temperatures ranging from 5 to 35 ∘C. Da Cunha et al. [33], pointed out that spraying should be avoided when temperature is above 30 ∘C and relative air humidity is below 55%. As a rule, if the relative humidity is above 70%, the conditions are ideal for spraying, and if the relative humidity is below 50%, it is quite critical and requires special attention [32]. Generally, spray drift can be significantly reduced by spraying at low wind speed, low temperature, with low turbulence, at times of low sun radiation and at high relative humidity [38].

Moreover, it should be noted that during rain or shortly before it occurs, PPP application should be avoided due to leaching risk from the crop canopy before the active substance is able to act [39], polluting the soil and the underground water resources.

Since the negative effect of spray drift has been recognized, there is a need for harmonized mitigation measures to reduce human health and environmental impact. Such measures have been developed and can be divided into three classes [40–42]: (i) The use of no-spray or even no-crop buffer zones; (ii) the reduction of exposure using vegetative or artificial windbreaks, and (iii) the application of drift-reducing technology, such as drift-reducing nozzles and spray additives to coarsen the droplet size distribution, as well as shielded and band sprayers. In addition, the EU legislation has been adjusted in this direction, with Directive 2009/128/EC [43] establishing a framework to achieve sustainable use of PPPs, while Directive 2009/127/EC [44] specifies that sprayers should be designed and constructed to ensure that PPPs are deposited on target areas, to minimize losses to other areas and to prevent drift. USA has also set certain measures for spray drift prevention that are in the same direction as the EU legislation [45].

In recent years, several attempts have been made to reduce spray drift through predicting weather conditions and creating automated spray drift reduction systems. Such systems may collect meteorological data from either ground meteorological stations or geostationary satellites and predict drift or provide information to the user on how to treat and regulate the sprayer. The frequency of meteorological data gathering is crucial to optimize weather prediction and develop a reliable spray drift model. As an example, Huang and Thomson [46] highlighted the importance of knowing meteorological data on 15 min basis than on 1 h basis, as climate conditions can change rapidly, making hourly forecasts unreliable. On-line applications are also used for spray drift prediction and reduction helping farmers make decisions for best practice implementation. For example, Nansen et al. [47] created an online decision support tool for farmers and agricultural advisors. This tool allows the prediction, measurement, and archiving of spraying coverage, which is quantified using water sensitive filter papers. Spray drift investigation and prediction under a wide range of conditions have been conducted [48,49]. A comprehensive model which accurately predicts the downwind movement of spray for given circumstances, including spray liquid characteristics, spray nozzle characteristics, and meteorological conditions, was developed by the US Spray Drift Task Force [50]. Hong et al. [51] presented a software application for spray drift estimation using an orchard air-assisted sprayer, through their research study on spray drift prediction. Another case of a spray drift prediction model, which simulates spray drift taking meteorological conditions into consideration, is the one produced by Nsibande et al. [52] in South Africa.

The aim of this research was to evaluate spray drift in vineyards using a drift risk assessment tool in order to improve the process of PPP application and mitigate spray drift. The spray drift risk assessment tool was developed taking into consideration a similar drift evaluation tool developed in the framework of the TOPPS-Prowadis [53]. The factors that the tool considers are wind speed and direction, air temperature, and relative humidity. The ultimate goal was to examine the reliability of the tool by measuring the meteorological conditions in the field and assessing ground and airborne spray drift. The evaluation used field trials in a vineyard under the standardized test methodology of ISO 22866:2005 [5]. The selected vineyard variety for spray drift trials was Savatiano, the most widespread winemaking variety in the Attica region, which has been cultivated for about 4000 years [54].

2. MATERIALS AND METHODS

2.1 Spay Drift Risk Assessment Tool

The tool evaluates the potential drift risk from PPP applications using air-assisted sprayers for different meteorological conditions under field conditions in vineyards.

The meteorological data input to the tool is obtained from a weather forecast website for the geographic coordinates of the field. The drift evaluation tool can help the farmers or spray contractors to make decisions for PPP applications with high spray efficiency and low spray drift risk to the environment.

The tool is based on the methodology developed from the TOPPS-Prowadis drift evaluation tool [53], and presents three categories depending on the risk of spray drift to occur:

- ☆ "**Low**" indicates that it is possible to apply spraying as the drift will have small/acceptable extent;
- ☆ "**Medium**" indicates that there is medium risk of spray drift due to conditions and, the use of drift-reducing technology and/or setting drift-reducing application parameters should be considered;
- ☆ "**High**" indicates that there is high risk of spray drift and; therefore, spraying should not be applied.

The tool takes into account limitations related to air temperature, relative humidity, wind speed, and wind direction. Considering the meteorological conditions in Greece and a series of experimental measurements carried out in Greek vineyards, the tool's limits on weather conditions varied from those used by the TOPPS-Prowadis drift evaluation tool as follows:

a. Air temperature: The limits between the three categories were set at 25 and 30 ○C, while the TOPPS-Prowadis tool uses, as limits, 15 and 25 ○C. This is because during the spraying period of vineyards in Greece temperatures vary from 20 ○C to above 30 ○C.
b. Relative humidity: The TOPPS-Prowadis tool's limits between the three categories were maintained and they are 40% and 60%.
c. Wind speed: The 5 classes of the TOPPS-Prowadis tool were made 3, with the limits between them set at 3 and 4.5 m s^{-1}. The classes of the TOPPS tool which refer to low and medium wind speeds were merged, because during the spraying period the high temperatures do not allow weak winds to occur.

In order to evaluate and categorize the drift risk, the tool combines the values of the meteorological parameters by giving priority to the wind speed. The above limits and categories apply with the following basic assumptions:

a. The wind direction is towards the sensitive area.
b. The canopy crop density is greater than 50%.
c. There is no rainfall.
d. The rows are sprayed from two sides and air is blown from two sides to each row.

2.2 Test Site and Crop Characteristics

The experimental field was an organic vineyard at the Agricultural University of Athens farm in Athens, Greece (37°59*06**N, 23°54*21**E).

The vineyard has 2.0 m row spacing with 1.6 m spacing of vines along the row to result in a density of 3125 vines ha^{-1}. The average vine height was about 1.1 m, with the leaves and grapes occupying the zone above ground between 0.3 and 1.2 m. At spraying times, the vines were in full leaf stage (BBCH 83 "Berries developing colors" and BBCH 91 "After harvest; end of wood maturation") [55], and the leaf area index (LAI) values were 1.58 and 1.46, respectively.

2.3 Experimental Design

The field experiments were carried out based on the ISO 22866:2005 standard [5], which set the criteria on the conditions for spray drift measurements. In accordance to this, the directly-sprayed area shall be at least 20 m wide upwind of the edge of the cropped area and the length of the spray track at least twice as the largest downwind sampling distance, and should be symmetrical to the axis of the sampling array. Therefore, every trial was carried out by spraying the ten outer downwind rows of the vineyard along a distance of 60 m, in order to treat a surface of 1200 m^2 (60 × 20 m) (Figure 1).

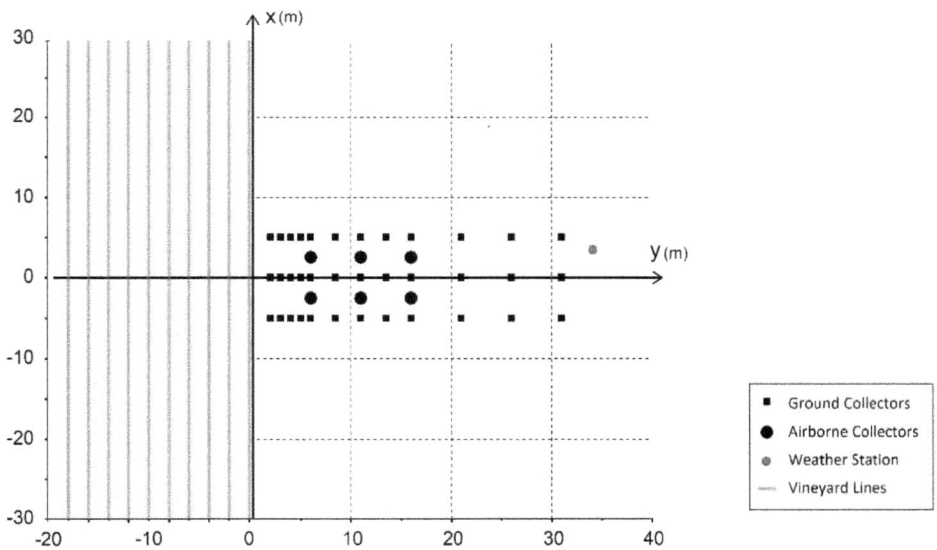

Figure 1. Test site layout according to the ISO 22866:2005 standard.

The experimental area was open and free of obstructions, other than the target crop (vineyard), as these may affect the airflow in the sampling area [5]. On the downwind side of the directly-sprayed area, there was bare soil or short vegetation (maximum height 7.5 cm), on which collectors were placed for the estimation of airborne spray drift and sedimenting spray drift [5] (Figure 2a).

In each trial, both ground sediment and airborne spray drift downwind to the directly-sprayed area were sampled. The ground collectors were placed at 12 different sampling distances in bare soil at 1, 2, 3, 4, 5, 7.5, 10, 12.5, 15, 20, 25, and 30 m from the edge of the directly-sprayed area (Figure 1). These distances started from the parallel straight line in front of the last plant row spaced 1 m (half of the row spacing). At each sampling distance, three wooden laths were placed with an upper surface covered in filter paper, Whatman Grade-1, 46 × 8 cm (Figure 2a,b), counting 36 soil samples per trial (Figure 1). So, each ground collector had a surface area of 368 cm2, summing up to a total of 1104 cm2 collector surface at each distance (the minimum collector area at any distance must be 1000 cm^2 [5]).

The airborne spray drift was monitored on cylindrical polyethylene lines with an external diameter of 2 mm, length 1 m, and collection area 62.8 cm2 (Figure 2c). The measurements were taken at 3 distances, 5, 10, and 15 m, downwind from the edge of the directly-sprayed area (Figure 1). At each sampling distance, two 6 m high columns were placed, each of which had support structures per meter for the polyethylene lines, thus forming an array of 6 sampling collectors (Figure 2a,c). Therefore, at each distance there were 12 collectors summing up to a total sampling area of 753.6 cm^2, resulting in 36 collectors per trial.

After each spraying repetition, the ground and air collectors were stored into plastic sachets and sealed in refrigerators for maintenance at the appropriate temperature (4°C) until their spectrophotometric analysis in the laboratory.

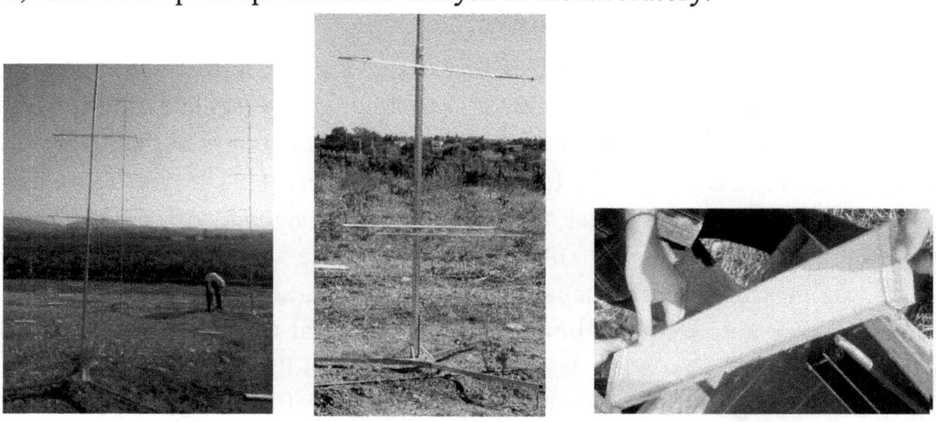

(a) *(b)* *(c)*

Figure 2. (a) The vineyard and sampling area; (b) detail showing the ground collector; (c) detail showing the pole holding the polythene lines.

2.4 Meteorological Conditions

Local weather conditions were measured following ISO 22866:2005 [5], but also to check the reliability of the weather forecast used by the drift risk assessment tool. Wind speed, wind direction, air temperature, and relative humidity were measured during the trials using a meteorological station placed at the edge of the downwind area in the center of the sampling area, 30 m from the sprayed area. Wind speed and direction were measured at a distance of 3 m from the ground, using an ultrasonic anemometer (Campbell Scientific WindSonic1 Gill 2D, Logan, UT, USA). Temperature and relative humidity were measured at two different heights, 2 and 3 m above ground, using two thermo-hygrometer probes (Rotronic HC2A-S3, Hauppauge, NY, USA). All measurements were taken at a frequency of 1 Hz sampling rate and all data were recorded automatically by a data logger (Campbell Scientific CR850, Logan, UT, USA).

The following parameters were calculated for each trial [5]:

a. Percentage of wind speed measurements less than 1 m s^{-1} (must be <10%);
b. Mean wind direction shall be at 90° ± 30° to the spray track (in this experiment between 0° and 60°) and no more than 30% of results shall be >90° ± 45° to the spray track;
c. Mean temperatures must be between 5 and 35 °C.

2.5 Treatments and Equipment Application Parameters

A total of six trials were carried out, two were related to the assessment of spray drift with a low risk tool indicator, two with medium, and two tests with high risk tool indicator.

Trials were performed using a trailed air-assisted sprayer for bush and tree crops Archimedes Turbo FS 1000 ("Archimedes" G. Roumeliotis, Aridaia Pellas, Greece), equipped with a 1000 L polyester tank, an axial fan of 800 mm in diameter with a two-speed gearbox and 7 nozzles for each side of the sprayer. The nozzles used were conventional hollow cone Teejet TXA 8002VK, yellow signed, with nominal nozzle flow rate of 1.40 L min^{-1} at 1.0 MPa. The real flow rate was closest to the nominal one. It was measured in the Department of Agricultural Engineering, Institute of Soil and Water Resources of the Hellenic Agricultural Organization "DEMETER", using an electronic measuring device (AAMS-Salvarani BVBA, Maldegem, Belgium) (Figure 3a). During testing, after adjusting the spray profile to target characteristics by means of water-sensitive paper spread on the vineyard canopy and poles, it was decided to activate a total of 6 nozzles (3 on each side of the sprayer). For all treatments, the working pressure was 1.0 MPa, driving speed 1.61 m s^{-1} (5.8 km h^{-1}), and volume application rate 434 L ha^{-1}. In all trials the PTO revolution speed was 56.55 rad s^{-1} (540 rev min^{-1}), fan speed 169.65 rad s^{-1} (1620 rev min^{-1}) (fan ratio 1:3), and fan airflow rate 10,000 m^3 h^{-1}. Fan air flow rate was measured in the Department of Agricultural Engineering, using a measuring tunnel (AAMS-Salvarani BVBA, Maldegem, Belgium) (Figure 3b).

(a) (b)

Figure 3. (a) Nozzles flow rate measurement; (b) fan air flow rate measurement.

2.6 Spray Liquid, Sample Extraction, and Spray Drift Estimation

The spray liquid was a solution of clean water and E-102 Tartrazine yellow dye tracer 85% (w/w) at a concentration of about 4 g L^{-1} [56,57].

Before each test, a blank sample of filter paper was placed in the sprayed area and collected just before the start of spraying. Two samples of the spray liquid were also collected from the spray tank directly from a nozzle, one at the beginning and one at the end of the test, to determine the precise tracer concentration at the nozzle outlet at each test.

After collecting the samples, they were transferred for analysis to the laboratory. Tartrazine's concentration in soil and air collectors was studied and quantified using a Shimadzu UV-1800 spectrophotometer, functioning at a wavelength of 426 nm.

Deposits of the spray tracer were extracted from samples using deionized water. For the respective samples the following volumes of deionized water were applied: A total of 40 mL for filter paper from 1 to 5 m distance, 20 mL for filter paper from 7.5 to 30 m distance, and 10 mL for polyethylene lines.

The reading of the spectrophotometer is related to the amount of tracer in solution through a calibration curve. From the reading of the spectrophotometer, the calibration factor, the collector surface area, the spray concentration and the volume of dilution liquid, and the amount of spray deposit per unit area were calculated as follows [5]:

$$drift_{dep} = \frac{\left(\rho_{smpl} - \rho_{blk}\right) \cdot F_{cal} \cdot V_{dil}}{\rho_{spray} \cdot A_{col}}, \quad (1)$$

driftdep = spray drift deposit (μL cm^{-2})

ρ smpl = spectrophotometer reading of the sample (Abs)

ρ blk = spectrophotometer reading of the blanks (collector + deionized water) (Abs)

F_{cal} = calibration factor (μg L^{-1}) Vdil = volume of dilution liquid (L)

ρ_{spray} = spray concentration of tracer (g L^{-1})

A_{col} = collection area of the spray drift collector (cm^2)

From this spray drift deposition figure, the percentage of spray drift on a collector can be calculated relating spray drift deposition to the amount applied in the field on the same unit of area, with the following formula:

$$drift_{\%} = \frac{drift_{dep} \cdot 10^4}{\beta_v}, \tag{2}$$

where βv is the spray application volume in liters per hectare (L ha-1) and given by the

$$\beta_v = \frac{Total\ nozzle\ flow\ rate \cdot Time}{Area} = \frac{Total\ nozzle\ flow\ rate \cdot 60}{Row\ Spacing \cdot Velocity/10} = \frac{Total\ nozzle\ flow\ rate \cdot 600}{Row\ Spacing \cdot Velocity},$$

For drift ground sediment, once the tracer amount on each collector was measured, the mean of values derived from the three samples placed at each downwind distance was calculated. For airborne drift, the mean tracer amount derived from the two samples placed at each sampling height above the ground was calculated separately for each of the three sampled downwind distances (5, 10, and 15 m from the sprayed area).

While the drift is precisely defined by ISO 22866:2005 [5], for mitigation estimate a 1% line was calculated, which is the distance from the sprayer where the drift equaled only 1% of the original applied rate. If the reference parameter is the distance of the 1% line from the sprayed area, the mitigation of sedimenting drift from one treatment to another is given in the following equation [58]:

$$M(\%) = \frac{D_i - D_j}{D_i} \cdot 100, \tag{3}$$

Where

M = mitigation of sedimenting drift from (i) to (j) treatment (%)

Di = distance of the 1% line from the sprayed area at (i) treatment

The 1% line was selected because its distance is detected with good precision in field trials and has been used by other researchers [58]; however, the results can be extended to other parameters (e.g., the 0.5% line).

2.7 Data Analysis

STATGRAPHICS Centurion XVI Version 16.1.15 software for Windows was used for all statistical analyses [59]. In all tests a confidence level of 95% was considered. The effect of each treatment on the sedimenting drift was studied using two-way analysis of variance (ANOVA) considering drift risk indication from the tool and distance from the sprayed area as sources of variation. Airborne drift was evaluated with three-way ANOVA considering drift risk indication, distance, and height above the ground as factors. Spearman's correlation was used to identify the correlations between the above parameters [60]. Fisher's least significant difference (LSD) procedure was applied for pair-by-pair comparison among the means of the three treatments for all distances and heights [61]. Previously, the homogeneity of variance of the studied variables was verified by the Levene's test [62].

3. RESULTS AND DISCUSSION

3.1 Weather Conditions

Trials were carried out during July, August, and October 2018 in order to achieve alignment with the weather conditions that respond to the three different categories of drift risk assessment tool. The weather forecasts used in the tool are shown in Table 1.

The weather conditions during trials were monitored (Table 2) to evaluate the tool and weather forecasts and to ensure following limitations set in ISO22866:2005 standard [5]. Tables 1 and 2 indicate the validity of the prediction tool, as the measurements from the weather station have very small differences compared to the forecasts used by the tool. This outcome is considered suffcient given that the drift risk assessment tool used data from an open data source which uses a weather station that is not located in the experimental site, in contrast to the local weather station.

During trials the mean temperature fell within ISO 22866 [5] requirements. The maximum differences (Δ) in air temperature and relative humidity measured for the two heights (2 and 3 m from ground) were 0.38°C and 0.46%, respectively. The mean wind speed was from 1.71 to 4.79 m s^{-1}, and the mean wind direction from 12.86° to 60.56○ (ideal direction was 30○ ± 30°). Wind direction deviation for treatment L1 was 7.30% over the limit prescribed in [5], meaning 24 records out of the 337 recorded during the trial. Considering the deviation slight, all data were used in the statistical analysis.

Table 1. Meteorological data used by the tool.

Date	Treatments	Replicates	Temperature	Relative Humidity	Wind Speed	Wind Direction	Rainfall
			°C	%	m s^{-1}		
07.10.2018	"Low"	L1	21.50	69.00	2.60	North-East	No
20.10.2018		L2	23.10	62.00	1.65	North-East	No
21.07.2018	"Medium"	M1	33.00	46.10	2.00	North-East	No
29.07.2018		M2	29.20	56.00	2.50	North-East	No
02.08.2018	"High"	H1	31.00	55.00	4.00	North/North-East	No
02.08.2018		H2	31.00	55.00	4.00	North/North-East	No

Table 2. Meteorological data collected during trials.

		Temperature		Relative Humidity		Wind Speed				Wind Direction		
Treatments	Replicates	Mean	Δ	Mean	Δ	Min	Max	Mean	Deviation [a]	Mean	Range	Deviation [b]
		°C	°C	%	%	m s^{-1}	m s^{-1}	m s^{-1}	%	° az	°	%
"Low"	L1	20.67	0.14	67.17	0.24	0.70	5.91	2.84	1.50	60.56	120	37.30
	L2	22.38	0.24	60.25	0.43	0.27	4.58	1.71	9.61	59.30	139	29.95
"Medium"	M1	32.67	0.38	44.76	0.46	0.26	4.40	2.12	10.19	47.80	281	29.21
	M2	28.91	0.16	54.92	0.40	0.77	5.30	2.77	0.24	31.27	110	0.72
"High"	H1	30.58	0.19	51.47	0.20	1.68	7.21	3.68	0.00	12.86	75	0.99
	H2	30.40	0.12	55.23	0.23	1.92	9.51	4.79	0.00	19.69	92	2.87

[a] Percentage of measurements <1 m s^{-1} (must be <10%); [b] percentage of measurements $\notin$ [30° − 45°, 30° + 45°] (must be <30%).

Temperature and relative humidity may vary much during the year, but they are stable during short periods of time (time of trial). This does not apply to wind speed and wind direction, which can drastically change during each trial, resulting in high influence on spray drift measurements [63]. For that reason, wind speed and wind direction correlation were analyzed indicating that high speed results in lower wind direction variance (Figure 4), in agreement with the result of previous research [64]. Studying the relationship between minimum wind speed, mean wind speed, and range of wind directions during trials, it appears that for minimum wind speeds less than 1 m s^{-1} and mean wind speeds less than 3 m s^{-1}, the range of wind directions was greater than 110○. Simultaneously, when the minimum wind speed was greater than 1.5 m s^{-1}, and the mean wind speed greater than 3.5 m s^{-1}, the wind direction was more uniform and range of the wind direction was less than about 90○ (Table 2).

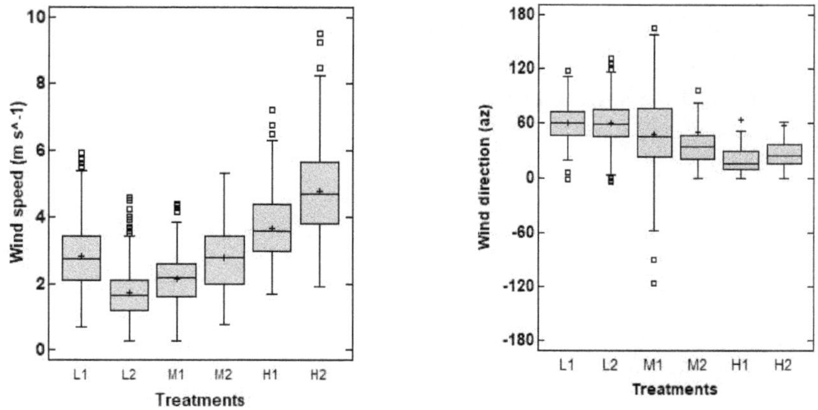

(a) (b)

Figure 4. Boxplots of all wind speed and wind direction measurements; (a) wind speed; (b) wind direction.

3.2 Sedimenting (Fallout) Spray Drift

The spray drift deposits on ground collectors measured at different distances downwind of the sprayed area for all treatments are presented in Table 3, and the mean data curves are shown in Figure 5.

Table 3. Ground deposit of spray drift.

Distance m	Treatments					
	"Low"		"Medium"		"High"	
	L1	L2	M1	M2	H1	H2
	% Application Rate	% Application Rate	% Application Rate	% Application Rate	% Application Rate	% Application Rate
1	13.07	13.88	14.44	14.95	14.31	23.93
2	9.86	10.50	10.04	11.79	12.85	15.02
3	7.70	9.63	9.79	9.21	10.82	13.06
4	5.33	7.16	6.07	6.97	7.87	8.61
5	3.59	5.13	3.35	6.47	7.47	6.97
7.5	1.80	2.25	2.42	3.21	4.19	3.64
10	1.05	1.12	1.17	2.13	2.62	2.73
12.5	0.54	0.67	0.77	1.54	1.63	2.02
15	0.52	0.52	0.41	1.30	1.36	1.65
20	0.29	0.36	0.30	0.97	0.89	1.39
25	0.16	0.15	0.19	0.62	0.46	0.89
30	0.14	0.14	0.17	0.44	0.36	0.69

The results indicate a significant amount of ground sediment at all sampled distances. In all treatments the greatest deposition was measured in the first few meters of the downwind area. Additionally, all curves showed continued decreased deposition as distance increased from 1 to 30 m, but the rate of decrease varied among the three treatments, keeping, though, a common trend for alltreatments. Similar results in both vineyard and orchard experiments have also been found by other researchers [65–68].

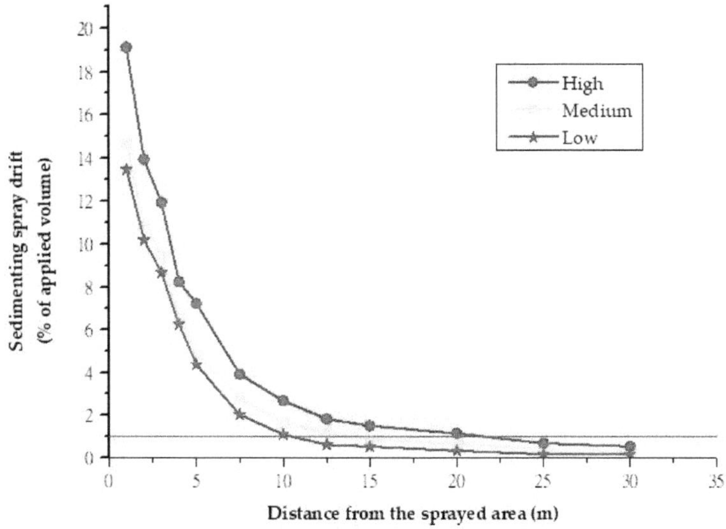

Figure 5. Spray drift deposit on the ground collectors.

It is also apparent that the three treatments (three categories of drift risk) generated different amounts of sedimenting drift over distance. As expected, the greatest amounts of ground sediment were observed when spraying was performed under worst-case conditions of high wind speed, high temperature, and low relative humidity ("High" treatment), while the lowest amounts were found under favorable weather conditions of low wind speed, low temperature, and high relative humidity ("Low" treatment). This finding comes in accordance with previous studies [69,70].

More precisely, spray drift deposit corresponding to about 1% of the spray volume was achieved at 10, 13.5, and 20 m from the sprayed area during "Low", "Medium", and "High" treatments, respectively (Figure 5).

Therefore, the mitigation of sedimenting drift, according to Equation (3), was 32.5% between "High" and "Medium" treatments, 50% between "High" and "Low" treatments, and 25.93% for "Medium" and "Low" treatments.

For the ground collectors the two-way ANOVA test showed that there is a statistically highly significant effect of the collector's placement distance, downwind of the sprayed area, and drift risk indication of the tool on the spray drift deposits ($p < 0.001$). However, there is no interaction between distance and drift risk indication ($p > 0.05$) (Table 4).

Table 4. Results of ANOVA for sedimenting spray drift as affected by distance and drift risk indication (Df: degrees of freedom).

Source	Df	Sum Sq	Mean Sq	F-Ratio	p-Value
Distance	11	5338.32	485.302	107.47	0.00001
Drift Risk	2	167.041	83.5206	18.50	0.00001
Distance × Drift Risk	22	92.4256	4.20116	0.93	0.5555
Residual	180	812.82	4.51566		

Spearman's rank correlation test indicated (Table 5) that there is very strong negative correlation between distance and drift ($r_s = -0.9392$), meaning that drift decreased drastically with increasing distance ($p < 0.001$), while drift risk category and drift have weak positive correlation ($r_s = 0.2047$), meaning that drift slightly increased when drift risk category changed following "Low"–"Medium"–"High" sequence ($p < 0.01$). These results are aligned with previous research trials, in which spray drift amounts are directly influenced by distance of the sprayed area and weather parameters, especially with the wind speed [48,64,71,72].

Table 5. Results of Spearman rank correlation test.

		Distance	Drift Risk
Drift (%)	Correlation	−0.9392	0.2047
	Sample size	216	216
	p-Value	0.00001	0.0027

Fisher's least significant difference (LSD) procedure was applied for pair-by-pair comparison among the means of the three treatments for all distances, showing statistically significant differences between "High"–"Medium" and "High"–"Low" treatments at downwind distances from 5 to 30 m, while no statistically significant differences were observed between "Medium"–"Low" treatments ($p < 0.05$) (Table 6). This finding enhances the fact, that the wind speed, which is less than 3 m s^{-1} in "Low" and "Medium" treatments, while in the "High" treatment is greater than 3.5 m s^{-1}, has the most significant effect on the spray drift from other meteorological parameters [30,73]. Additionally, the effect of meteorological parameters on ground drift is shown in Figure 6, where the means of sedimenting drift indicate that "High" treatment gave the highest ground drift deposition and the other treatments decreased from "Medium" to "Low".

Table 6. Multiple range LSD tests for ground collectors (LSD: least significant difference; Sig: significant).

Distance	+/−Limits	Low–Medium		Medium–High		Low–High	
		Sig.	Difference	Sig.	Difference	Sig.	Difference
1	5.44014		−1.22583		−4.4235	*	−5.64933
2	3.95954		−0.736167		−3.02083		−3.757
3	4.70071		−0.832333		−2.446		−3.27833
4	2.77943		−0.277167		−1.7155		−1.99267
5	2.12408		−0.546667	*	−2.31417	*	−2.86083
7.5	1.03509		−0.7955	*	−1.0965	*	−1.892
10	0.718107		−0.5655	*	−1.02733	*	−1.59283
12.5	0.541351	*	−0.554	*	−0.670167	*	−1.22417
15	0.506432		−0.338667	*	−0.644167	*	−0.982833
20	0.466114		−0.31	*	−0.507	*	−0.817
25	0.264212		−0.247667	*	−0.2675	*	−0.515167
30	0.172946		−0.162667	*	−0.216833	*	−0.3795

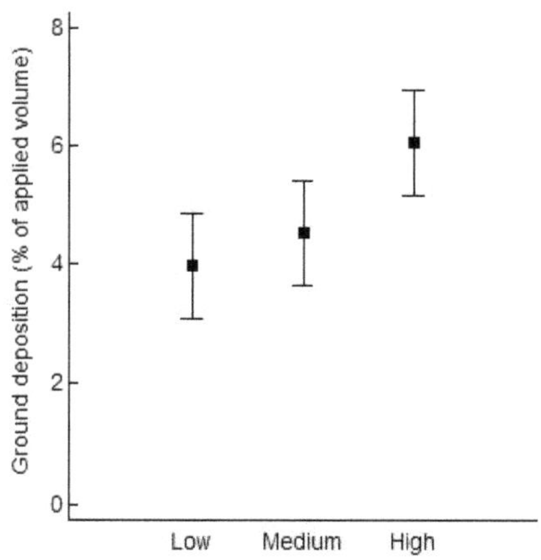

Figure 6. Sedimenting spray drift mean ± SE of the mean.

3.3 Airborne Spray Drift

The spray drift deposits on vertical samplers measured at six different heights for three different distances, downwind of the sprayed area, are presented in Table 7, and the mean data curves are shown in Figure 7.

Table 7. Airborne spray drift deposition.

Distance m	Height m	Treatments "Low" L1 % Application Rate	"Low" L2 % Application Rate	"Medium" M1 % Application Rate	"Medium" M2 % Application Rate	"High" H1 % Application Rate	"High" H2 % Application Rate
5	1	5.42	6.71	9.93	11.69	14.54	17.93
	2	5.53	5.39	9.63	10.41	14.98	15.55
	3	4.73	4.41	8.29	9.14	13.40	12.10
	4	3.61	2.72	7.43	6.31	9.83	9.83
	5	2.81	2.09	5.66	4.21	6.50	5.35
	6	2.40	1.48	3.83	3.51	4.33	3.75
10	1	3.70	2.34	3.39	8.81	10.31	10.16
	2	3.25	2.24	3.25	8.56	9.78	10.60
	3	2.77	1.72	3.89	7.12	10.78	8.50
	4	2.20	1.87	3.86	5.27	7.75	7.12
	5	2.06	1.24	3.07	3.59	7.04	5.39
	6	1.91	1.76	3.03	3.07	4.85	3.47
15	1	2.39	1.42	2.32	7.88	8.35	8.26
	2	1.82	2.31	1.84	7.21	7.80	7.99
	3	1.68	1.38	2.35	5.99	7.36	7.22
	4	1.62	1.63	2.02	5.01	6.60	6.35
	5	1.82	1.04	2.54	4.08	5.29	4.67
	6	1.45	0.91	1.78	3.58	3.95	3.41

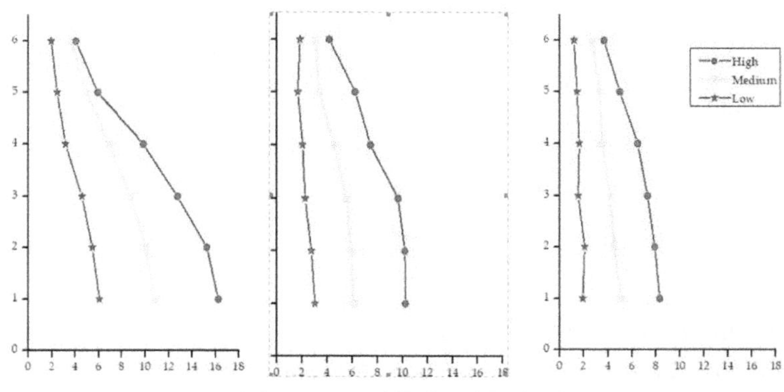

Figure 7. Airborne spray drift deposition profile at three distances from the sprayed area.

The results indicate that, in all treatments, the greatest deposition was measured at 5 m distance from the sprayed area, followed by 10 and 15 m distances. The three treatments (three categories of drift risk) generated different amounts of airborne drift over distance, having the greatest amount for "High" treatment and the lowest for "Low" treatment. The meteorological parameters had more influence on airborne spray drift than ground spray drift, since at 5, 10, and 15 m distances the amounts of airborne drift were greatest than the corresponding ground sediment.

Furthermore, in all treatments, spray drift deposition decreased with increasing height above ground, especially at 5 m distance from the sprayed area. This shape of the airborne drift profile is similar to the profile described by other researches in orchards [67,74], while there are also studies in vineyards showing that the airborne spray drift deposition increased with increasing height [64]. This discrepancy is justified by the fact that spray drift amount depends heavily on the architecture and geometry of the canopy [75,76]. The experimental vineyard had particularly poor canopy and small height, resulting in larger drift depositions in lower heights.

For the vertical samplers the three-way ANOVA test showed that there is a statistically significant effect of the collector's placement distance, downwind of the sprayed area, height from the ground, and drift risk indication of the tool on the spray drift deposits at level $p < 0.001$. Moreover, it was found that there are statistically highly significant interactions between distance and height ($p < 0.001$), height and drift risk indication ($p < 0.001$), and distance and drift risk indication ($p < 0.05$) (Table 8).

Table 8. Results of ANOVA for airborne spray drift as affected by distance, height, and drift risk indication.

Source	Df	Sum Sq	Mean Sq	F-Ratio	*p*-Value
Distance	2	436.733	218.366	91.09	0.00001
Height	5	614.543	122.909	51.27	0.00001
Drift Risk	2	1194.12	597.058	249.05	0.00001
Distance × Height	10	161.672	16.1672	6.74	0.00001
Distance × Drift Risk	4	24.7385	6.18463	2.58	0.0393
Height × Drift Risk	10	158.014	15.8014	6.59	0.00001
Distance × Height × Drift Risk	20	23.0276	1.15138	0.48	0.9709
Residual	162	388.369	2.39734		

Spearman's rank correlation test indicated that there is weak negative correlation between distance and drift (r_s = -0.3599), moderate negative correlation between height and drift (r_s = -0.4130), while drift risk category and drift have a strong positive correlation (r_s = 0.6993), meaning that drift strongly increased when wind speed and air temperature increased and relative humidity decreased, and the drift risk category followed the "Low"–"Medium"–"High" succession ($p < 0.001$) (Table 9). These results are largely consistent with other field studies. Fox et al. [77], in experiments on apple trees, found that deposits on floss decreased with height at 7.5 m distance downwind, but were more uniform across all heights at 15, 30, and 60 m downwind. Kasner et al. [78] reported that the distance from sprayed area, the height above ground, and the wind speed were significantly associated with drift level. Grella et al. [64] found a good significant relationship between airborne drift and wind speed variables, especially for the maximum and mean wind speed.

Table 9. Results of Spearman rank correlation test.

		Distance	Height	Drift Risk
	Correlation	-0.3599	-0.4130	0.6993
Drift (%)	Sample size	216	216	216
	p-Value	0.00001	0.00001	0.00001

Fisher's least significant difference (LSD) procedure was applied for pair-by-pair comparison among the means of the three treatments for the distances of 5, 10, and 15 m from the sprayed area and for all heights above the ground ($p < 0.05$). The results showed statistically significant differences between all treatment pairs, except for "Medium"–"High" pair at the heights of 5 and 6 m, at 5 m distance, and 6 m high at 15 m distance (Table 10). Furthermore, in Figure 8 the means of airborne drift for all treatments at distances 5, 10, and 15 m are presented. The highest airborne drift was registered in the distance closest to the sprayed area, and in each distance airborne drift was reduced following the "High"–"Medium"–"Low" order.

Table 10. Multiple-range LSD tests for vertical samplers.

Distance	Height	+/−Limits	Low-Medium		Medium-High		Low-High	
			Sig.	Difference	Sig.	Difference	Sig.	Difference
5	1	4.56578	*	−4.7505	*	−5.42075	*	−10.1713
	2	3.73621	*	−4.56775	*	−5.2415	*	−9.80925
	3	2.71095	*	-4.148	*	−4.03025	*	−8.17825
	4	2.08707	*	−3.70825	*	−2.95925	*	−6.6675
	5	1.65814	*	−2.48625		−0.98875	*	−3.475
	6	0.989128	*	−1.73475		−0.369	*	−2.10375
10	1	3.07183	*	−3.087	*	−4.13	*	−7.217
	2	3.10751	*	−3.1645	*	−4.2845	*	−7.449
	3	2.82998	*	−3.25975	*	−4.134	*	−7.39375
	4	1.44807	*	−2.5295	*	−2.86675	*	−5.39625
	5	1.27078	*	−1.67875	*	−2.8825	*	−4.56125
	6	0.897631	*	−1.2185	*	−1.10525	*	−2.32375
15	1	3.06148	*	−3.19225	*	-3.209	*	−6.40125
	2	2.949	*	−3.156	*	-3.3745	*	−5.8305
	3	2.069	*	−2.64175	*	−3.1235	*	−5.76525
	4	1.86667	*	−1.89125	*	−2.9625	*	−4.85375
	5	1.11037	*	−1.88125	*	−1.66925	*	−3.5505
	6	1.21974	*	−1.49475		−1.002	*	−2.49675

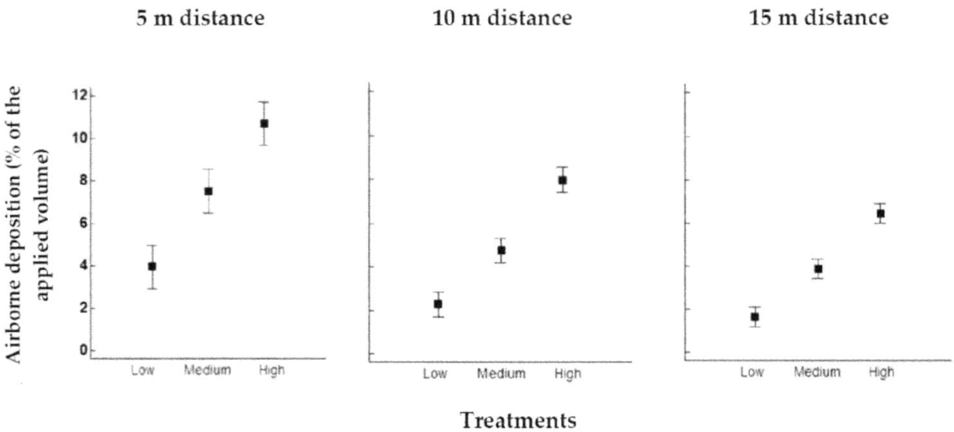

Figure 8. Airborne spray drift mean + - SE of the mean.

4. CONCLUSIONS

It is undeniable that there is great need to develop measures for spray drift reduction. Weather conditions during spraying application have considerable impact on sediment and airborne spray drift. Therefore, this work has developed a drift risk assessment tool based on the methodology derived from the TOPPS-Prowadis drift evaluation tool, which uses weather forecast data to predict spray drift extend. This tool is adapted to the Greek meteorological conditions and has three classification categories (low, medium, high) based on air temperature, relative humidity, and wind speed.

This tool and its limits between categories were evaluated successfully in real conditions, by conducting spray drift measurements in the vineyard of the Agricultural University of Athens, according to the ISO 22866:2005 methodology. Results showed that there are significant differences in airborne as well as in sediment drift among the treatments that correspond to the three drift risk classifications. The highest amount of spray drift deposits was observed within "High risk" treatments, which relate to unfavorable weather conditions; and the lowest within "Low risk" treatments, which are related to ideal weather conditions for spraying application.

The experimental results make evident that fine-tuning of the limits provides an optimized tool for Greek conditions that allows farmers to spray their vineyards with limited spray drift, resulting in higher spraying efficacy and minimized environmental impact. This tool shows that the limits classifying spray drift risk can be adjusted based on local experience for optimized results in spraying, but at the same time strengthens the importance of the TOPPS-Prowadis tool as a basis for such local optimization. Such a tool could be converted to software to assist farmers to predict and plan future spraying activities avoiding PPP application in "High" spray drift risk days.

This research produced the first set of data on spray drift amounts in vineyards when working with a conventional air-assisted sprayer and proved the efficiency of the tool. However, since the assessment of the effect of uncontrolled environmental conditions is objectively very difficult, the developed drift assessment tool will have to be tested in a wider range of environmental conditions and under different spraying techniques and crop characteristics.

References

1. Oerke, E.C. Crop losses to pests. *J. Agric. Sci.* **2006**, *144*, 31–43. [CrossRef]
2. Cooper, J.; Dobson, H. The benefits of pesticides to mankind and the environment. *Crop Prot.* **2007**, *26*, 1337–1348. [CrossRef]
3. Giles, D.K.; Akesson, N.B.; Yates, W.E. Pesticide application technology: Research and development and the growth of the industry. *Trans. ASABE* **2008**, *51*, 397–403. [CrossRef]
4. Garcera, C.; Roman, C.; Molto, E.; Abad, R.; Insa, J.A.; Torrent, X.; Planas, S.; Chueca, P. Comparison between standard and drift-reducing nozzles for pesticide application in citrus: Part II. Effects on canopy spray distribution, control efficacy of Aonidiella aurantii (Maskell), beneficial parasitoids and pesticide residues on fruit. *Crop Prot.* **2017**, 94, 83–96. [CrossRef]
5. ISO 22866:2005. *Equipment for Crop Protection-Methods for Field Measurement of Spray Drift*; International Organization for Standardization: Geneva, Switzerland, 2005; pp. 1–17.
6. Nuyttens, D.; De Schampheleire, M.; Baetens, K.; Sonck, B. The influence of operator- controlled variables on spray drift from field crop sprayers. *Trans. ASABE* **2007**, 50, 1129–1140. [CrossRef]
7. Butler Ellis, M.C.; Lane, A.G.; O'Sullivan, C.M.; Miller, P.C.H.; Glass, C.R. Bystander exposure to pesticide spray drift: new data for model development and validation. *Biosyst. Eng.* **2010**, 107, 162–168. [CrossRef]
8. Felsot, A.S.; Unsworth, J.B.; Linders, J.B.H.J.; Roberts, G. Agrochemical spray drift; assessment and mitigation-a review. *J. Environ. Sci. Health Part B* **2011**, 46, 1–23. [CrossRef]
9. Hilz, E.; Vermeer, A.W.P. Spray drift review: the extent to which a formulation can contribute to spray drift reduction. *Crop Prot.* **2013**, 44, 75–83. [CrossRef]

10. Benbrook, C.M.; Baker, B.P. Perspective on dietary risk assessment of pesticide residues in organic food. *Sustainability* **2014**, 6, 3552–3570. [CrossRef]
11. Kruger, R.G.; Klein, N.R.; Ogg, L.C. *Spray Drift of Pesticides*; University of Nebraska-Lincoln Extension: Nebraska, NE, USA, 2013; p. G1773.
12. Balsari, P.; Grella, M.; Marucco, P.; Matta, F.; Miranda-Fuentes, A. Assessing the influence of air speed and liquid flow rate on the droplet size and homogeneity in pneumatic spraying. *Pest Manag. Sci.* **2019**, 75, 366–379. [CrossRef]
13. Ozkan, H.E. *New Nozzles for Spray Drift Reduction;* AEX-523-98; Ohio State University Extension Fact Sheet Food Agricultural and Biological Engineering: Columbus, OH, USA, 1998.
14. Hofman, V.; Solseng, E. *Reducing Spray Drift;* North Dakota State University NDSU Extension Service AE-1210: Fargo, ND, USA, 2001.
15. Farooq, M.; Salyani, M. Modeling of spray penetration and deposition on citrus tree canopies. *Trans. ASABE* **2004**, *47*, 619–627. [CrossRef]
16. Da Silva, A.; Sinfort, C.; Tinet, C.; Pierrot, D.; Huberson, S. A lagrangian model for spray behaviour within vine canopies. *Aerosol Sci.* **2006**, *37*, 658–674. [CrossRef]
17. Yi, C. Momentum transfer within canopies. *J. Appl. Meteorol. Climatol.* **2008**, 47, 262–275. [CrossRef]
18. Take, M.E.; Barry, J.W.; Richardson, B. *An FSCBG Sensitivity Study for Decision Support Systems;* ASAE Annual Meeting: Phoenix, AZ, USA, 1996; p. 961037.
19. Arvidsson, T.; Bergström, L.; Kreuger, J. Spray drift as influenced by meteorological and technical factors. *Pest. Manag. Sci.* **2011**, 67, 586–598. [CrossRef] [PubMed]
20. Nuyttens, D.; Schampheleire, M.D.; Verboven, P.; Sonck, B. Comparison between indirect and direct spray drift assessment methods. *Biosyst. Eng.* **2010**, 105, 2–12. [CrossRef]
21. Nuyttens, D.; de Schampheleire, M.; Baetens, K.; Brusselman, E.; Dekeyser, D.; Verboven, P. Drift from field crop sprayers using an integrated approach: results of a five-year study. **Trans. ASABE 2011**, *54*, 403–408. [CrossRef]
22. Miranda-Fuentes, A.; Marucco, P.; Gonzalez-Sanchez, E.J.; Gil, E.; Grella, M.; Balsari, P. Developing strategies to reduce spray drift in pneumatic spraying vineyards: Assessment of the parameters affecting droplet size in pneumatic spraying. *Sci. Total Environ.* **2018**, 616–617, 805–815. [CrossRef]
23. Miller, P.C.H.; Butler Ellis, M.C. Effects of formulation on spray nozzle performance for applications from ground-based boom sprayers. *Crop Prot.* **2000**, 19, 609–615. [CrossRef]

24. Stainier, C.; Destain, M.F.; Schiffers, B.; Lebeau, F. Effect of the entrained air and initial droplet velocity on the release height parameter of a Gaussian spray drift model. *Commun. Agric. Appl. Biolog. Sci.* **2006**, 71, 197–200.

25. De Ruiter, H.; Holterman, H.J.; Kempenaar, C.; Mol, H.G.J.; de Vlieger, J.J.; van de Zande, J. Influence of Adjuvants and Formulations on the Emission of Pesticides to the Atmosphere. In *A Literature Study for the Dutch Research Programme Pesticides and the Environment (DWK) Theme C-2*; Plant Research International B.V.: Wageningen, The Netherlands, 2003; Report 59.

26. Hobson, P.A.; Miller, P.C.H.; Walklate, P.J.; Tuck, C.R.; Western, N.M. Spray drift from hydraulic spray nozzles: the use of a computer simulation model to examine factors influencing drift. *J. Agric. Eng. Res.* **1993**, *54*, 293–305. [CrossRef]

27. Miller, P.C.H. The measurement of spray drift. *Pestic. Outlook* **2003**, *14*, 205–209. [CrossRef]

28. van de Zande, J.C.; Stallinga, H.; Michielsen, J.M.G.P.; van Velde, P. Effect of sprayer speed on spray drift. *Annu. Rev. Agric. Eng.* **2005**, 4, 129–142.

29. Grella, M.; Marucco, P.; Manzone, M.; Gallart, M.; Balsari, P. Effect of sprayer settings on spray drift during pesticide application in poplar plantations (Populus spp.). *Sci. Total Environ.* **2017**, 578, 427–439. [CrossRef] [PubMed]

30. Arvidsson, T. *Spray Drift as Influenced by Meteorological and Technical Factors. A Methodological Study*; Swedish University of Agricultural Sciences, Acta Universitatis Agriculturae Sueciae: Agraria Sweden, 1997; Volume 71, p. 144.

31. Nuyttens, D.; Sonck, B.; De Schampheleire, M.; Steurbaut, W.; Baetens, K.; Verboven, P.; Nicolai, B.; Ramon, H. Spray drift as affected by meteorological conditions. *Commun. Agric. Appl. Biol. Sci.* **2005**, 70, 947–959. [PubMed]

32. Sumner, P.E. Reducing Spray Drift. In *Cooperative Extension Service*; The University of Georgia College of Agricultural and Environmental Sciences: Athens, GA, USA, 1997.

33. da Cunha, J.P.A.R.; Pereira, J.N.P.; Barbosa, L.A.; da Silva, C.R. Pesticide Application Windows in the Region of Uberlândia-MG, Brazil. *Biosci. J. Uberlândia* **2016**, 32, 403–411.

34. Maciel, C.F.S.; Teixeira, M.M.; Fernandes, H.C.; Zolnier, S.; Cecon, P.R. Droplet Spectrum of a Spray Nozzle under Different Weather Conditions. *Revista Ciência Agronômica* **2018**, 49, 430–436. [CrossRef]

35. Al-Jumaili, A.; Salyani, M. *Wind Effect on the Deposition of an Air-Assisted Sprayer*; University of Florida: Gainesville, FL, USA, 2014.

36. Thistle, H. Meteorological concepts in the drift of pesticide. In Proceedings of the International Conference on Pesticide Application for Drift Management, Washington State University, Waikoloa, HI, USA, 27–29 October 2004; pp. 156–162.
37. Holterman, H.J. *Kinetics and Evaporation of Water Drops in Air*; IMAG Report 2003-2012; Institute of Agricultural and Environmental Engendering: Wageningen, The Netherlands, 2003.
38. Carlsen, S.C.K.; Spliid, N.H.; Svensmark, B. Drift of 10 herbicides after tractor spray application. Primary drift (droplet drift). *Chemosphere* **2006**, 64, 778–786. [CrossRef]
39. Fishel, F.M. *When a Pesticide Doesn't Work*; Agronomy Department: Florida, FL, USA, 2008.
40. FOCUS, 2004. *Focus Surface Water Scenarios in the EU evaluation process under 91/414/EEC*; Report prepared by the FOCUS working group on Surface Water Scenarios; European Commission: Brussels, Belgium, March 2004; p. 238.
41. FOCUS, 2007a. Landscape and mitigation factors. In *Aquatic Risk Assessment. Extended Summary and Recommendations, vol. 1, Report of the FOCUS Working Group on Landscape and Mitigation Factors in Ecological Risk Assessment*; EC Document Reference SANCO/10422/2005 V.2.0; European Commission: Brussels, Belgium, 2007; p. 169.
42. FOCUS, 2007b. Landscape and mitigation factors. In *Aquatic Risk Assessment. Detailed Technical Reviews, vol. 2, Report of the FOCUS Working Group on Landscape and Mitigation Factors in Ecological Risk Assessment*; EC Document Reference: SANCO/10422/2005 V.2.0; European Commission: Brussels, Belgium, September 2007; p. 436.
43. Directive 2009/128/EC of the European parliament and the council of 21 October 2009 establishing a framework for community action to achieve the sustainable use of pesticides. Off. *J. Eur. Union* **2009**, *309*, 71–86.
44. Directive 2009/127/EC of the European parliament and of the council of 21 October 2009 amending Directive 2006/42/EC with regard to machinery for pesticide application. Off. *J. Eur. Union L* **2009**, *310*, 29–33.
45. EPA-United States Environmental Protection Agency. 2015; Reducing Pesticide Drift. Available online: http://www.epa.gov/reducing-pesticide-drift (accessed on 22 December 2015).
46. Huang, Y.; Thomson, S.J. Atmospheric Stability Determination Using Fine Time-Step Intervals for Timing of Aerial Application. In Proceedings of the ASABE, 2016 Annual International Meeting 162461162, Orlando, FL, USA, 17–20 July 2016. [CrossRef]

47. Nansen, C.; Ferguson, J.C.; Moore, J.; Groves, L.; Emery, R.; Garel, N.; Hewitt, A. Optimizing Pesticide Spray Coverage Using a Novel Web and smartphone Tool, SnapCard. *Agron. Sustain. Dev.* **2015**, 35, 1075–1085, INRA and Springer Verlag France, 2015. [CrossRef]
48. Baetens, K.; Nuyttens, D.; Verbovena, P.; De Schampheleire, M.; Nicolai, B.; Ramona, H. Predicting drift from field spraying by means of a 3D computational fluid dynamics model. *Comput. Electron. Agric.* **2007**, 56, 161–173. [CrossRef]
49. Kruckeberg, J.; Hanna, M.; Darr, M.; Steward, B. An interactive spray drift simulator. In Proceedings of the American Society of Agricultural and Biological Engineers Annual International Meeting, ASABE, Dallas, TX, USA, 13–16 September 2010; Volume 3, pp. 2297–2311.
50. Maber, J.; Dewar, P.; Praat, J.P.; Hewitt, A.J. Real Time Spray Drift Prediction. *Acta Hortic.* **2001**, *566*, 493–498. [CrossRef]
51. Hong, S.W.; Zhaoa, L.; Zhuc, H. SAAS, a computer program for estimating pesticide spray efficiency and drift of air-assisted pesticide applications. *Comput. Electron. Agric.* **2018**, *155*, 58–68. [CrossRef]
52. Nsibande, S.A.; Dabrowski, J.M.; van der Walt, E.; Venter, A.; Forbes, P.B.C. Validation of the AGDISP model for predicting airborne atrazine spray drift: A South African ground application case study. *Chemosphere* **2015**, *138*, 454–461. [CrossRef] [PubMed]
53. TOPPS-Prowadis Project. Best Management Practices to Reduce Spray Drift. 2014. Available online: http://www.topps-life.org/ (accessed on 24 June 2019).
54. Stavrakas, E.D. *Ampelographia*; Ziti Publications: Thessaloniki, Greece, 2010.
55. Meier, U. *Growth Stages of Mono-and Dicotyledonous Plants*: BBCH Monograph, 2nd ed.; Uwe Meier Federal Biological Research Centre for Agriculture and Forestry: Braunschweigh, Germany, 2001; pp. 1–158.
56. Pergher, G. Recovery rate of tracer dyes used for spray deposit assessment. *Trans. ASABE* **2001**, *44*, 787–794. [CrossRef]
57. Gil, E.; Balsari, P.; Gallart, M.; Llorens, J.; Marucco, P.; Andersen, P.G.; Fàbregas, X.; Llop, J. Determination of drift potential of different flat fan nozzles on a boom sprayer using a test bench. *Crop Prot.* **2014**, 56, 58–68. [CrossRef]
58. Otto, S.; Loddo, D.; Baldoin, C.; Zanin, G. Spray drift reduction techniques for vineyards in fragmented landscapes. *J. Environ. Manag.* **2015**, 162, 290–298. [CrossRef] [PubMed]
59. StatPoint Technologies Inc. *STATGRAPHICS Centurion XVI Version 16.1.15*; StatPoint Technologies Inc.: Warrenton, VA, USA, 1982–2011.

60. Spearman, C. The Proof and Measurement of Association between Two Things. *Am. J. Psychol.* **1904**, 15, 72–101. [CrossRef]

61. Fisher, R.A. *The Design of Experiments;* Oliver and Boyd: Edinburg, TX, USA; London, UK, 1935.

62. Levene, H. Robust tests for equality of variances. In *Contributions to Probability and Statistics: Essays in Honor of Harold Hotelling*; Olkin, I., Ghyrye, S.G., Hoeffding, W., Madow, W.G., Mann, H.B., Eds.; Stanford University Press: Palo Alto, CA, USA, 1960; pp. 278–292.

63. Bode, L.E.; Butler, B.J.; Goering, C.E. Spray drift and recovery as affected by spray thickener, nozzle type, and nozzle pressure. *Trans. ASAE* **1976**, 19, 213–218. [CrossRef]

64. Grella, M.; Gallart, M.; Marucco, P.; Balsari, P.; Gil, E. Ground Deposition and Airborne Spray Drift Assessment in Vineyard and Orchard: The Influence of Environmental Variables and Sprayer Settings. *Sustainability* **2017**, 9, 728. [CrossRef]

65. Balsari, P.; Marucco, P.; Grella, M.; Savoia, S. Spray drift measurements in Italian vineyards and orchards. In Proceedings of the 13th Workshop on Spray Application in Fruit Growing (SuproFruit 2015), Lindau/Lake Costance, Germany, 15–18 July 2015; pp. 30–31.

66. Balsari, P.; Marucco, P. *Sprayer Adjustment and Vine Canopy Parameters Affecting Spray Drift: The Italian Experience;* DEIAFA-University of Turin: Torino, Italy, 2004.

67. Torrent, X.; Garcera, C.; Molto, E.; Chueca, P.; Abad, R.; Grafulla, C.; Roman, C.; Planas, S. Comparison between standard and drift-reducing nozzles for pesticide application in citrus: Part, I. Effects on wind tunnel and field spray drift. *Crop Prot.* **2017**, 96, 130–143. [CrossRef]

68. Grella, M.; Marucco, P.; Balsari, P. Toward a new method to classify the airblast sprayers according to their potential drift reduction: comparison of direct and new indirect measurement methods. *Pest Manag. Sci.* **2019**, 75, 2219–2235. [CrossRef] [PubMed]

69. Threadgill, E.D.; Smith, D.B. Effects of physical and meteorological parameters on the drift of controlled-size droplets. *Trans. ASAE* **1975**, 18, 51–56.

70. De Schampheleire, M.; Baetens, K.; Nuyttens, D.; Spanoghe, P. Spray drift measurements to evaluate the Belgian drift mitigation measures in field crops. *Crop Prot.* **2008**, 27, 577–589. [CrossRef]

71. Nuyttens, D.; Zwertvaegher, I.; Dekeyser, D. Comparison between drift test bench results and other drift assessment techniques. *Asp. Appl. Biol. Int. Adv. Pestic. Appl.* **2014**, *122*, 293–302.

72. Combellack, J.H.; Westen, N.M.; Richardson, R.G. A comparison of the drift potential of a novel twin fluid nozzle with conventional low volume flat fan nozzles when using a range of adjuvants. *Crop Prot.* **1996**, 15, 147–152. [CrossRef]

73. Huang, Y.; Zhan, W.; Fritz, B.; Thomson, S.; Fang, A. Analysis of impact of various factors on downwind deposition using a simulation method. *J. ASTM Int.* **2010**, 7, 1–11. [CrossRef]

74. van de Zande, J.C.; Butler Ellis, M.C.; Wenneker, M.; Walklate, P.J.; Kennedy, M. Spray drift and bystander risk from fruit crop spraying. *Asp. Appl. Biol.* **2014**, *122*, 177–185.

75. Bird, S.L.; Esterly, D.M.; Perry, S.G. Atmospheric pollutants and trace gases. Off-target deposition of pesticides from agricultural aerial spray applications. *J. Environ. Qual.* **1996**, *25*, 1095–1104. [CrossRef]

76. Duga, A.T.; Ruysen, K.; Dekeyser, D.; Nuyttens, D.; Bylemans, D.; Nicolai, B.M.; Verboven, P. Spray deposition profiles in pome fruit trees: Effects of sprayer design, training system and tree canopy characteristics. *Crop Prot.* **2015**, 67, 200–213. [CrossRef]

77. Fox, R.; Hall, F.; Reichard, D.; Brazee, R.D.; Krueger, H.R. Pesticide tracers for measuring orchard spray drift. *Appl. Eng. Agric.* **1993**, *9*, 501–505. [CrossRef]

78. Kasner, E.J.; Fenske, R.A.; Hoheisel, G.A.; Galvin, K.; Blanco, M.N.; Seto, E.Y.W.; Yost, M.G. Spray Drift from a Conventional Axial Fan Airblast Sprayer in a Modern Orchard Work Environment. *Ann. Work Expos. Health* **2018**, 62, 1134–1146. [CrossRef] [PubMed]

Chapter 7

DEVELOPMENT OF AN AUTONOMOUS ELECTRIC ROBOT IMPLEMENT FOR INTRA-ROW WEEDING IN VINEYARDS

1. INTRODUCTION: AN OVERVIEW

The impact of chemical pesticides became an important global issue for the sustainability of the food production system. More and more organic food is produced and has become popular all over the world. One reason for the use of chemical pesticides is to reduce weeds, as they can be responsible for high yield losses [1]. However, to leave them untreated is not an option, as weeds could be responsible for a decrease of up to 40% of yield [2,3]. The importance of mulching and weeding in vineyards was already conducted in different studies. It was proclaimed that organic mulching was the most sustainable practice for yearly vine production [4]. In order to reduce chemical weed control, mechanical weeding approaches are promising alternatives. Mechanical weed control can be conducted between the tree/crop rows (inter-row) and within the tree/crop rows (intra-row). The main challenge of mechanical weed control is the realization at the intra-row area [5]. Weeding close to the crop/trunk and the use of intra-row weeding tools needs a very accurate steering for not damaging the crop/trunk [6]. Therefore, accurate guidance systems are needed. During the last decades, navigation has been improved by the use of new automatic row guidance systems using feelers, Global Navigation Satellite Systems (GNSS), distance sensors and cameras [7–9]. The development of these technologies has created wide opportunities for weed management and may become a key element of modernweed control [10]. Anyhow, in order to increase the selectivity of the weed control and reduce the crop plants damage, intelligent autonomous weeding systems are needed. Identifying individual weeds and crops is possible by the use of mechanical feelers, GNSS coordinates, sonar sensors, laser scanners, or cameras [11–14]. Automation offers the possibility to determine and differentiate crops from weeds and at the same time, to remove the weeds with a precisely controlled device. Automated weeding implements developed [15–18], provide examples of how the control of mechanical weeding can be performed. Real-time weeding robots have the potential to control intra-row weeds plus reduce the reliance on herbicides and hand weeding [19,20].

For reducing labor costs of mechanical weed control the whole task has to be automated. This approach could result in self-guided, self-propelled and autonomous machines that could cultivate crops with minimal operator intervention. Weeding robots can improve labor productivity, solve labor shortage, improve the environment of agricultural production, improve work quality, reduce energy input, improve resource management, and help farmers to change their traditional working methods and conditions [21]. Introducing self-governing mobile robots in agriculture, forestry and landscape conservation is dependent on the natural environment and the structure of the facilities. The solution proposed is the use of little and light machines, like unmanned vehicles that are self-propelled, autonomous and low powered [22]. Mobile robots must always be aware of its surroundings such as unpredicted obstacles [23]. To this scope, real-time environment detection could be performed using different sensor systems.

There were several autonomous machines and robots developed in the past [24]. Zhang et al. investigated the performance of a mechanical weeding robot [25]. They used machine vision to locate crop plants and a side shift mechanism to move along the crop rows. The intra-row weeds were controlled by moving hoes that could open and close to prevent damage to crop plants. Another study by Nørremark et al. developed an automatic GNSS-based intra-row weeding machine using a cycloid hoe [13]. The weeder used eight rotating tines, which could be released to tolerate crop plants. This automated weeding system mainly used Real Time Kinematic (RTK)-GNSS to navigate and control the side-shifting frame plus a second RTK-GNSS to control the rotating tines. Astrand and Baerveldt investigated the performance of a vision-based intra-row robotic weed control system [16]. The robotic system implemented two vision systems to guide the robot along the crop row and to discriminate between crops and weed plants. The weeding tool consisted of a rotating wheel connected to a pneumatic actuator that lifted and lowered the weeding tool to tolerate crop plants. Robots for weeding in vineyards were also developed, mainly focusing on trunk detection [26,27]. Weeds in vineyards can cause significant reductions in vine growth and grape yields [11]. Conventional control methods rely on herbicide applications in the vine rows and the area between them (middles), or a combination of herbicide strip application in the vine row and mowing or disking of the middles [28]. For a mobile robot to navigate in between the vine or tree rows, it must detect the position of the rows first, by means of detecting the trunks. The complete integration of an under-vine rotary weeder in an autonomous robot system for active intra-row weeding in vineyards and orchards was not done until now.

The overall goal of the current research was to develop and test the performance of a rotating electrical tiller weeder mechanism, built for automated intra-row weeding for vineyards. The system should be integrated in an autonomous robot platform and should be evaluated under controlled soil bin and outdoor conditions. The power requirement of the rotary blades mechanism and the total power for the autonomous robotic machine were of interest. The second interest was to test and compare two control strategies for detecting trunks, using a sonar sensor and a feeler.

2. MATERIALS AND METHODS

2.1 Mechanics and Electronics

An electric tiller head rotary weeder cultivator was manufactured and designed in the laboratory of the Institute of Agricultural Engineering, Hohenheim University and was mounted to an autonomous caterpillar robot called "phoenix" [14] (see Figure 1). The prototype electric rotary weeder was built up using a rotary weeder from Humus Co. called "Humus Planet" (Humus, Bermatingen, Germany). This tool originally operated with a hydraulic motor and was redesigned to get driven by an electric DC motor. A 1 kW 48DC brushed motor model ZY1020 (Ningbo Jirun Electric Machine Co.,Ltd., Nigbo, China) was used. The motor had a nominal torque of 2 Nm and a rated speed of 3200 rpm. The motor was mounted using a gearbox reducing with 30:1 ratio to produce a nominal torque of 60 Nm and a motor speed of 106 rpm at the rotary weeder. The gearbox was directly connected with a drive shaft to the rotary weeder. The motor and the gearbox were connected using a v-belt. The developed rotary weeder was fixed on a metal bar together with the drive unit.

The whole implement was attached to the robot using a coupling triangle. The implement could be moved up, down, left, right and could be tilted with the help of three Linak LA36 actuators (Linak, Nordborg, Denmark). They were capable of creating a pull/push force between 1700 N and 4500 N. This was sufficient to move the implement under all circumstances. The robot system was driven by two brushless motors HBL5000 (Golden Motor Technology Co., Ltd., Jiangsu, China), creating a maximum driving power of 10 kW, sufficient to move the 500 kg weight of the robot plus implement even in harsh terrain. The two times 80 × 15 cm belt system of the phoenix minimized soil compaction and provided enough grip and pull force for mechanical weeding. The width of the robot was 1.70 m. The implement could move 30 cm side wards, to move the tool inside the crop row. The power of the whole robot system was provided by batteries, with a nominal voltage of 48 V and 304 Ah. The implement motor was powered by a motor controller Model LB57 (Yongkang YIYUN Elektronic Co., Ltd., Zhejiang, China). The drive motors were controlled by two ACS48S motor controllers (Inmotion Technologies AB, Stockholm, Sweden). The whole robot was controlled and programmed with an embedded computer with i5 processor, 4 GB Ram and 320 GB disk space.

Figure 1. Autonomous robot called "phoenix" with attached rotary weeder implement and sensors.

2.2 Sensors

For following the tree and vine rows, a LMS 111 (Sick, Waldkirch, Germany) 2D laser scanner was used [29]. The mounting of the scanner was in the front of the vehicle to enable a security stop if an obstacle appeared in front of the robot. Additionally, there were four security switches attached to the robot, disabling the drive and implement motors in case of an emergency. For gaining the heading and traveled distance, an Inertial Measure Unit VN100 (VectorNav, Dallas, TX, USA) was fused with the hall sensor information of the drive motors [30]. The traveling distance was defined by the hall sensors, and the heading was estimated with the IMU. For guiding the implement, two different methods and sensors were tested. The first method used a standard feeler, a modified version of the original feeler of the rotary weeder. As the second method, a sonar sensor pico + 100/U (microsonic, Dortmund, Germany) was used to detect the trunks [31]. This sensor was used to detect the trunks, because it could be placed at any position at the robot. The signal could directly replace the input signal of the feeler without any additional software changes.

The specifications of the different sensor systems can be found in Table 1. The data of the feeler and the sonar sensor were digitalized using an analog digital converter from RedLab 1208LS (Meilhaus Electronic, Alling, Germany). This information could be used to send control signals to the implement and the linear motors.

Table 1. Sensor specifications of the used sensors of the robot.

Sensor	Specification	Value
LMS 111 [29]:	Operating range:	0.5 m to 20 m
	Field of view/scanning angle:	270°
	Data rate:	25 Hz
	Angular resolution:	0.5°
	Systematic error: Statistical error (1σ):	± 30 mm 12 mm (0.5–10 m)
Pico + 100/U [31]:	Operating range:	0.12 to 1.0 m
	Field of view/scanning angle:	22°
	Data rate:	10 Hz
	Distance measurement accuracy:	± 1%
VN100 [30]:	Data rate:	40 Hz
	Angular resolution:	0.05°
	Accuracy heading:	2.0°
	Accuracy pitch/roll:	1.0°

2.3 Software

The control and data logging of the whole robot system and implement was programmed using ROS Indigo-middleware [32]. The embedded computer ran under Ubuntu 14.04. All parts of the guidance software were programmed using C/C++. All data created were pushed to a so called "topic" which could be time stamped and recorded with the ROS system. This data could be analyzed for evaluating the power consumption, the movement of the robot and the implement behavior. The movement and the transformations between the robot base and the implement was performed using the Transform Library of ROS [33].

The guidance of the robot was performed with the use of 2D laser scanner data. First the data were converted to Cartesian coordinates and were filtered with a range filter of 5 m. The points were separated into two clusters, one for the left and one for the right side of the row. This was done by separating all points exactly at the laser scanner position into two point clouds (left and right). Afterwards the points were filtered using a Euclidean Clustering method using a distance threshold of 1.5 m. The resulting cluster with the highest point cloud number was estimated as the row point cluster. This clustering helped to get the row filtering more robust. The resulting points of each side were afterwards used to estimate a best fit line with the use of a RANSAC algorithm [34]. This resulted in two estimated lines, one for each row side. The median of the two resulting lines was used to define the row direction and the next goal point. When one line was missing, or a fixed distance to the row was needed, like in this weeding application, a fixed offset to one side could be set.

In this application a side offset of 0.15 m to the right side detected row was used. Therefore, the implement could move 0.15 m inside the crop row, as the maximum side shift was limited to 0.3 m. The forward speed was controlled by the software and just an active angle adjustment was done based on the line following algorithm of the laser scanner. The robot drive motor controller and the implement motors were controlled with CAN bus signals. The update rate of the motor control was fixed to 10 Hz.

The control of the implement was realized in a separate ROS node, analyzing the values of the switch and the sonar sensor. As soon as the feeler or the sonar detected an obstacle like a trunk, the linear actuator was shifted “inside”. When there was no obstacle detected, the actor moved “outside” until the maximum outer position was reached. The robot program could be started and stopped using a basic joystick. There were two different programs available, one using just the sonar value and one the feeler signals. The feeler signal just included a digital signal. The output of the sonar sensor was an analog signal between 0 and 10 V. This sensor output correlated with the object distance, detected by the sensor. As soon as the limit distance of 5V was read in the ADC Redlab, a digital input signal was triggered, shifting the actuator “inside”.

The height of the implement could be fixed to one depth, or could use the actual motor current of the implement to control the depth dependent on the motor torque. This could help to minimize the necessary force and prevent the rotary tiller weeder to get stuck. In the indoor soil bin, the height was fixed, as the surface was even and without any huge disturbances. In the outdoor tests the adaptive height adjustment variance was used to adjust the implement to the soil surface.

The following Figure 2 shows the ROS-visualization tool “rviz” used to visualize the live data, acquired at the robot system. The robot pose was visualized in real time together with the implement pose, the laser scanner data, the sonar data and the next waypoint the robot wanted to follow. This helped to debug the system while programming and provided feedback to the user.

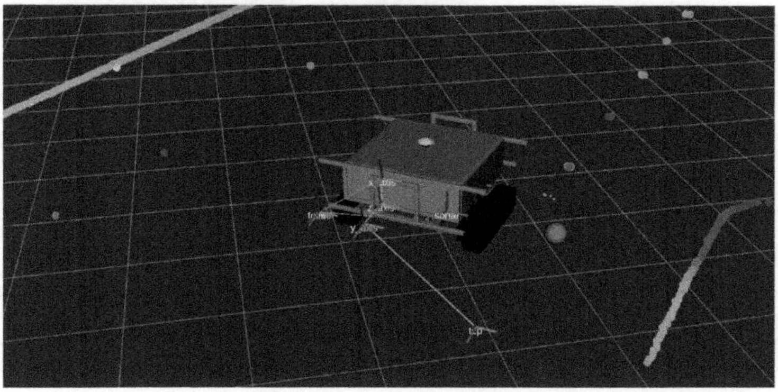

Figure 2. ROS-Software environment visualization "rviz" of robot position, implement position and the laser scanner points (small dots) and sonar data (purple ball).

The tilled area in the vineyard was analyzed by using the open source software ImageJ version 5.22 [35]. For further processing of the gained data, the software Microsoft Excel (Microsoft Corporation, Redmond, WA, USA) was used.

2.4 Test Environments

The indoor experiment was performed at the Hohenheim University soil bin laboratory. The indoor soil bin had 46 m length, 5 m width and 1.2 m depth. The mixture of the soil content was 73% sand, 16% silt and 11% clay. The soil was prepared by a rotary harrow, first loosening the soil and afterwards re-compacting the upper soil layer using a flat roller. 22 different plots were prepared and the soil was marked and oriented by using poles of wood with the size of 3 × 3 × 160 cm. The poles were separated with 1.5 m distance and simulated the trees/trunks for the indoor test. The total area evaluated for the treatment quality between every two poles was 0.45 m2 (1.5 m length, 0.3 m row width). This area was divided into 18 rectangles marked on the soil by using chalk. After the experiment, the rectangles with no interaction with the tiller were defined as non-tilled area. Three main soil parameters were recorded for each individual plot. The penetration resistance, soil moisture content and soil shear strength conditions were measured with an H-60 Hand-Held Vane Tester (GEONOR, Inc., Augusta, GA, USA). The specific soil moisture was logged using a TRIME-PICo 64 Time-Domain-Reflectometry sensor (IMKO Micromodultechnik GmbH, Ettlingen, Germany). The penetration force was measured with an Eijkelkamp penetrometer logger (Eijkelkamp, Giesbeek, The Netherlands).

It was decided to use the absolute rotational speed for the implement and a forward speed for the robot of 0.16 m s^{-1}. 11 plots were treated using a feeler and the remaining 11 plots were treated while detecting the poles with the sonar sensor. The position of the sonar and the feeler was placed m before the weeder, so that the implement could move out of the way even without stopping the robots forward speed. The goal distance between the robot and the row was fixed to 0.15 m in the row following software. When a pole was detected, the electric actuator shifted the rotary out of the row, to prevent damage to the poles. As soon as the pole was passed, the actuator got shifted back inside the row. The linear motor used for the side shift had a maximum speed of 68 mm/s with no load and 52 mm/s at maximum load. On average this resulted in a travel speed of 60 mm/s of the linear LA36 actuator. As the implement was set up to move 0.15 m to the intra row area, at least a time difference between pole detection and arrival of the tool of 2.5 s were needed. As the traveling speed of the robot was fixed to 0.16 m/s, it was necessary to detect the tree at least 0.4 m before the implement. For gaining some buffer, the pole detection was set up at a distance of 0.5 m. The time delay for the shifting mechanism was fixed to 2.5 s in the software to get as close as possible to the poles and therefore to potential trunks. The following Figure 3 describes the mechanical and electrical components of the two control methods using the sonar sensor and the feeler.

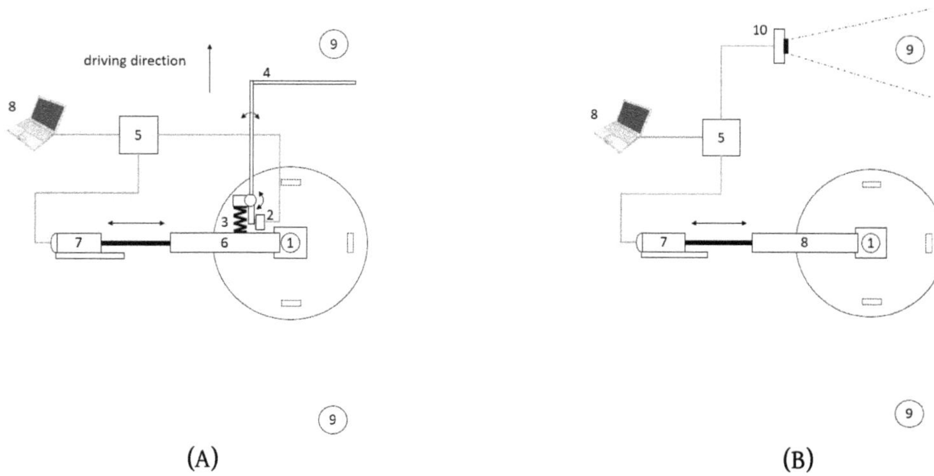

Figure 3. The two different control methods: (A) Feeler controlled unit and (B) Sonar sensor controlled actuator for the developed autonomous electric weeder. 1-Electric weeder; 2-Switch; 3-Spring; 4-Feeler; 5-Embedded controller; 6-Bar for implement; 7-Linear actuator; 8-PC; 9-Trunks (poles); 10-Sonar sensor.

For each treatment, the power consumption and the accuracy of the processing procedure were evaluated estimating the treated and the non-treated area. The better control method tested in the indoor laboratory was evaluated at the vineyard of Hohenheim University (48.710115 N, 9.212913 E) on 9 October 2018 (see Figure 4). The following figure shows the two different test areas for the autonomous weeding robot.

(A)

(B)

Figure 4. The two test environments- (A) indoor laboratory and (B) outdoor vineyard.

3. RESULTS AND DISCUSSION

3.1 Comparison of Feeler and Sonar

Over the plots of the indoor trial, the soil moisture varied between 13.1% and 7.4% by volume. The shear stress varied between 59 and 82 MPa. The penetration resistance varied between 2 and 4 MPa for the depths of the first 10 cm. The detailed results of the soil physical properties for the laboratory test can be found in Table 2.

Table 2. Soil physical properties for the evaluation of the sonar and the feeler method in the soil bin and the resulting tilled area (indoor laboratory test).

Plot	Sensor	Soil Penetration, MPa (0–10 cm Depth)	Moisture Vol. %	Soil Temp., °C	Shear Stress, MPa	Tilled Area, %
t1	Sonar	2	13.8	20.1	61.5	94
t2	Sonar	2	12.7	20.4	61	83
t3	Sonar	3.8	9.09	20.8	61	89
t4	Sonar	4	9.7	20.9	59	100
t5	Sonar	3	8.8	21	60.5	89
t6	Sonar	2	7.4	21.3	56	94
t7	Sonar	2	13.2	21.9	60.5	72
t8	Sonar	2.5	9.8	22	69.5	89
t9	Sonar	2.5	11.5	22.2	66.5	61
t10	Sonar	2.2	8.8	22.4	67.5	83
t11	Sonar	2.4	9.5	22.5	68	50
t1	Feeler	3.8	10.1	22.6	62	50
t2	Feeler	3	11.9	22.7	66	61
t3	Feeler	3.5	12.07	22.9	79.5	61
t4	Feeler	2.5	13.1	23	82	94
t5	Feeler	4	9.8	23.6	70	78
t6	Feeler	4	8.8	23.7	81	78
t7	Feeler	4	9.5	24	70	56
t8	Feeler	4	9.7	24	73	44
t9	Feeler	3.5	10.07	24	75	78
t10	Feeler	4	9.7	24.8	79	56
t11	Feeler	3.8	9.7	24.8	79	61

3.2 Efficiency of the Control Algorithm

The percentages of the tilled area between the poles (trunks) were used as parameters to indicate the efficiency of weed control in the current study. Table 2 indicates that the sonar treatment performed better than the feeler. Only at the plots t9 and t11 the feeler performed better than the sonar. In all other plots sonar performed equal or better compared to the feeler solution. The overall performance of both methods was at least over 50% of the intra-row area, at some plots reaching a tilled area of almost 100%. The average of the tilled area for the feeler was 65% and for the sonar it was 82%.

The results of the tilled area indicated a high success rate of both methods. However, the sonar control method outperformed the feeler control. One reason for this was that the sonar signals of the poles could directly be converted to a control message. The feeler sometimes got stuck at the poles for some seconds, causing wrong signals to the actuator and made it hard for the software to define the exact pole position. Additionally, the long lever of the feeler caused vibrations, sometimes triggering the switch even when there was no pole. With no other disturbances e.g., of weeds and branches, the sonar worked reliably and without any failure. Therefore, it was suggested, to test this control method under outdoor conditions later on.

The signals from the sonar sensor over the travelled distance in the soil bin can be seen in the following Figure 5. The gained signals showed a peak at the pole positions. At the beginning the sonar values hit the edges of the poles, causing wrong distance signals. This effect could be seen at both sides of the poles. However, the minimum value of the sticks was quite accurate in distance and position. The signals always formed small peaks with similar values. This shape could be used for further filtering of the exact trunk positions in the future and separate high weed spots from trunks. All poles triggered at least one signal of less than 0.5 m distance, which triggered the side shift of the implement. Therefore, the sonar could detect all poles at the indoor test.

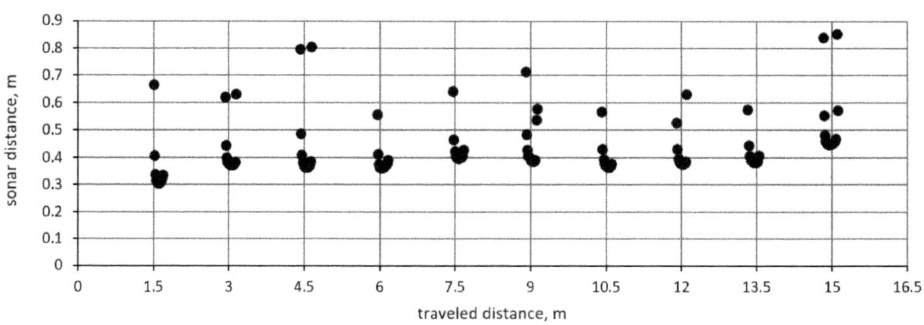

Figure 5. Sonar values detected at the indoor test at the laboratory (soil bin).

In the following outdoor test just the sonar detection was evaluated. Figure 6 shows the sonar detections based on the test in the vineyard of Hohenheim University. In this case the trunks were not spaced out evenly like in the indoor tests. Sometimes individual trunks were missing, or were closer together (less than 0.5 m), causing that in this case the intra-row area could not be entered by the tilling device. Additionally, there were shoots and weeds causing disturbances.

However, with small height adjustments of the sonar sensor, the system worked well in the outdoor vineyard. The linearity of the row was not given, because of curved trunks, as they were not in a straight line as the wooden poles. However, all vines were successfully detected, causing no issues with hitting the vines basal area with the implement or similar. But the within-row spacing of the vines was not always suitable for the 0.3 m diameter electric weeder head, as there was not enough space to enter the intra row area between some trunks. Some of them were close to 0.5 m distance, causing the system to stay outside of the crop row. Using different sizes of weeder heads would solve this issue. At the spots with bigger gaps, the system performed well and did not cause any problems at all. When analyzing the tilled area of the outdoor test using the ImageJ software, the min value reached 57%, the max values 98% of the tilled area. Areas where the tiller head could not enter were excluded from the analysis.

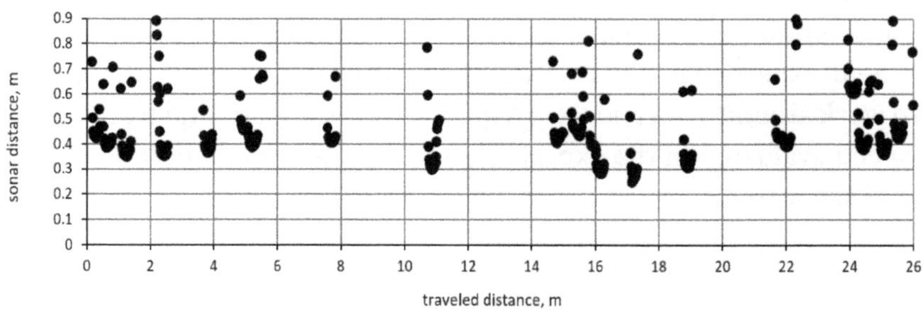

Figure 6. Sonar values detected at the outdoor test (vineyard).

It could be shown, that the outdoor test did perform well under the given circumstances. However, the environment was not sloped, which could bring more issues to the trunk detection. At high angles, shoots and leaves could hang more into the area of the sonar sensor. The automatic trunk detection would work in ordered and well cultivated vineyards. More complex examples would need an additional tree detection system like a laser scanner or camera, which could filter out the sonar values depending on the environment and context. Additionally, the trajectory of the implement could be planned dependent on the robot forward speed. This could make a more flexible path planning possible. However, the tilled area and the detection rate of the system were sufficient for the evaluated test area.

3.3 Power Requirement for the Autonomous Robot

The maximum and average power requirement for the implement guidance, actuator motor and belt drive of the autonomous robot is shown in the following Figure 7. In the indoor test, the power consumption increased with higher soil penetration resistance and shear stress. The overall power consumption for the implement was smaller than in the outdoor test. Here, an additional factor was the vegetation layer of weeds, which caused higher power consumption. Because of the implemented algorithm for limiting the working depth depending on the implement motor current, the power consumption peaks in the outdoor test were smaller than in the indoor test. When the soil in the indoor experiment would be compacted more, this difference between indoor and outdoor power consumption for the implement could be equalized.

In all tests the right drive motor of the robot needed more power than the left one. This was quite logical, as the implement was attached to the right side of the robot. The inertial forces of the tiller caused the robot to move to the right. To overcome this momentum, the row follower sent correction signals to the drive motors, forcing the right motor to consume more current. This effect could be seen at all performed tests. In the outdoor experiment, the slope of the tested area was not completely horizontal, causing higher power consumption for the drive motors. The overall power consumption of the machine was 1.8 KW in the soil bin and 2.2 kW in the outdoor field experiment. This would allow this machine with the given battery power to work around 8.1 h and 6.6 h depending on the slope and weed density. With a fixed speed evaluated in the experiments, the system could perform an intra-row weeding of 500 m/h.

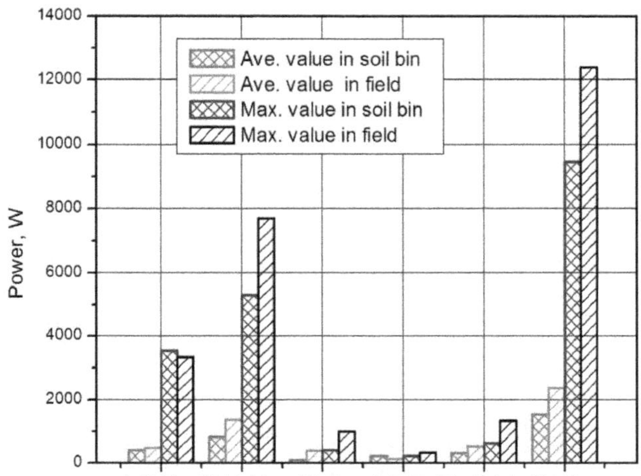

Figure 7. Overall power consumption of the robot for the weeding task.

When comparing the power consumption of the whole machine with the official power requirement provided by the rotary tiller manufacturer, the energy reduction is immense. The official power requirement for the tiller was set up with 8.1 kW by the company. Comparing the power consumption of the whole robot, the savings are more than 5.9 kW, meaning a 73% energy reduction. When we just compare the attachment, a 1 kW motor was enough to drive the implement in the performed experiments. When driving the robot system in a sloped environment, the power consumption of the drive motors would increase. However, it is expected, that the power consumption of the actuator would not differ in sloped environments.

3.4 Outlook

It could be shown that with an electrical implement a high power reduction could be possible, even when using commercial weeding tools. Combined with autonomous robots, this system could be highly energy efficient and sustainable. However, the system could be improved in the future. When using a higher speed for the linear side shift control, the overall speed of the machine could be increased, depending on the rotary speed of the rotary tiller and the side shift performance. Additionally, the tillage quality of the system could be increased, when the robot would stop at each trunk, to move the rotary weeder out of the row. Afterwards the robot could move a little bit forward and move the tool back into the row as soon as the trunk has passed. This would result in a rectangular trajectory, just excluding the trunk. Between the trunks, the speed could be increased until the next trunk is detected.

A different solution for detecting the trunks could be to use the information of the navigation laser scanner in front of the robot. Together with the odometry value, the software could increase the accuracy and plan the best trajectory for the implement based on speed, trunk spacing and weed density. This could help optimize path and work quality for weeding tasks in vineyards and orchards. In combination with a feeler or a sonar sensor this information could be filtered and optimized. This could help to increase the robustness and to not be disturbed from shoots and higher weeds.

4. CONCLUSIONS

An autonomous tiller weeder for intra-row weeding in vineyards was developed and tested at the University of Hohenheim. The system was built up using a commercial rotary tiller weeder. It was mounted on a mobile robot platform. For the side shift of the system, linear electrical motors were used. The trunk detection was performed with two different methods, using a feeler and a sonar. Both methods performed well and did not harm any trunks. The overall tilled area between the trunks with the sonar was higher than with the feeler, caused by vibrations and the design of the feeler. The autonomous row following based on the laser scanner enabled the machine to follow the rows accurately. The energy consumption of the whole system was evaluated. It could be shown that electrical driven machines could perform autonomous weeding in vineyards, which is energy efficient. Therefore, autonomous intra-row weeding machines could save energy and time for workers.

References

1. Peteinatos, G.G.; Weis, M.; Andújar, D.; Rueda Ayala, V.; Gerhards, R. Potential use of ground-based sensor technologies for weed detection. *Pest Manag. Sci.* **2014**, 70, 190–199. [CrossRef] [PubMed]

2. Andújar, D.; Dorado, J.; Fernández-Quintanilla, C.; Ribeiro, A. An approach to the use of depth cameras for weed volume estimation. *Sensors* **2016**, *16*, 972. [CrossRef] [PubMed]

3. Oerke, E.-C. Crop losses to pests. *J. Agric. Sci.* **2006,** 144, 31. [CrossRef]

4. Susaj, L.; Susaj, E.; Belegu, M.; Mustafa, S.; Dervishi, B.; Ferraj, B. Effects of different weed management practices on production and quality of wine grape cultivar Kallmet in North-Western Albania. *J. Food Agric. Environ.* **2013**, 11, 379–382.

5. Bowman, G. *Steel in the Field—A Farmer's Guide to Weed Management Tools*; SARE Outreach: College Park, MD, USA, 1997; ISBN 188862602X.

6. Van Der Weide, R.Y.; Bleeker, P.O.; Achten, V.T.J.M.; Lotz, L.A.P.; Fogelberg, F.; Melander, B. Innovation in mechanical weed control in crop rows. *Weed Res.* **2008**, 48, 215–224. [CrossRef]

7. Barawid, O.C.; Mizushima, A.; Ishii, K.; Noguchi, N. Development of an autonomous navigation system using a two-dimensional laser scanner in an orchard application. *Biosyst. Eng.* **2007**, 96, 139–149. [CrossRef]

8. Reiser, D.; Paraforos, D.S.; Khan, M.T.; Griepentrog, H.W.; Vázquez-Arellano, M. Autonomous field navigation, data acquisition and node location in wireless sensor networks. *Precis. Agric.* **2017**, *18*, 279–292. [CrossRef]

9. Andújar, D.; Escolà, A.; Rosell-Polo, J.R.; Fernández-Quintanilla, C.; Dorado, J. Potential of a terrestrial LiDAR-based system to characterise weed vegetation in maize crops. *Comput. Electron. Agric.* **2013**, *92*, 11–15. [CrossRef]

10. Bajwa, A.A.; Mahajan, G.; Chauhan, B.S. Nonconventional weed management strategies for modern agriculture. *Weed Sci.* **2015**, 63, 723–747. [CrossRef]

11. Gerhards, R.; Oebel, H. Practical experiences with a system of site-specific weed control in arable crops using real-time image analysis and GPS-controlled patch spraying. *Weed Res.* **2006**, 46, 185–193. [CrossRef]

12. Weiss, U.; Biber, P.; Laible, S.; Bohlmann, K.; Zell, A. Plant species classification using a 3D LIDAR sensor and machine learning. *Proc. Int. Conf. Mach. Learn. Appl. ICMLA* **2010**, 339–345. [CrossRef]
13. Nørremark, M.; Griepentrog, H.W.; Nielsen, J.; Søgaard, H.T. The development and assessment of the accuracy of an autonomous GPS-based system for intra-row mechanical weed control in row crops. *Biosyst. Eng.* **2008**, *101*, 396–410. [CrossRef]
14. Reiser, D.; Martín-López, J.; Memic, E.; Vázquez-Arellano, M.; Brandner, S.; Griepentrog, H. 3D imaging with a sonar sensor and an automated 3-axes frame for selective spraying in controlled conditions. *J. Imaging* **2017**, 3, 9. [CrossRef]
15. Tillett, N.D.; Hague, T.; Grundy, A.C.; Dedousis, A.P. Mechanical within-row weed control for transplanted crops using computer vision. *Biosyst. Eng.* **2008**, 99, 171–178. [CrossRef]
16. Astrand, B.; Baerveldt, A.J. An agricultural mobile robot with vision-based perception for mechanical weed control. *Auton. Robots* **2002**, 13, 21–35. [CrossRef]
17. Cloutier, D.C.; Van der Weide, R.Y.; Peruzzi, A.; Leblanc, M.L. Mechanical weed management. *Non Chem. Weed Manag.* **2007**, 111–134. [CrossRef]
18. Griepentrog, H.W.; Nørremark, M.; Nielsen, J. Autonomous intra-row rotor weeding based on GPS. In Proceedings of the 2006 CIGR World Congress Agricultural Engineering for a Better World, Bonn, Germany, 3–7 September 2006; pp. 2–6.
19. Lee, W.S.; Slaughter, D.C.; Giles, D.K. Robotic weed control system for tomatoes. *Precis. Agric.* **1999**, 1, 95–113. [CrossRef]
20. Slaughter, D.C.; Giles, D.K.; Downey, D. Autonomous robotic weed control systems: A review. *Comput. Electron. Agric.* **2008**, 61, 63–78. [CrossRef]
21. Ge, Z.; Wu, W.; Yu, Y.; Zhang, R. Design of mechanical arm for laser weeding robot. In Proceedings of the 2nd International Conference on Computer Science and Electronics Engineering (CSE 2013), Los Angeles, CA, USA, 1–2 July 2013; pp. 2340–2343.
22. Blackmore, B.S.; Griepentrog, H.W.; Fountas, S.; Gemtos, T.A. A specification for an autonomous crop production mechanization system. *Agric. Eng. Int. CIGR Ej.* **2007**, IX. [CrossRef]
23. Bechar, A.; Vigneault, C. Agricultural robots for field operations: Concepts and components. *Biosyst. Eng.* **2016**, 149, 94–111. [CrossRef]
24. Anatis The Robotic in Support of the Environmental Friendly Agriculture! The Only Robot for Hoeing and Aid with Decision Making. Available online: http://

www.trp.uk.com/trp/uploads/carrebrochures/023_Anatis%20-%20GB.pdf (accessed on 9 January 2019).

25. Zhang, C.; Li, N. *System Integration Design of Intra-Row Weeding Robot*; ASABE: Saint Joseph, MI, USA, 2013.
26. Contente, O.M.D.S.; Lau, J.N.P.N.; Morgado, J.F.M.; Santos, R.M.P.M. Dos vineyard skeletonization for autonomous robot navigation. In Proceedings of the 2015 IEEE International Conference on Autonomous Robot Systems and Competitions (ICARSC 2015), Villa Real, Portugal, 8–10 April 2015; pp. 50–55. [CrossRef]
27. Igawa, H.; Tanaka, T.; Kaneko, S.; Tada, T.; Suzuki, S. Visual and tactual recognition of trunk of grape for weeding robot in vineyards. In Proceedings of the 35th Annual Conference of the Industrial Electronics Society (IECON 2009), Porto, Portugal, 3–5 November 2009; pp. 4274–4279. [CrossRef]
28. Velicka, R.; Mockeviciene, R.; Marcinkeviciene, A.; Pupaliene, R.; Kriauciuniene, Z.; Butkeviciene, L.M.; Kosteckas, R.; Cekanauskas, S. The effect of non-chemical weed control on soil biological properties in a spring oilseed rape crop. *Zemdirb. Agric.* **2017**, 104, 107–114. [CrossRef]
29. Sick AG Waldkirch Operating Instructions LMS1xx. Available online: https://mysick.com/saqqara/ im0031331.pdf (accessed on 11 January 2017).
30. Vectornav VN-100 IMU/AHRS High-Performance Embedded Navigation. Available online: https://www.vectornav.com/docs/default-source/documentation/vn-100-documentation/PB-12-0002.pdf?sfvrsn=9f9fe6b9_16 (accessed on 1 December 2017).
31. Microsonic Datasheet pico+100/U. Available online: https://www.microsonic.de/DWD/_111327/pdf/ 1033/microsonic_pico+100_U.pdf (accessed on 9 January 2019).
32. Quigley, M.; Conley, K.; Gerkey, B.; Faust, J.; Foote, T.; Leibs, J.; Berger, E.; Wheeler, R.; Mg, A. ROS: An open-source robot operating system. In Proceedings of the 2009 ICRA Workshop on Open Source Software, Kobe, Japan, 12–17 May 2009; Volume III, p. 5.
33. Foote, T. Tf: The transform library. In Proceedings of the 2013 IEEE Conference On Tehnologies for Practical Robot Applications (TePRA), Woburn, MA, USA, 22–23 April 2013.
34. Fischler, M.A.; Bolles, R.C. Random sample consensus: A paradigm for model fitting with applications to image analysis and automated cartography. *Commun. ACM* **1981**, 24, 381–395. [CrossRef]
35. Software, I. Open Source Software ImageJ. Available online: https://imagej.net/ Welcome (accessed on 29 October 2018).

Chapter 8

INFRARED THERMOGRAPHY APPLIED TO TREE HEALTH ASSESSMENT: A REVIEW

1. INTRODUCTION: AN OVERVIEW

The tree is a fundamental living being of ecosystems [1–3]. It contributes to regulating nature, climate, and urban greening [1]. It is also a source of wealth and employment [4]. Some trees are classified as natural and cultural heritage [5]. Trees provide enormous benefits, such as the ecosystems balance, prevention of desertification, and global warming, as well as the well-being of human populations and urban ecosystems [1]. However, trees are also prone to the risk of damage to people and assets associated with the falling of branches/trees, above all in urban spaces [6,7]. Risk increases when they have defects because it compromises their health and stability. Consequently, the safety of people and goods are compromised. To support decision making about trees, it is essential to monitor their health condition to understand their biological viability, cost-effectiveness, and associated risks.

The first discovery of the infrared spectrum was made by Herschel in 1800 during his study about optical filters to reduce the brightness of the sun in telescopes [8]. In his findings, Herschel also observed that these new rays had similar behaviour to the visible ones, i.e., they were absorbed, transmitted, refracted, and reflected [9]. The first thermal image (thermograph) was obtained by Herschel in 1840, by experiments with differential evaporation of a film of oil submitted to a heat pattern [8]. However, the crucial mark in the sensitivity of infrared detection was achieved by Langley in 1880 through the invention of the bolometer. The first infrared camera was developed by Tihanyi in 1929 and was applied for anti-aircraft defense by the British army [10]. The first uses of Infrared thermography (IRT) in trees was the assessment by infrared imaging of the crown condition, using monochrome and false colour films [11]. IRT is a non-contact technology that allows detection of damaged tissue in real-time. It is based on the detection of thermal differences between healthy and damaged zones.

This review paper concerns identifying the IRT effectiveness for inspection of the health of trees. For that purpose, this review was structured as follows. It starts with the research methods and then the importance of trees and their classification. The mechanism that trees use to protect themselves, as well as risks associated with them, and the way they should be managed, is referred to. Some methods and inspection techniques are also mentioned. The importance of sustainable techniques of inspection is highlighted. Principles of IRT, usefulness, and relevance are mentioned. A description of the IRT application for inspection of trees is done, and several experiments are described. The main differences between wood and tree properties are reviewed. Finally, the advantages and limitations associated with the use of IRT are described.

2. MATERIALS AND METHODS

A database query research paper was carried out in B-on, DOAJ, Google Scholar, Microsoft Academic, ProQuest, SciELO, ScienceDirect, Scopus, and Web of Science. The review was done in Portuguese, English and Spanish. It was carried out between October 2018 and February 2019. The keywords were: Infrared thermography, inspection, inspection techniques, non-destructive, non-invasive, sustainable, monitoring, classification, deterioration detection, relevance, benefits, risks, management, tree, tree heritage, and timber. The keywords were combined with each other. Articles from the reference list of reviewed papers were also taken into account. Whenever combinations of keywords were found, the abstracts and keywords were read. When the abstract proved to be insufficient for classification, the full article was read. The selected articles were classified according to sections, subsections, and theoretical references. The gathered data was organized by type of item, title, authors, abstract, keywords, and classification. These procedures facilitated the identification and analysis of the main references used by the authors, as well as citation frequency. All possible contributions to the application of IRT for health tree monitoring were considered. Duplicate references were withdrawn. A total of 239 papers were collected, of which only 81 papers were selected.

3. TREE RELEVANCE AND RISKS

In 1950, the world's urban population was 30%; nowadays, the urban population reached 55%, and by 2050, 68% of the world population is to probjected to live in urban areas [12]. That is, the urbanization in the world population is increasing. Trees and green spaces play important roles in integrating the inhabitants into their urban communities. Trees provide well-being and quality of life; they establish the connection of inhabitants with the natural world and contribute to the mitigation of climate change. Urban trees provide numerous benefits for their inhabitants, but they are also prone to risk [6,13,14].

Trees interact with the surrounding environment, making it more pleasant. The basic function of trees is the improvement of air quality. Trees reduce the rainwater flow, allowing the infiltration into underground aquifer environments. Trees increase biodiversity by providing shelter for wildlife. Trees provide wind protection to people and animals since their leaves alter the direction and the speed of the wind. The tree roots hold the soil, and the tree canopy protects the soil from falling rain. Trees reduce urban noises and promote a sense of calm and rest. Trees create spaces for recreation and tourism. Consequently, trees favour the reduction of stress and even crime, since public open space with trees tends to be used much more than space without trees and this increases surveillance [15]. They influence on the decrease of energy consumption in the summer season, providing shade and a low incidence of solar radiation [16]. Trees have influence in the increasing the property value, if they are in or even outside the property. Trees produce aesthetic benefits because they decorate the urban landscape and enhance the season' landscape [14,17–20].

Trees interact in the ecosystem. They are exposed to aggression caused by birds, insects, animals, fire, atmospheric conditions, or even human beings during their activities, and these interactions could result in wounds, which act as entrance doors for pathogenic microorganisms. Trees have developed a system of compartmentalisation known as CODIT (Compartmentalization of Decay in Trees), described by Shigo and Marx [21]. In this system, chemical and physical barriers are created; cells undergo changes to form four types of walls around the wound; the walls are organized in axial, radial, and tangential directions. They act against the proliferation of pathogens by isolating the wounded tissues and thus impeding their dissemination to prevent/mitigate tree deterioration [22]. Wood deterioration and trees ageing are natural processes, but they carry on risks to the tree itself, as well as people and property [23].

On the other hand, some trees are of special interest because of age, size, physical characteristics, as well as cultural and historical connections [5]. Some examples of classification are:

- ☆ Heritage: trees that are linked to history and culture; trees that are rare and/or botanically relevant (see for instance Reference [24]).

- ☆ Notable: trees that have reached maturity stand out from their surroundings because they are larger than the trees around them (see for instance Reference [25]).
- ☆ Ancient: trees that have overpassed the maturity and despite the hollow trunk they continue to be healthy. Their value is intrinsically linked to their age (see for instance Reference [26]).
- ☆ Veterans: trees that survived wounds and deterioration (decay); young trees that developed old trees characteristics (see for instance Reference [27]).

Nevertheless, trees are also a risk factor for people and property. Large numbers of imposing trees characterize most urban gardens and avenues. These trees are subjected to undergoing stress hostile environment characteristic of urban centres, and very often under poor management. As a consequence, many are unhealthy, and a damaged tree poses a threat to people safety [11]. Three people die each year as a result of falling trees in public spaces in the United Kingdom. It is a low risk when considered that one in 10 million die as a result of it [6,7]. While there is a low rate of tree incidents, it ends up altering the risk perception. The risk perception of the falling tree has led tree managers (owners and managers of urban green spaces) to knock down trees at the slightest sign of danger. Tree felling is being aggravated, as who is responsible for tree management bear the costs and legal duties related to trees, meanwhile the public benefits of trees without knowing about their management issues [6]. In order to continue providing benefits, it is crucial to managing trees in such a way as to minimize risks and conflicts they may cause [20]. However, the safety of people and goods are not the only factor to take into account. Environmental and aesthetic factors are also considered when assessing costs and benefits [6].

As described above, trees represent enormous environmental and social benefits, but they are also a source of risk. The aim of tree health monitoring is, consequently, multifunctional. There are techniques to tree health inspection that support decision making related to trees management, in which the IRT presents several specific advantages [11].

4. METHODS AND TECHNIQUES OF INSPECTION

VTA (Visual Tree Assessment) is an inspection method for tree diagnosis. It was developed by Mattheck and Breloer (1994), and according to VTA, interprets the body language of trees and provides the expert with failure criteria [28]. The method consists of three steps: a) External visual inspection in search of evidence of internal damage; b) the use of diagnostic tools for confirmation and measurement of damage; and c) assessment of damage extent and failure risk.

4.1 Instruments for Detecting Deterioration

Visual inspection alone does not provide enough evidence of internal damage extent. Then, additional diagnostic tools are required. According to several authors, those tools are classified as:

- ☆ Invasive and non-invasive: invasive instruments require drilling for deep penetration in the sapwood through one or more holes; the sapwood is living wood. Non-invasive instruments do not need contact, or they penetrate superficially in the sapwood. It is noted that any type of wound is an entry for pathogens into the tree [29].
- ☆ Destructive and non-destructive: Unlike the destructive instruments, non-destructive ones allow the identification of damage presence or not in the trees without causing harm [30]. Then, non-contact instruments are considered as non-destructive. Table 1 shows only non-contact instruments since they were considered non-destructive.
- ☆ Screening, diagnostic and evaluation: Screening tools allow a quick assessment to identify healthy and non-healthy trees. Diagnostic instruments allow a more accurate assessment, but require more time to identify the extent and type of damage in the tree. An intermediate method is an evaluation; it is a combination of screening speed and diagnostic accuracy methods [31].

Table 1. Classification of instruments for detecting deterioration in trees (adapted from Reference [31]).

	Unidentified Method in [31]	Screening	Evaluation	Diagnostic
Invasive	Increment borer, Boroscope	Resistograph	Shigometer, Fractometer	
Non-Invasive	-	Stress wave velocity	Electrical resistance	Stress wave tomography, Electromagnetic tomography
Non-contact	Nuclear magnetic resonance (NMR)	IRT, Electronic nose	-	Gamma-ray computed Tomography

Table 1 summarizes the main instruments for deterioration detection in trees, allowing to relate the purposes of the instruments to the invasive or destructive nature.

4.2 Sustainable Techniques Relevance

Besides IRT, the equipment performance evaluation and its advantages/disadvantages for detecting tree deterioration are briefly described in Table 2, and more details can be found in References [29,31]. It is noted that invasive techniques can increase tree damage [22], and radiation exposure is life detrimental for any living thing [11]. Inspection equipment must be safe for users and people around. It is therefore relevant to choose sustainable equipment. This type of equipment maintains a good relationship between portability, use speed, and they are not invasive or ionising. Besides, they have low running costs due to low energy consumption and low use of consumables [32].

Table 2. A short summary of the techniques for detecting trees deterioration.

Technique	Principle and Brief Description	Key Highlights
Increment borer [29]	Visual inspection: a sample of tree core is extracted for visual inspection.	Measures tree growth rate, age and soundness. Requires experience of decay potential causes. Invasive method (it may itself be a decay factor).
Boroscope [29]	Remote visual inspection: the tree trunk is drilled and it is used a small video camera to observe inside.	Enables visual confirmation from inside. Same disadvantages of the technique "Increment borer".
Resistograph [33]	Penetration resistance: a small drill/needle is inserted into the tree; drilling resistance is measured and registered.	Fast and easy to execute as well as interpret the graph. Does not detect early to intermediate decay stages; Requires comparison with known patterns (samples without decomposition).
Shigometer [34,35]	Electrical resistivity: a pulsed direct electric current generated by electrodes is applied into a drilled hole; it goes into the wood or the tree bark; electrical resistance is measured and registered.	Detects deterioration in early stages. The information is limited by the probe length.
Fractometer [28,35]	Strength and stiffness: it measures radial bending fracture strength and stiffness value.	Small device and easy to carry. Portable compression meter has depth limited.
Stress wave velocity [35,36]	Single-path acoustic wave velocity: detects cavities by measuring the acoustic wave velocity as it passes through the tree stem.	Quickly performed; defines the location and extent of internal decay. Difficult to determine early stages of decay.
Electrical resistance [29,37,38]	Electrical resistivity: resistivity is determined from the voltage difference between electrodes when the electricity is injected.	Effective to detect advanced stages of tree decay.
Stress wave tomography [38]	Multiple path acoustic wave velocity: the tomogram is obtained by measuring and registering the sound waves travel time generated by acoustic transducers positioned around the circumference of the tree stem.	Detects internal decay; accurately locates the anomalies; sensitive to early stages of decay. High cost and difficult to operate.

Electromagnetic tomography [31,38,39]	Electromagnetic wave permittivity: the tomogram is obtained by variations in the return signal. The variation is reflected by the receiving antenna when the electromagnetic wave emitted by the transmitting antenna encounters a boundary with a different dielectric constant.	Higher frequencies provide better resolution but penetration depth decreases. High cost and difficult to operate.
Nuclear magnetic resonance (NMR) [29,40]	Magnetic properties: uses the magnetic properties of certain atomic nuclei (as the hydrogen nucleus). They are aligned and oscillated using a strong magnetic field in the scanner.	Non-ionising radiation; delivers very detailed images that facilitate the analysis of the structure and function of the tissues. High cost and difficult to operate.
Electronic nose [41]	Odour: it distinguishes healthy and decayed wood through changes in the volatile organic compounds released by wood decay fungi.	Provides high levels of accuracy and reliability. Difficult to determine early stages of decay.
Gamma-ray computed Tomography [31,42]	Gamma-ray transmissivity: the tomogram is obtained from radiation absorption after directing gamma rays in multiple directions on a tree trunk thin slice.	Reliable and non-invasive; evaluates fungal decay and its extension. Ionizing radiation; high cost and difficult to operate; difficult to carry.

5. INFRARED THERMOGRAPHY APPLIED TO TREES

5.1 Principles of IRT

Infrared thermography (IRT) is a non-contact, non-destructive, and non-invasive technique that allows the detection of radiated heat energy from objects and bodies in the infrared range of the electromagnetic spectrum (wavelength range between 0.8–14 μm [43]). The conversion of infrared energy into visible imaging is possible through instruments that produce false-colour images, such as an infrared camera [44]. The principle underlying IRT is a method or equipment, which detects IR energy emitted from a surface, converts it to temperature, and displays the image of temperature field [43,45–50]. The fundamental concept behind it is that all bodies (alive and non-alive) have temperatures above absolute zero degrees (0 K) and emit infrared radiation that is captured by equipment capable of transforming that energy into pictures. The internal structure of the body shows different thermal behaviour depending on the health conditions of its parts. The differences in thermal behaviour result in differences in the colour pattern of the body surface image.

Figure 1 illustrates the heat transfer processes involved in a thermographic measurement. When the camera views the object, it receives radiation from the object, from the ambient between the camera and the object's surface, and from the atmosphere itself [8].

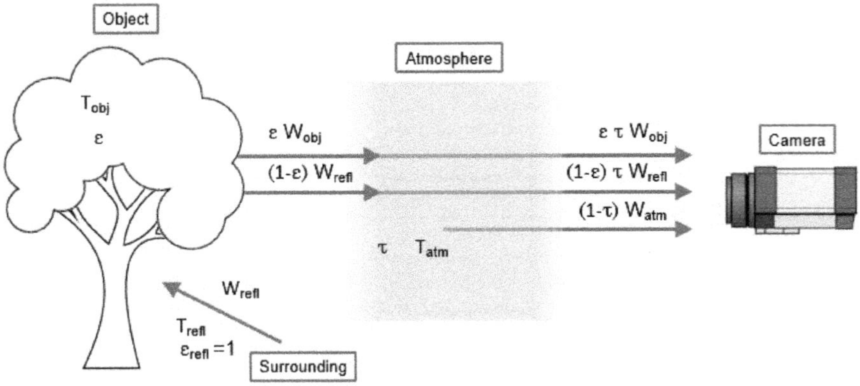

Figure 1. Schematic representation of a generic thermographic measurement (adapted from [8]).

The total received radiation power that the camera receives is equal to the sum of the emission of the object ($\epsilon \tau W_{obj}$), the reflected emission of environmental sources ($(1 - \epsilon) \tau W_{refl}$), and the emission from the atmosphere ($(1 - \tau) W_{atm}$). This is represented in the following equation (Equation (1)):

$$W_{tot} = \varepsilon \tau W_{obj} + (1 - \varepsilon) \tau W_{refl} + (1 - \tau) W_{atm} \tag{1}$$

where:

ϵ Object emissivity

τ Air transmittance

W_{obj} Emission from the object

W_{refl} Reflected emission from ambient sources

W_{atm} Emission from the atmosphere

The appropriate infrared camera selection depends on the characteristics of the object to be inspected, the atmospheric conditions, and the observation distance. Depending on the wavelength in the infrared range of the electromagnetic spectrum, much of the radiation is absorbed by the atmosphere. Lower absorption allows more radiation to reach the camera sensor. It happens at bands: Mid-wavelength infrared (MWIR) between 2 μm and 5 μm, and long-wavelength infrared (LWIR) between 8 μm and 14 μm. Limitations: Atmospheric humidity influences MWIR cameras, and LWIR cameras may capture optical and electronic noise [43].

The object emissivity depends on several factors such as wavelength. The emissivity is also highly sensitive to the nature and type of the material surface. As the emissivity increases, the influence of parasitic reflections diminishes. Metallic materials have low emissivity as they are highly reflective, whereas nonmetallic materials are high emissive and low reflection, leading to a better measurement [43,51]. Handling and interpretation of IRT information need cameras equipped with screens to visualise the energy emitted in the image format. The recording is an accessory that allows comparison between thermograms to follow the evolution of cases/pathologies [11]. It is also advantageous when the camera is equipped to capture thermograms and photographs simultaneously. It is preferably an integrated illumination accessory in the case of low illumination; this feature facilitates data interpretation [52].

There are two main thermographic procedures: Passive and active modes. In the active procedure, an external energy source is used to obtain the thermal contrast into the object of study. Different processes can trigger the heat flow, for example, thermal sources like lamps or heaters. The defects and damages on or near the object surface cause thermal discontinuities producing thermal contrast. It is detected during the thermographic inspection. In the passive procedure, the thermal contrast is generated by natural sources such as sunlight [44,50]. In most of the thermographic application scenarios, it is necessary to introduce an emissivity factor to calibrate the camera temperature measurement. The determination of the correct emissivity value of a specific wood remains a relevant issue [8].

IRT can be applied to both wood and tree plantlets to identify deterioration and voids that may compromise the structure, stability, and durability. Defects within the object disrupt the flow of energy and cause temperature differences on its surface. The thermal contrast results from differences in radiation emission that are registered by the thermal camera [43].

5.2 Wood and Trees

Wood and trees have significantly different thermal characteristics. That is, they differ in the variables that determine the temperature, density, humidity, thermal conductivity and thermal diffusivity, specific heat, convective coefficient, and emissivity [53–55].

After determining the density and the remaining properties of a species of wood, it is possible to predict its thermal behaviour. Therefore, when biodegradation changes the density and humidity of the wood, it also changes the thermal behaviour and temperature distribution along the surface [55–58]. Figure 2 shows the conventional photograph of a wood sample surface and its thermogram obtained in the active IRT procedure. By the active mode, this type of nodes and cracks show colour heterogeneity in relation to the general pattern denoting higher temperature. As shown in Figure 2, the cracks are barely visible in the photograph, but they are well seen in the thermogram [50].

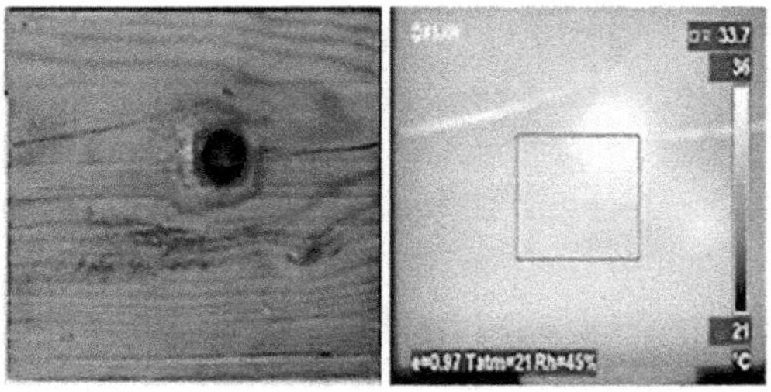

(a) *(b)*

Figure 2. Pine sample: (a) photograph and (b) IRT thermogram (active mode) [50].

The sap circulation causes the temperature gradient to vary along the trunk, and trees with more available water have better sap circulation [59]. This characteristic allows the differentiation between functional and dysfunctional tissue since the transport is done through the functional tissue. Therefore this feature allows that health and vitality can be verified, however, does not allow extrapolation of its thermal behaviour to the other trees, even of the same species, in order to identify damages. In fact, even for the same type of pathology, trees must be analyzed on a case-by-case basis [52,60].

Actually, two damaged trees of similar species, even when they have the same pathology, can generate different thermal patterns, because the availability of water to which the tree is subject is different and varies the temperature gradient along the trunk/tree. Thus, the temperature pattern, which allows the identification of functional or dysfunctional tissues, is unique for each tree.

Figure 3 shows the difference in surface temperature between a tree and a wooden stake [2]. The tree analysed was a specimen Prunus domestica L. (common name plum-tree) that was observed about 4 h after the sunset in the summer. The authors used an emissivity of 0.97 and a Rainbow colour pallet. During the observation, the atmospheric temperature was 22.5 ∘C, and the relative humidity was 55%.

Most of the wooden stake presented a lower temperature than the tree, except its lower part. The lower part of the wood stake showed a slightly higher temperature compared to the lower trunk because the tree was irrigated before observation. The tree keeps a balanced relationship with the environment temperature, so tree temperatures are usually lower than the atmospheric temperature when the sun heating effect is over [61]. Therefore, the tree in Figure 3 showed higher temperatures in the healthier parts and lower temperatures in the deteriorated parts, as well as in parts where the tree was recently water wet [46].

IRT is commonly applied to wood in laboratory ambient. In fact, the lab is the first selected scenario because it is possible to control several environmental parameters (humidity, temperature, and luminosity), the thermal energy of the material, as well as the easy manipulation. Some studies in the lab have induced damage in the samples to verify the ability of IRT to detect wood degradation [55,56,62]. In the case of trees, most thermographic studies are performed on trees that are already programmed to be felled or when it is possible to observe the anomalies without causing damage [60,61].

(a) (b)

Figure 3. A tree (Prunus domestica) and a wooden stake that supports it: (a) photograph; (b) IRT thermogram (passive mode). Temperature values (○C): Sp1 = 20.0; Sp2 = 20.5; Sp3 = 21.5; Sp4 = 20.5; and Sp5 = 19.5. Adapted from Reference [2].

The application of IRT to wood has been used for diverse purposes. Inspection of wood pieces quality in a production line is presented by Reference [63]. The results showed that TIV can be used to mark the debonded areas or to completely remove materials from the assembly line. IRT associated with ultrasonic equipment in the inspection of historic buildings is given by Reference [64]. Oratorio San Felipe Neri in Cadiz, Spain, was analyzed, and the authors conclude that the union of the ultrasound technique and the thermography represents a good tool for wooden structure inspection. IRT associated with ultrasonic equipment, supported by laboratory analyses of timber samples in the inspection of the Aslanhane Mosque in Ankara, Turkey, which can be found in Reference [65]. The combined analysis of these techniques was done to evaluate the condition of the structural elements of wood in terms of their state of preservation, moisture problems, and recent incompatible repairs that affect them. This combination of techniques was reported as useful in assessing the reliability of timber, increasing the accuracy and effectiveness of the survey, facilitating and distinguishing the work of urgent intervention from long-term conservation programs. The detection of termite pests in wood or wood-clad structures is analyzed by Reference [66]. The author points out that TIV enables finding two basic elements that an inspector looks to identify the presence of pest infestation: Areas exhibiting anomalous elements that could be associated with the presence of moisture or hot spots and the presence of subsurface defects. Additionally, in large wooden structures such as a railway bridge where, in addition to other techniques, the IRT was used to identify a rapidly significant amount of structural damage that was not identified by other techniques [67].

5.3 Analysis of Tree Health

The thermal contrast captured by the thermal camera only represents the temperature on the surface of the tree bark. However, the tree bark near deteriorated tissue and voids show a lower temperature than the area around it. As in wood, tree deteriorated tissue and voids undergo a change in their thermal properties. Whenever there are significant differences in the thermal properties of the wood and tree bark, they can be detected by IRT [60,68]. IRT applied to trees and offers a means to differentiate damaged and deteriorated tissue from healthy tissues. However, due to the specific characteristics of each species, whenever a species is evaluated for the first time, it is advisable to observe the representation of the thermal pattern of the tree bark in the thermal image for better results interpretation [11].

The first use of IRT in trees was for aerial surveillance of the canopies to detect the distribution and spread of forest damage [60]. Later, IRT was applied to tree bark (trunks and branches), and it is now possible to evaluate some types of damage in the lower trunk and to deduce the cause that affects roots (root system) [60,69].

Several factors affect tree health, and consequently, can change the thermal properties of the tree trunk and branches. Some of the conditions that IRT can identify are described below, namely, diseases, pest attack, water stress, and formation of new functional tissue.

Regarding diseases, the application of IRT in the detection of tree bark lesions such as hemorrhagic cancers caused by *Phytophthora* spp., even before its consequences can be detected with the naked eye [60,70]. Fungi such as *Inonotus hispidus* (Bull.) P. Karst., *Phellinus punctatus* (Fr.) Pilát, *Coriolus pubescens* (Schumach.) and Corticium sp. were found in samples taken from the woody material of platan previously analyzed by IRT [71].

IRT can also be used for detection of insect pests on trees. Insects attack both leaves for feeding as well as trunks for their habitat, feeding, and egg-laying. While trees have a certain tolerance to pest attack on their leaves, it repeated attacks and factors such as poor irrigation, flooding, and inadequate trimming generate greater stress on the tree and consequently compromise its ability to regenerate or resist. On the other hand, larvae and insects create wounds, and these wounds make the trees susceptible to other pathogens. Some larvae can dig deep into the trunk creating tunnels that ultimately affect the transportation of nutrients and water causing rapid tree deterioration. Some of the insects recently detected by IRT were: White Pine Cone Beetle, Australian 'fire-beetle', Citrus Long-horned Beetle, Woodworm, and Red Palm Weevil (RPW) [72]. However, the IRT was not able to detect larval stage insects in some young tree species, as in the case of goat moth larvae. A hypothesis was the thermal adaptation of the larvae to the environment, that even with the elevation of temperature due to the effect of the activity of the larvae, it was not sufficient to detect them [73].

IRT is also applied for trees water stress assessment. The temperature in the canopy is an indicator of the amount of water available in the soil because the leaves transpiration is a mechanism to dissipate excess energy (heat). The tree uses surplus energy to transform water into water vapour through the leaves to cool it. However, when the soil is in water shortage, the tree perspires less, and as a consequence, there is an increase in leaf temperature when compared to trees in soil adequately wet. Water stress is a relevant indicator because the monitoring provides information for the adequate water supply of each tree species [74–77].

IRT detects the formations of new functional tissue. It was confirmed after dissection that the thermogram showed a band of higher wood formation in the area of the thermogram. This happens as a consequence of the adaptive growth of the change, because as Shigo (1984) has defined, the vascular cambium is the cell generator, and the new cells have different thermal properties from the other tissues [22]. This formation of new cells has different thermal properties than the other tissues.

This tree had emptiness along the trunk that could also be observed in the thermogram. Catena and Catena [60] suggest that an adequate prognosis of the actual state of the tree requires the identification of the cause for tissue forming, that is, as result of damage and/or mechanical stress.

5.4 Advantages and Limitations of IRT Applicated to Trees

It is an advantage that IRT allows the tree to be observed as a whole, then the damage is early identified, even in trees that do not yet have external visual signs [52]. IRT requires little time to perform the inspection and is relatively easy to interpret the results. It does not require contact with the tree, and observations can be made at a distance of up to 25 m (depending on the camera and lens) safely. It is a safe technology for both the trees and the examiner as it does not emit harmful radiation. It is not necessary to use ladders or lifting platforms to observe the high parts of the tree; it reduces time and costs [60,68,78]. IRT also allows real-time assessment of root damage by observation of the lower trunk. It is a tool for stability and safety evaluation [69,79]. It eases to monitor the health status of trees over time, following up the evolution of pathologies previously identified as well as the identification of new patterns that indicate the development of new pathologies [69]. However, the major advantage of IRT compared to other inspection methods is the ability to differentiate functional tissue from dysfunctional tissue. In fact, IRT provides information for analysing the vitality and health status of a tree in a non-destructive, rapid, and cost-effective manner [60]. In essence, agriculture has been mechanized by the green revolution and now, like any other industry, agriculture is being digitalized [80]. According to this study, most farmers are ready to accept technology if it is profitable, less complex, and makes their life easier. IRT may be one of the relevant components of smart farming and agriculture, as can better manage and help to carry out real-time events.

As in other non-invasive methods, the main limitation is that IRT does not identify whether the damage detected is a void or a deteriorated tissue, nor its causative agent. Nor can it give precise indications of the magnitude of the damage. However, the fact that IRT identifies damaged areas early, it reduces the tests time and thus the damage progression. The acquired data leads to a more precise investigation of the pathologies, and identify the locations where invasive techniques are required [11,81]. Interferences are observed in thermograms when the surface of the tree is obscured by mosses or other vegetation, and when the tree is wet, or when IRT has been carried out directly to sunlight. Some of these limitations can be overpassed by avoiding inspections after rainy days or carry them out before the trees are watered. In the case of sunlight, the tests should preferably be performed at night. Another commonly noted limitation is the cost of the equipment when compared to simpler equipment.

Table 3. A short summary of infrared thermography applied to tree health analysis.

Study	Main Focus	Findings
Al-doski et al. [72]	Pest detection.	Pest infestation detected.
Ballester et al. [74]	Water stress detection on citrus and persimmon trees.	Water stress detected.
Bellett-Travers & Morris [61]	Relationship between surface temperature and radial wood thickness.	No apparent relationship in most of the trees; strong relationship when there was a gradual change in radial wood thickness caused by a cavity.
Burcham et al. [59]	Effect of mechanically induced internal voids on *Dracaena fragrans* L. stem temperature.	Only able to identify reductions temperature in internal defects with at least 76% of the stem cross-sectional area.
Burcham et al. [70]	Evaluate the relationship between the internal defects and trunk surface temperature in *Casuarina equisetifolia* L.	Does not provide accurate results about the internal condition of trees.
Catena & Catena [60]	Review in order to assess the accuracy, reliability, and costs.	Does not automatically distinguish between different kinds of alteration; does not accurately provide the extent of the damages found; provides enough information to decide regarding the need for remedial action or a more detailed kind of assessment; non-invasive, fast, reasonable prices, in real time.

Catena, G. [68,71]	Internal cavities in trees detection.	Enables the detection of cavities.
Catena, A. [79] and Catena et al. [69]	Damages in the roots.	Enables that damages in the roots can be deduced in real time.
Crisóstomo et al. [32]	Considerations over IRT as applied to the state of the tree healthiness.	Enables healthiness evaluation.
Crisóstomo et al. [52]	Quercus pygmentosis tree Analysis.	Evaluating its healthiness status.
García-Tejero et al [76]	Water stress detection in almond trees.	Water stress detected.
Giuliani & Flore [77]	Water stress detection in apple trees.	Water stress detected.
Goh et al. [29]	Review of the current sensing methods used for decay detection in trees.	Comparing methods concerning the fundamental of measurements, hardware implementation, damage caused to the tree and the ease of use.
Hoffmann et al. [73]	Detecting the larval stage of goat moth´s larvae in young tree species.	Was not able to detect
Jones et al. [75]	Water stress detection on grapevine.	Water stress detected.
Leong et al. [31]	Evaluating the current tree decay detection tools.	Classifying the tree decay detection tools in terms of measurement speed, resolution and accuracy.

This is not the case when compared to more sophisticated equipment, or when multiple applications beyond the inspection of tree health are taken into account [60,68,78].

A brief search result summary of IRT applied to tree health analysis is presented in Table 3.

6. CONCLUSIONS

The IRT was presented as a tool to inspect tree health, and it has proved to be an efficient tool in the early detection of damages even though it is not possible to identify the type of damage. As well as other non-invasive and non-destructive techniques, the causative agent can be identified when resorting to invasive methods. However, the comparison of inspection methods has shown that IRT has great advantages in terms of the capacity to differentiate functional tissue from dysfunctional tissue, and thus to inspect the vitality and health status of trees. It allows monitoring the evolution of pathologies in a fast, economical, and non-destructive way. Result improvement can be achieved with the evolution and increased application of IRT.

7. RECOMMENDATIONS

Despite its merits, IRT is still a relatively new technique in assessing tree health and remains residually implemented in agriculture, but it needs more detailed studies to establish a solid application basis, which can guide practitioners. This would allow developing the IRT potential for applicability on a larger scale. Some challenges such as the complexity of the technique, the new and atypical aspects of the problem, and several knowledge gaps related to technical issues and applicability specificities the affect of the technique performance. The latest high-definition thermal cameras record thermal images of high resolution and sensitivity will also contribute to overcoming these challenges, which will help to turn the IRT technique as a decision-making tool to assess the health status of trees.

References

1. Pitarma, R.; Crisóstomo, J.; Ferreira, M.E. Learning About Trees in Primary Education: Potentiality of IRT Technology in Science Teaching. In Proceedings of the EDULEARN18 Conference, Palma, Spain, 2–4 July 2018; pp. 208–213.
2. Ferreira, M.; Crisóstomo, J.; Pitarma, R. Infrared thermography technology to support science teaching-meaningful learning about trees with university students. In Proceedings of the 13th International Technology, Education and Development Conference (INTED2019), Valencia, Spain, 11–13 March 2019.
3. Ferreira, M.E.; André, A.C.; Pitarma, R. Potentialities of Thermography in Ecocentric Education of Children: An Experience on Training of Future Primary Teachers. *Sustainability* **2019**, 11, 2668. [CrossRef]
4. Lier, M.; Parviainen, J. *Integration of Nature Protection in Forest Policy in Finland;* INTEGRATE Country Report; EFICENT-OEF: Freiburg, Germany, 2013.
5. Ancient Tree Forum & The Woodland Trust. *Ancient Tree Guide No. 4: What Are Ancient, Veteran and Other Trees of Special Interest;* The Woodland Trust: Grantham, UK, 2008.
6. National Tree Safety Group. *Common Sense Risk Management of Trees: Guidance on Trees and Public Safety in the UK for Owners, Managers and Advisers;* Forestry Commission: Edinburgh, Scotland, 2011; ISBN 978-0-85538-840-9.
7. Health and Safety Executive. *Management of the Risk from Falling Trees;* Health & Safety Executive/Local Authorities Enforcement Liaison Committee (HELA): Bootle, UK, 2007.
8. *FLIR Tools+ User´s Guide*; Flir Systems, Inc.: Wilsonville, OR, USA, 2016.

9. Barr, E.S. Historical Survey of the Early Development of the Infrared Spectral Region. *Am. J. Phys.* **1960**, 28, 42–54. [CrossRef]

10. Kylili, A.; Fokaides, P.A.; Christou, P.; Kalogirou, S.A. Infrared thermography (IRT) applications for building diagnostics: A review. *Appl. Energy* **2014**, *134*, 531–549. [CrossRef]

11. Catena, A. Thermography Reveals Hidden Tree Decay. *Arboric. J.* **2003**, 27, 27–42. [CrossRef]

12. United Nations. World *Urbanization Prospects: The 2018 Revision;* United Nations, Department of Economic and Social Affairs, Population Division: New York, NY, USA, 2018.

13. Pokorny, J.; O'Brien, J.; Hauer, R.; Johnson, G.; Albers, J.; Bedker, P.; Mielke, M. *Urban Tree Risk Management: A Community Guide to Program Design and Implementation;* USDA Forest Service Northeastern Area State and Private Forestr: St. Paul, MN, USA, 2003.

14. Rotherham, I.D. Editorial: Trees In A Changing World. *Arboric J.* **2010**, 33, 1–2. [CrossRef]

15. Kuo, F.E.; Sullivan, W.C. Environment and Crime in the Inner City: Does Vegetation Reduce Crime? *Environ. Behav.* **2001**, 33, 343–367. [CrossRef]

16. Coder, K.D. *Identified Benefits of Community Trees and Forests;* University of Georgia School of Forest Resources: Athens, GA, USA, 1996.

17. *International Society of Arboriculture Benefits of Trees*; International Society of Arboriculture: Champaign, IL, USA, 2011.

18. Roy, S.; Byrne, J.; Pickering, C. A systematic quantitative review of urban tree benefits, costs, and assessment methods across cities in different climatic zones. *Urban For. Urban Green.* **2012**, *11*, 351–363. [CrossRef]

19. Song, X.P.; Tan, P.Y.; Edwards, P.; Richards, D. The economic benefits and costs of trees in urban forest stewardship: A systematic review. *Urban For. Urban Green.* **2018**, 29, 162–170. [CrossRef]

20. Johnston, M.; Hirons, A. Urban Trees. *In Horticulture: Plants for People and Places;* Dixon, G.R., Aldous, D.E., Eds.; Springer: Dordrecht, The Netherlands, 2014; Volume 2, ISBN 978-94-017-8580-8.

21. Shigo, A.L.; Marx, H.G. *Compartmentalization of Decay in Trees;* U. S. Government Printing Office: Washington, DC, USA, 1977; pp. 4–15.

22. Shigo, A.L. Compartmentalization: A Conceptual Framework for Understanding How Trees Grow and Defend Themselves. *Annu. Rev. Phytopathol.* **1984**, 22, 189–214. [CrossRef]

23. Shortle, W.C.; Dudzik, K.R. *Wood Decay in Living and Dead Trees: A Pictorial Overview;* U.S. Department of Agriculture, Forest Service, Northern Research Station: Newtown Square, PA, USA, 2012.

24. Sherwood Forest. Available online: https://www.visitsherwood.co.uk/things-to-do/the-major-oak/ (accessed on 18 June 2019).

25. Undiscovered Scotland. Available online: https://www.undiscoveredscotland.co.uk/blairgowrie/ meikleourhedge/index.html (accessed on 18 June 2019).

26. Woodland Trust. Available online: https://www.woodlandtrust.org.uk/visiting-woods/trees-woods-and- wildlife/woodland-habitats/ancient-trees/ (accessed on 18 June 2019).

27. National Trust. Available online: https://www.nationaltrust.org.uk/ashridge-estate/features/looking-after-our-veteran-trees-at-ashridge-estate (accessed on 18 June 2019).

28. Mattheck, C.; Breloer, H. Field guide for visual tree assessment (Vta). *Arboric. J.* **1994**, 18, 1–23. [CrossRef]

29. Goh, C.L.; Abdul Rahim, R.; Fazalul Rahiman, M.H.; Mohamad Talib, M.T.; Tee, Z.C. Sensing wood decay in standing trees: A review. *Sens. Actuators A Phys.* **2018**, 269, 276–282. [CrossRef]

30. Hellier, C. Introduction to Nondestructive Testisng. In *Handbook of Nondestructive Evaluation*; Hellier, C., Ed.; McGraw-Hill: New York, NY, USA, 2003; ISBN 978-0-07-139947-0.

31. Leong, E.-C.; Burcham, D.C.; Fong, Y.-K. A purposeful classification of tree decay detection tools. *Arboric. J.* **2012**, 34, 91–115. [CrossRef]

32. Crisóstomo, J.; Pereira, C.; Roque, E.; Jorge, L.; Pitarma, R. Considerações na Observação do Estado de Salubridade de Árvores Através da Termografia por Infravermelhos. In Proceedings of the 1st Iberic Conference on Theoretical and Experimental Mechanics and Materials/11th National Congress on Experimental Mechanics, Porto, Portugal, 4–7 November 2018; pp. 745–748.

33. Mattheck, C.; Bethge, K.; Albrecht, W. How To Read The Results Of Resistograth M. *Arboric. J.* **1997**, 21, 331–346. [CrossRef]

34. Shigo, A.L.; Shortle, W.C. *Spruce Budworms Handbook: Shigometry–A Reference Guide*; Agric. Handb.; U.S. Department of Agriculture, Forest Service, Cooperative State Research Service: Washington, DC, USA, 1985.

35. Monk, B. Evaluation of Decay Detection Equipment in Standing Trees. Available online: https://www.fs.fed.us/t-d/programs/im/tree_decay/tree_decay_detect_equip.shtml (accessed on 17 June 2019).

36. Ross, R.J.; Pellerin, R.F. Inspection of Timber Structures Using Stress Wave Timing Nondestructive Evaluation Tools. In *Wood and Timber Condition Assessment Manual: Second Edition;* White, R.H., Ross, R.J., Eds.; U.S. Department of Agriculture, Forest Service, Forest Products Laboratory: Madison, WI, USA, 2014.

37. Oliva, J.; Romeralo, C.; Stenlid, J. Accuracy of the Rotfinder instrument in detecting decay on Norway spruce (Picea abies) trees. *For. Ecol. Manag.* **2011**, *262*, 1378–1386. [CrossRef]

38. Nicolotti, G.; Socco, L.V.; Martinis, R.; Godio, A.; Sambuelli, L. Application And Comparison Of Three Tomographic Techniques For Detection Of Decay In Trees. *J. Arboric.* **2003**, 29, 66–78.

39. Bogosanovic, M.; Al Anbuky, A.; Emms, G.W. Overview and comparison of microwave noncontact wood measurement techniques. *J. Wood Sci.* **2010**, 56, 357–365. [CrossRef]

40. Wang, P.C.; Chang, S.J. Nuclear Magnetic Resonance Imaging of Wood. *Wood Fiber Sci. 1986*, *18*, 308–314.

41. Baietto, M.; Wilson, A.; Bassi, D.; Ferrini, F. Evaluation of Three Electronic Noses for Detecting Incipient Wood Decay. *Sensors* **2010**, *10*, 1062–1092. [CrossRef]

42. Habermehl, A.; Ridder, H.-W. Computerised Tomographic Investigationa Of Street And Park Trees. *Arboric. J.* **1995**, 19, 419–437. [CrossRef]

43. Usamentiaga, R.; Venegas, P.; Guerediaga, J.; Vega, L.; Molleda, J.; Bulnes, F. Infrared Thermography for Temperature Measurement and Non-Destructive Testing. *Sensors* **2014**, 14, 12305–12348. [CrossRef]

44. Maldague, X.P.V.; Streckert, H.H.; Trimm, M.W. Introduction to *Infrared and Thermal Testing.* In Infrared and Thermal Testing; Maldague, X.P.V., Moore, P.O., Eds.; Nondestructive Testing Handbook; American Society for Nondestructive Testing: Columbus, OH, USA, 2001; ISBN 978-1-57117-044-6.

45. Snell, J.R., Jr. Thermal Infrared Testing. In *Handbook of Nondestructive Evaluation;* Hellier, C., Ed.; McGraw-Hill: New York, NY, USA, 2003; ISBN 978-0-07-139947-0.

46. Meola, C. (Ed.) Carosena Origin and Theory of *Infrared Thermography. In Infrared Thermography Recent Advances and Future Trends;* Bentham Science Publishers: New York, NY, USA, 2012; pp. 3–28. ISBN 978-1-60805-143-4.

47. Ibarra-Castanedo, C.; Maldague, X.P.V. Infrared Thermography. In *Handbook of Technical Diagnostics*; Czichos, H., Ed.; Springer: Berlin/Heidelberg, Germany, 2013; pp. 175–220. ISBN 978-3-642-25849-7.

48. Holst, G.C. *Common Sense Approach to Thermal Imaging*; JCD Pub.; Co-Published by SPIE Optical Engineering Press: Winter Park, FL, USA; Bellingham, WA, USA, 2000; ISBN 978-0-9640000-7-0.

49. Pitarma, R.; Crisóstomo, J.; Jorge, L. Analysis of Materials Emissivity Based on Image Software. In *New Advances in Information Systems and Technologies;* Rocha, Á., Correia, A.M., Adeli, H., Reis, L.P., Mendonça Teixeira, M., Eds.; Springer International Publishing: Cham, Switzerland, 2016; Volume 444, pp. 749–757. ISBN 978-3-319-31231-6.

50. Crisóstomo, J.; Pitarma, R. The Importance of Emissivity on Monitoring and Conservation of Wooden Structures Using Infrared Thermography. In *Advances in Structural Health Monitoring;* Hassan, M., Ed.; IntechOpen: London, UK, 2019. [CrossRef]

51. Maldague, X.P.V. *Nondestructive Evaluation of Materials by Infrared Thermography*; Springer: London, UK, 1993; ISBN 978-1-4471-1997-5.

52. Crisóstomo, J.; Pereira, C.; Roque, E.; Jorge, L.; Pitarma, R. *Análise da Salubridade de Árvores Através da Termografia por Infravermelhos*; Gomes, J.F.S., Ed.; INEGI/FEUP: Porto, Portugal, 2018; pp. 749–758.

53. Meola, C.; Carlomagno, G.M. Recent advances in the use of infrared thermography. *Meas. Sci. Technol.* **2004**, 15, R27–R58. [CrossRef]

54. Wyckhuyse, A.; Maldague, X. A Study of Wood Inspection by Infrared Thermography, Part I: Wood Pole Inspection by Infrared Thermography. *Res. Nondestruct. Eval.* **2001**, 13, 1–12. [CrossRef]

55. Conde, M.J.M.; Liñán, C.R.; de Hita, P.R.; Gálvez, F.P. Infrared Thermography Applied to Wood. *Res. Nondestruct. Eval.* **2012**, 23, 32–45. [CrossRef]

56. Rodríguez-Liñán, C.; Morales-Conde, M.J.; Rubio-de Hita, P.; Pérez-Gálvez, F. Análisis sobre la influencia de la densidad en la termografía de infrarrojos y el alcance de esta técnica en la detección de defectos internos en la madera. *Mater. de Construcción* **2012**, 62, 99–113. [CrossRef]

57. Tanaka, T.; Divós, F. Wood Inspection by Thermography. In Proceedings of the 12th International Symposium on Nondestructive Testing, Sopron, Hungary, 13–15 September 2000.

58. Pereira, J.C.A. *Contribuição para a Análise de Manifestações Patológicas em Madeira na Construção com Recurso à Termografia;* Instituto Politécnico de Castelo Branco: Castelo Branco, Portugal, 2014.

59. Burcham, D.C.; Leong, E.-C.; Fong, Y.-K. Passive infrared camera measurements demonstrate modest effect of mechanically induced internal voids on Dracaena fragrans stem temperature. *Urban For. Urban Green.* **2012**, 11, 169–178. [CrossRef]

60. Catena, A.; Catena, G. Overview of Thermal Imaging For Tree Assessment. *Arboric. J.* **2008**, 30, 259–270. [CrossRef]

61. Bellett-Travers, M.; Morris, S. The Relationship Between Surface Temperature And Radial Wood Thickness Of Twelve Trees Harvested In Nottinghamshire. *Arboric. J.* **2010**, 33, 15–26. [CrossRef]

62. López, G.; Basterra, L.-A.; Ramón-Cueto, G.; Diego, A. de Detection of Singularities and Subsurface Defects in Wood by Infrared Thermography. *Int. J. Archit. Herit.* **2014**, 8, 517–536. [CrossRef]

63. Meinlschmidt, P. Thermographic Detection of Defects in Wood and Wood-based Materials. In Proceedings of the 14th International Symposium of Nondestructive Testing of Wood, Hannover, Germany, 2–4 May 2005.

64. Rodríguez Liñán, C.; Morales Conde, M.J.; Rubio de Hita, P.; Pérez Gálvez, F. Inspección mediante técnicas no destructivas de un edificio histórico: Oratorio San Felipe Neri (Cádiz). *Inf. de la Construcción* **2011**, 63, 13–22. [CrossRef]

65. Kandemir-Yucel, A.; Tavukcuoglu, A.; Caner-Saltik, E.N. In situ assessment of structural timber elements of a historic building by infrared thermography and ultrasonic velocity. *Infrared Phys. Technol.* **2007**, 49, 243–248. [CrossRef]

66. Grossman, J.L. Advanced techniques in IR thermography as a tool for the pest management professional. In Proceedings of the SPIE, Orlando, FL, USA, 18 April 2006; Volume 6205.

67. Grossman, J.L. Trestles anyone? A Thermographic Nightmare. In Proceedings of the SPIE, Orlando, FL, USA, 9 April 2007; Volume 6541.

68. Catena, G. A new application of thermography. *Atti Della Fond. Giorgio Ronchi* **1990**, 6, 947–952.

69. Catena, A.; Catena, G.; Lugaresi, D.; Gasperoni, R. La Termografia rivela la presenza di danni anche nell'apparato radicale degli alberi. *Agric. Ric.* **2002**, 81–100.

70. Burcham, D.C.; Leong, E.-C.; Fong, Y.-K.; Tan, P.Y. An Evaluation of Internal Defects and Their Effect on Trunk Surface Temperature in *Casuarina equisetifolia* L. (Casuarinaceae). *Arboric. Urban For.* **2012**, 38, 277–286.

71. Catena, G. Une Appication De La Thermographie En Phytopathologie. *Phytoma-La Défense Des Végétaux* **1992**, 439, 46–48.

72. Al-doski, J.; Mansor, S.B.; Shafri, H.Z.B.M. Thermal Imaging For Pests Detecting-A Review. *Int. J. Agric. For. Plant.* **2016**, 2, 10–30.

73. Hoffmann, N.; Schröder, T.; Schlüter, F.; Meinlschmidt, P. Potenzial von Infrarotthermographie zur Detektion von Insektenstadien und -schäden in Jungbäumen. *J. Für Kult.* **2013**, 65, 2013.

74. Ballester, C.; Jiménez-Bello, M.A.; Castel, J.R.; Intrigliolo, D.S. Usefulness of thermography for plant water stress detection in citrus and persimmon trees. *Agric. For. Meteorol.* **2013**, *168*, 120–129. [CrossRef]

75. Jones, H.G.; Serraj, R.; Loveys, B.R.; Xiong, L.; Wheaton, A.; Price, A.H. Thermal infrared imaging of crop canopies for the remote diagnosis and quantification of plant responses to water stress in the field. *Funct. Plant Biol.* **2009**, 36, 978. [CrossRef]

76. García-Tejero, I.; Durán-Zuazo, V.H.; Arriaga, J.; Hernández, A.; Vélez, L.M.; Muriel-Fernández, J.L. Approach to assess infrared thermal imaging of almond trees under water-stress conditions. *Fruits* **2012**, 67, 463–474. [CrossRef]

77. Giuliani, R.; Flore, J.A. Potential Use Of Infra-Red Thermometry For The Detection Of Water Stress In Apple Trees. *Acta Hortic.* **2000**, 537, 383–392. [CrossRef]

78. Ibarra-Castanedo, C.; Tarpani, J.R.; Maldague, X.P.V. Nondestructive testing with thermography. *Eur. J. Phys.* **2013**, 34, S91–S109. [CrossRef]

79. Catena, A. Thermography Shows Damaged Tissue and Cavities Present in Trees. In *Nondestructive Characterization of Materials XI.*; Green, R.E., Djordjevic, B.B., Hentschel, M.P., Eds.; Springer: Berlin/Heidelberg, Germany, 2002; pp. 515–522. ISBN 978-3-540-40154-4.

80. Sharma, S.; Kaushik, A. Views of Irish Farmers on Smart Farming Technologies: An Observational Study. *AgriEngineering* **2019**, 1, 164–187.

81. Bellett-Travers, M. A Risk Assessment Methodology For Trees In Parkland Based On Comparative Population Analysis. *Arboric. J.* **2010**, 33, 3–14. [CrossRef]

Chapter 9

CONTRIBUTION TO TREES HEALTH ASSESSMENT USING INFRARED THERMOGRAPHY

1. INTRODUCTION: AN OVERVIEW

The tree is an essential natural resource for ecosystem balance. It regulates nature, climate and urban greening. Trees play a key role in local biodiversity as they release oxygen and reduce global warming. Trees regulate climate by mitigating urban heat islands. Trees behave as barriers for noise pollution and wind. They provide moisture to the atmosphere, which favours precipitation. They facilitate water infiltration into the soil contributing to the formation and maintenance of groundwater aquifers. They are also essential for soil building as their roots fix the soil preventing erosion. All trees are of importance, whether by their age, type, size and shape. Some of them are classified as remarkable or even monumental trees, and the law protects them [1]. Therefore, it is fundamental to act for the preservation and sustainability of these living beings. This is a current social challenge [2–5]. It is decisive to move from the anthropocentric conception of trees to ecocentric conception, that is, all living beings including humankind are interrelated and interdependent to keep the ecosphere equilibrium.

Contrary to what people thought for many years, trees react to physical and biological damages that lead to deterioration. The compartmentalization of decay in trees

(CODIT) is a good example of this [6]. Therefore, trees must be cared for in a way that enhances their own defence systems [6]. In order to verify the health status of trees especially when more detailed information is required, invasive and even destructive techniques [7] are used. Unfortunately, they interfere with the structural integrity of this living being. Therefore, it is urgent to disseminate and implement non-aggressive inspection techniques that preserve biological integrity and functionality [3,4]. The basic rule must always be to start with less invasive techniques and only if required, to utilise the most aggressive ones, so that the damage produced in the tree [4,5,8] is minimized.

Infrared thermography (IRT) is a non-invasive non-contact technique that relies on the detection of body heat emission [9–11]. It measures continuously surface temperature in real-time [9]. Aerial surveillance of the canopies to detect the distribution and spread of forest damage was the first use of IRT in trees [4]. Later, this technique was applied to tree bark (branches and trunks). Now, it allows the evaluation of some types of damage in the lower trunk and to estimate the presumable cause that affects roots (the root system) [4]. The latest thermal cameras capture thermal images of high resolution and sensitivity. It has been demonstrated that IRT in conjunction with visual inspection when applied to assessment, and monitoring of tree health provides reliable data [2–4,12]. The IRT capabilities for tree inspection have not been exploited sufficiently yet, since the elaborate sap conductive tissues are practically on the surface of the living structure. The reading of tree surface temperature reveals specific variations when there is deterioration and voids inside the tree. The IRT observes the tree as a functional whole body and thermograms provide information to differentiate deteriorated tissues from healthy tissues. Other methods do not have this assessment capacity, much less the naked eye inspection. Other non-invasive methods survey the body by points, and then extrapolation applies to have an idea of the functional whole body. Through the naked eye, only advanced deterioration is detected, then a corrective measure is hardly effective, and the solution is tree felling.

Thermography allows the early detection of damage, while it is still not visually noticeable. Even more, it monitors the progress of pathology. The IRT is expeditious and non-invasive [2–5,12]. It is, therefore, a powerful, fast and efficient tool to detect changes in the integrity of trees and branches, identifying if one of them should be removed. The differences in the thermal patterns of the tree surface indicate the deteriorated areas of the tree. The greater the differences in the thermal patterns of the trunk and branches, the worse the tree health condition [2–5].

The methods different from IRT that acquire data for diagnosis are time-consuming, and require more hand labour, especially if the part of the tree to be examined cannot be reached from the ground [2–4]. Some techniques, such as the resistograph, require perforation, and these holes may become pathways for pathogens [2–4]. Other methods use X-rays or γ. A short summary of the other main methods for detecting tree deterioration is presented in Table 1.

Table 1. A brief summary of the other main methods for detecting tree deterioration (adapted from [19]).

Method	Principle	Key Highlights
Increment borer	Visual inspection	Measures growth rate, age and soundness. Invasive method (it may itself be a decay factor) and needs experience on decay causes.
Boroscope	Remote visual inspection	Visual analysis from inside. Same disadvantages of the "Increment borer".
Resistograph	Penetration resistance	Fast, easy to execute and interpret. Does not detect early to intermediate decay stages; requires comparison with known patterns.
Shigometer	Electrical resistivity	Detects deterioration in early stages. Information is limited by the probe length.
Fractometer	Strength and stiffness	Small device and easy to carry. Portable compression meter has depth limited.
Stress wave velocity	Single-path acoustic wave velocity	Quickly performed; defines the location and extent of internal decay. Difficult to determine early stages of decay.
Electrical resistance	Electrical.	Effective to detect advanced decay stages.
Stress wave tomography	Multiple path acoustic wave velocity	Detects internal decay; accurately locates the anomalies; sensitive to early stages of decay. High cost and difficult to operate.
Electromagnetic tomography	Electromagnetic wave permittivity	Higher frequencies provide better resolution but penetration depth decreases. High cost and difficult to operate.
Nuclear magnetic resonance (NMR)	Magnetic properties	Non-ionising radiation; very detailed images that facilitate the analysis. High cost and difficult to operate.
Electronic nose	Odour	High levels of accuracy and reliability. Difficult to determine early stages of decay.
Gamma-ray computed Tomography	Gamma-ray transmissivity	Reliable and non-invasive. Ionizing radiation; high cost, and difficult to carry and operate.

The ionising radiation used in some methods conveys the perception that they are not safe for the health of living beings [3]. Moreover, when IRT is compared to more sophisticated techniques, such as X-ray and γ, as well as tomographic acoustic techniques such as Picus and ArborSonic 3D, or even nuclear magnetic resonance [7,13–15], IRT is the only one able to evaluate the health condition and functionality of tree tissues. That is, the assessment of the structural integrity to detect voids and deterioration inside the tree is possible because the tree is analysed as a functional whole body [4,5]. IRT analyses the tree as a whole, in a holistic way, while other techniques provide information only on specified points, and the whole is obtained by extrapolation after a series of investigations [2–4]. The authors raise concerns about avoiding errors when selecting the emissivity value for temperature reading calibration. However, more parameters are required for IRT quantitative readings. It is necessary to assess the value of reflected (or reflective) apparent temperature, which varies according to the angle between the camera and the tree surface, and the direction of radiation from the environment and sunrays. In some studies, the tree is cut into logs, and then the logs are perforated to simulate deterioration after the logs are sealed at both sides to maintain the water content. When observing the logs in thermograms, it is difficult to detect the holes and the larvae introduced inside them. Even with the same water content, the temperature of the logs does not behave like the tree because there is no flow of the sap, as this happens when the tree structure is alive [16,17]. Some studies analyse trees of the same species affected by several pathologies. The main aim was to look for similar temperature patterns on the surface all along the trees applying IRT [18]. The statistical analysis showed that there was no correlation [18].

Considering the above, IRT is a well-established technique in a range of fields, such as industrial and building maintenance. Despite its merits, it is still a relatively recent technique in assessing tree health [4,5] and remains residually implemented in agriculture. Due to the relatively scarce studies in this field [5], more research to ensure its potential and applicability on a large scale is required. Thus, it is crucial more studies focus on contributing to show and analyse the complexity of the technique applied to trees, the new and atypical aspects of the problem and respond to some knowledge gaps. Accordingly, the present paper intends to detail some relevant features related to the applicability of the technology, thus contributing to turning the IRT technique as a decision-making tool to assess the health status of trees. To accomplish this, two sample trees, one of species *Quercus pyrenaica* Willd and another of species *Olea europaea* L., are qualitatively analysed from their thermal images.

2. MATERIALS AND METHODS

2.1 Background

The fundamental aim of thermographic surveying is the heat transfer process shown in Figure 1. When the IRT camera aims at the target, it receives radiation from the object itself, radiation reflected on the surface of the object coming from the emissions of neighbouring bodies, and radiation emitted by the atmosphere. In fact, the atmosphere results in interfering in some way with radiation that arrives at the camera [20]. The total power of the radiation arriving at the IRT (W_{tot}) camera is equal to the sum of the emission of the body ($\epsilon\tau W_{obj}$), the emission reflected in the object coming from sources of the near ambient ($(1 - \epsilon)\tau W_{refl}$), and the emission of

$$W_{tot} = \varepsilon\tau W_{obj} + (1-\varepsilon)\tau W_{refl} + (1-\tau)W_{atm}, \quad (1)$$

where:

ϵ	Emissivity
τ	Coefficient of atmosphere transmission
W_{obj}	Energy radiation emitted by the object
W_{refl}	Reflected energy from surrounding bodies
W_{atm}	Atmospheric energy

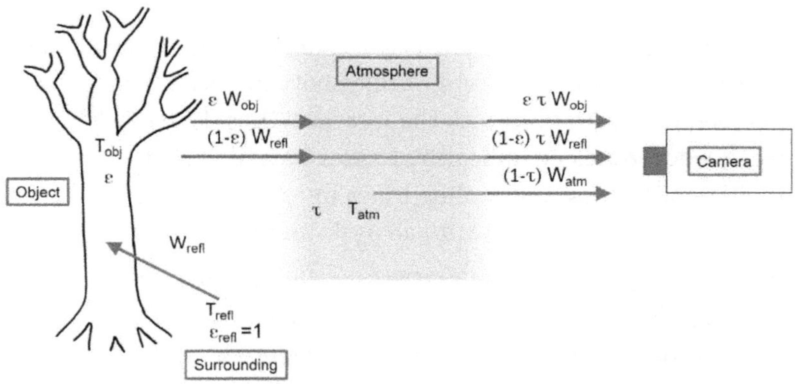

Figure 1. A schematic representation of infrared thermography (IRT) basics (adapted from [19]).

The thermal regime of a tree is determined by the heating of the surface by solar radiation and the transport, by conduction, of sensible heat to the interior. During the day, the surface heats up, generating a sensible heat flow towards the tree (a relatively slow process). The transportation of the elaborated sap promotes the convective transport of heat and ensures more tree thermal uniformity (a faster process of energy exchange). At night, the surface cooling by the emission of radiation (long waves) reverses the flow direction that changes from the interior to the surface leading to the cooling of the entire structure. The sap circulation causes the temperature gradient to vary along the trunk, and trees with more water available have better sap circulation [16]. This characteristic allows the differentiation between functional and dysfunctional tissue since the transport is done through the functional tissue. This feature allows health and vitality to be verified by IRT [16].

The density, the specific heat and the thermal conductivity play an important role in these processes. High specific heat leads to a smaller thermal stimulation for the same amount of heat. Thermal conductivity expresses the ability of the material to allow the heat to pass through it. It is commonly used in steady-state heat transfer analysis. Thermal diffusivity is much more relevant for transient heat transfer processes because it shows how well the heat could diffuse through the material. Thus, the diffusivity is a more important variable for the thermal characterization of a body than the conductivity (k), because it expresses how quickly the body adjusts completely to the temperature of its environment. In fact, the material thermal diffusivity (α) expresses how much easier the heat moves through its volume, that depends on the velocity conduction of the heat (k) and the amount of heat needed to increase temperature (ρC_p). It is given by the following equation:

$$\propto = \frac{k}{\rho C_p}, \tag{2}$$

where k is the thermal conductivity, ρ is the volumetric mass density, and C_p is the material specific heat. For instance, low diffusivity material delays the transfer of external temperature variations into the interior. Thus, it derives from the above, that damaged wood zones or unhealthy tissues have different thermal properties, leading to different heating and cooling times in relation to healthy zones. It is this principle that is explored in tree health analysis using IRT technology.

As mentioned above, wood (deteriorated/non-live tissues) and trees (functional tissues) have significantly different thermal characteristics, and as a consequence, different temperatures. This fact can be seen in Figure 2, that shows the difference in surface temperatures between a tree and a wooden stake supporting it [21]. The analysed tree was a specimen Prunus domestica L. (common name plum-tree) and the thermogram was obtained in the summer (air temperature 22.5 ○C, relative humidity 55%), 4 h after the sunset, with an emissivity of 0.97 and a rainbow colour pallet. The wooden stake displayed a lower temperature as compared to the tree (except its lower part). The lower part shows slightly higher temperature values than the lower trunk because the tree was irrigated before the observation. The tree keeps a balanced relationship with the environment temperature, so the tree temperatures usually are lower than the atmospheric temperature when the sun heating effect is over [5]. Therefore, in Figure 2, the tree shows higher temperatures in the healthier parts and lower temperatures in the deteriorated parts, as well as in parts where the tree was recently wet with water [22]. The greater the differences in the thermal pattern of trunk and branches, the worse the condition of the tree health [2–5].

(a) (b)

Figure 2. A tree (Prunus domestica L.) and a wooden stake supporting it: (a) photograph; (b) IRT thermogram (passive mode). The temperature values (○C): Sp1 = 20.0; Sp2 = 20.5; Sp3 = 21.5; Sp4 = 20.5; Sp5 = 19.5. (Adapted from [21]).

Equipment

A thermohygrometer, FLIR MR 176 (Figure 3A) was used to measure the atmospheric temperature and relative humidity. The thermographic camera used in this study is the FLIR T1030sc (Figure 3B).

Figure 3. (A) Humidity meter FLIR MR 176; (B) Thermal camera FLIR T1030sc.

It is a HD imaging and measurement camera recording 1024 × 768 pixels. It has a focal plane array (FPA) detector type, uncooled microbolometer with a detector pitch of 17 µm. The T1030sc operates in a spectral range of 7.5–14 µm and offers a thermal sensitivity < 20 mK, at 30°C (86°F). The accuracy, at 25°C, is +1°C or +1%, for body temperatures from 5°C to 150°C. The camera was equipped with 28° lens, FOV 28° × 21° (36 mm) and number F 1.2. The thermal camera is equipped with a compass and GPS functions to locate the data on the thermogram [20]. The minimum focusing distance is 0.4 m for the standard 28○ lens. The focus is manual, automatic or continuous. The continuous digital zoom from 1 to 8× can be used. It has a tactile screen of 800 × 480 pixels that facilitates the introduction of instructions. The software used to treat the thermograms were FLIR Tools + and FLIR Research IR Max 4 [23,24].

2.2 Sample Trees

Two trees were analysed: *Quercus pyrenaica* Willd (Figure 4), and *Olea europaea* L. (Figure 5). Table 2 shows their general characteristics. The common name of species *Quercus pyrenaica* Willd is Pyrenean oak. It is an autochthonous species of the Iberian Peninsula, from the mountain regions of continental climates [25]. This deciduous tree sprouts in acidic substrates preferably of granitic and schist origin. The immature subjects are marcescent with the capacity to sprout at the base of the trunk, as well as from the roots. The trunk is straight with opaque grey bark cracked in the form of panels. The specimen under study (Figure 4) is approximately 10 m in height. It consists of three trunks that arise from the base. The diameter at breast height (DBH) measured at 1.3 m in height, from south to north (south at the left of the photo) was: 0.20 m, 0.17 m and 0.17 m, respectively. The second tree is a perennial of the species *Olea europaea* L. It is native to the Mediterranean region (Southern Europe, North Africa and the Middle East) and the common name is olive (Figure 5). The olive tree is of exceptional longevity, being able to surpass 1500 years of age, and reaches a height of up to 8 m. It has greyish bark and branches. The specimen studied is approximately 3.5 m high. The branching begins at 1.75 m from the base, and the DBH measured at 1.30 m height is 0.35 m. Both specimens are located in the interior region of Portugal.

Figure 4. A photograph of the Quercus pyrenaica Willd tree taken at 2 p.m. under direct sunlight.

Table 2. Characteristics of sample trees.

Scientific Name	Common Name	Height (m)	HGB (m)	DBH (m)	Picture
Quercus pyrenaica Willd	Pyrenean oak	10.0	1.50	0.20/0.17/0.17	Figure 4
Olea europaea L.	Olive	3.5	1.75	0.35	Figure 5

HGB: Height from the ground (collar) to where branches begin (bifurcation); DBH: diameter at breast height.

Figure 5. A photograph of the Olea europaea L. tree taken at 5:30 p.m. under direct sunlight.

3. METHODOLOGY

The tree inspection was carried out applying qualitative IRT at the passive mode, that is, the heat source was the environment through solar radiation. The heat flows from the hotter to the colder zones of the tree. The irregularities of the thermal pattern on the tree surface can indicate the existence of defects, voids and deteriorated tissue. When it is found, deterioration as voids and defects occur in both the thermal properties of the constituents and the heat transfer process [2–5,7].

Thermograms were recorded at different times along the day, from when direct solar radiation was hitting the trees until after sunset. Besides the thermograms, photographs were taken to support the visual inspection and thermogram interpretation. The thermograms were processed running FLIR software [23,24]. The authors marked the spots at different locations over the tree trunk. The thermal patterns were analysed in order to correlate the temperature distribution with tree health. The atmospheric temperature and the relative humidity were measured at the time of the recording pf each thermogram, as well as the observation distance (between the thermal camera and the tree).

From a qualitative approach, although important, the value of emissivity is not preponderant. In fact, what is at issue is not the exact value of the measured temperature, but the temperature differences among the spots. The temperature values are comparable because the camera measured them with the same emissivity and reflected temperature.

Nevertheless, the emissivity value introduced was the one appropriate to the surface of the element under study. As the wood and bark are little reflective materials, the emissivity is high. Then, a high emissivity value was set in the camera. Thus, for both specimens, the emissivity value was 0.95. For each thermogram, the atmospheric temperature at the time of the capture was set in the camera. As it was a qualitative analysis, this study did not consider the reflected temperature. Further, it did not rain more than one week before surveying, then the thermograms did not register noises due to the humidity factor.

From the qualitative analysis, a more quantitative approach was taken, in particular, as regards to the comparison of recorded temperature values. It is important to understand thermal pattern differences among the trunks of similar diameter. The thermal patterns resulted from differences in the temperature along the surface. Naturally, this process introduces systematic errors that affect all points equally in each thermogram, whereby do not affect the result.

4. RESULTS AND DISCUSSION

Figure 6 shows the thermogram of the *Quercus pyrenaica* Willd tree recorded from the same point of view and at the same time as the photo shown in Figure 4 (on 21 January 2018, 2 p.m., winter in the northern hemisphere). The southern part of the tree (left side of the thermogram) is exposed to sunlight. Figure 7 shows another thermogram of the same tree at the same day but it was taken at 9:45 p.m. (4 h after sunset) from the same point of view. Table 3 shows the observation conditions, and the parameters set in the camera to capture the thermograms. Table 4 contains the temperatures of the spots marked in the thermograms.

Figure 6. Thermogram of Quercus pyrenaica Willd tree taken at 2 p.m. under direct sunlight. (Spot's size out of scale for better visualization).

Figure 7. Thermogram of the Quercus pyrenaica Willd tree taken at 9:45 p.m., 4 h after sunset. (Spot's size out of scale for better visualization).

Table 3. The ambient conditions and parameters for the thermograms of *Quercus pyrenaica* (Figures 6 and 7).

Thermogram Image	Tree (Specimen)	Air Temperature (°C)	Relative Humidity (%)	Emissivity	Reflected Temperature (°C)	Observation Distance (m)	Colour Palette	Temperature Range (°C)	Daytime
Figure 6	Quercus pyrenaica	20	50	0.95	20	5	Rainbow	10–35	sunlight
Figure 7	Quercus pyrenaica	10	70	0.95	10	5	Rainbow	0–10	night

Table 4. The temperature (○C) of spots on the *Quercus pyrenaica* thermograms (Figures 6 and 7).

Thermogram Image	Sp1	Sp2	Sp3	Sp4	Sp5
Figure 6	32.5	18.0	12.0	15.0	23.5
Figure 7	8.5	8.0	3.0	6.0	4.5

Figure 6 shows areas under direct sun exposure (left zone of the tree) and areas in the shadow (right of the tree). On the left of the trunk is Sp1 under direct sun exposure at 32.5°C, while Sp2 is in the shadow at 18°C. The Sp3, on the bottom right, with 12°C, point out an old cut trunk. Sp4 is in the shadow at 15°C on a branch of larger diameter than the branch where Sp5 is located. Sp5 is under direct sun exposure at 23.5°C.

Figure 7 shows the thermogram of the same tree captured 4 h after sunset. It can be seen that Sp1 is on the left trunk at 8.5°C and Sp2 is on the central trunk at 8°C. Sp3 is at 3°C on the stump. Sp4 at 6°C is located on a branch of larger diameter than the branch of Sp5. Sp5 is at 4.5°C. The observation conditions and the parameters assumed to capture the thermogram are in Table 3. Table 4 shows the spot temperatures in the thermogram.

Figure 8 shows a photograph (A) and a thermogram (B) of the tree *Quercus pyrenaica* Willd taken at the same time. The capture angle (west) is different from previous images. A red arrow (B) shows a stump that remained from an old cut trunk identified as Sp3 in the thermograms of Figures 6 and 7.

Figure 8. Photo (A) and thermogram (B) of the Quercus pyrenaica Willd tree. The white arrow shows an enlarged image of the CODIT development wall; the green arrow shows the same wall of CODIT on the thermogram; the red arrow indicates a stump. CODIT: compartmentalization of decay in trees.

Figure 9. Thermogram of the Olea europaea L. tree recorded in passive mode at 5:50 p.m., 15 min after sunset. (Spot's size out of scale for better visualization).

Figure 9 shows the thermogram of the tree of species *Olea europaea* L. described in Figure 5 and Table 2. Table 5 shows the environmental conditions during the observation, and the parameters assumed. The spot temperatures in Figure 9 are detailed in Table 6. The thermogram was recorded after sunset on a winter day (on 21 January 2018, at 5:50 p.m.; sunset at 5:35 p.m.). The atmospheric temperature was 18.0°C and the relative humidity was 50%. The atmospheric temperature was 18.0°C and the relative humidity was 50%. The thermogram was taken a distance of 5 m, using an emissivity of 0.95. A rainbow colour palette and a temperature range of 5.0°C to 20.0°C were used.

Table 5. The ambient conditions and parameters assumed for recording thermogram of *Olea europaea* L.

Thermogram Image	Tree (Specimen)	Air Temperature (°C)	Relative Humidity (%)	Emissivity	Reflected Temperature (°C)	Observation Distance (m)	Colour Palette	Temperature Range (°C)	Daytime
Figure 9	*Olea europaea* L.	18	50	0.95	18	5	Rainbow	5–20	after sunset

Table 6. The temperature (○C) of the spots on the Olea europaea L. thermogram (Figure 9).

Thermogram Image	Sp1	Sp2	Sp3	Sp4	Sp5	Sp6Sp7
Figure 9	18.5	8.5	14.0	10.0	9.0	12.011.0

The *Olea europaea* L. tree shows a large crack identified as Sp1 in the thermogram of Figure 9. Sp1 temperature is slightly higher than the atmospheric temperature. This spot represents a zone of the tree that has been exposed to the sun for a longer time. Sp2 temperature is much lower than Sp1 temperature—the difference between them is 10 ○C. The differences in the temperature at the same diameter trunk is an indicator of possible deterioration. Then, it is possible that the Sp2 zone has deteriorated. However, this zone was in the shadow for a longer time. In addition, Sp1 is in a crack area. Thus, the comparison of these two spots is not conclusive, but it reveals a clue of possible deterioration to take into account.

In the same Figure 9, Sp2 is at a much lower temperature than Sp3. These spots are comparable because they are located on trunks of similar calibre. On the other hand, Sp2 was in the shadow for some time. Thus, the comparison is not conclusive, but it reveals a new clue to take into account for possible deterioration. Then, this study analysed the temperature of Sp2 in relation to the atmospheric temperature. The Sp2 temperature of 8.5°C is much lower than the atmospheric temperature of 18°C.

This condition strengthens the possibility of deterioration at Sp2. Note that Sp3 is in the middle of the main trunk receiving sunlight. Although it already lost temperature, it was at 14°C.

Sp4 in Figure 9 is at 10°C, even exposed to direct sunlight, and it is on the same trunk as Sp3, which is at 14°C. The two of them are located in trunks of the same diameter. The difference of 4°C between the two points is a strong indicator of deterioration in Sp4. When comparing the temperature of Sp4 with the atmospheric temperature, the difference is 8°C, which is relevant since Sp4 was exposed to direct solar radiation until just before the thermogram was recorded. This is another reason that indicates deterioration. Sp4 is approximately at the opening of an orifice. Then, it was expected the temperature to be higher than the registered temperature, and still higher than the values of the boundary zones, as is the case at the slot of Sp1. A probable cause for unhealthy conditions in that area is the hole at Sp4 that stores water from precipitation.

Sp5, Sp6 and Sp7 (Figure 9) are on branches of approximately the same diameter. They are at lower temperatures than the main trunk, because they are of smaller calibre. The trunks of smaller calibre heat and cool faster than larger ones. Smaller trunks have a larger surface per volume unit, so they heat and cool faster. The thermal inertia of the trunks of larger diameter is greater because they have more mass. Thus, from the point of view of health, Sp5, Sp6 and Sp7 should be at approximately the same temperature if subjected to the same solar exposure. Sp7 should have a higher temperature since it received directly more sunlight. This did not happen. Sp6 temperature is 1°C higher than Sp7, which is not significant. The most disturbing is Sp5, because it is at 9 ∘ C. This value is too low for a healthy area even if exposed to less time in the sun. Sp5 is at a much lower temperature than the atmospheric temperature, which is 18 ∘C. This is a strong indicator of possible deterioration.

Sp2, Sp4 and Sp5 conditions (Figure 9) strongly indicate possible deterioration. This strong evidence was confirmed with observations of an enormous interior cavity linked to the crack (Figure 10).

Figure 10. Felling the tree has shown the presence and size of the concealed damage (red arrow) detected by the IRT method.

For doubtful cases, more thermography analysis would be recommended, e.g., at night, and if doubt remained, these spots become flags for the application of other diagnostic techniques and methods. Note that, these other diagnostic techniques do not need to be applied to the whole tree, but only in the identified points.

Epiphytic vegetation, such as lichens, is distributed more or less homogeneously throughout the tree and should not be responsible for the differences in the thermal pattern presented.

In IRT, as in all existing techniques, there are several conditions to take into account in order to achieve reliable results on tree analysis. That is: Exposure to sunlight and shadow; thermal contrast between the environment and the object targeted; the absence of water like rainfall; vegetation covers such as mosses and lichens; typical bark patterns of each species; the thermal comparison between the trunks of different calibre within the same tree [2–4,9].

Thermographic observations to trees are often made against the sun, or to tree surfaces when they are under the incidence of solar radiation. Thermographic recordings of trees when the sun is in front of the camera and sunlight directly hitting the tree surface introduces recording noise. In these cases, the temperature differences are more influenced by reading errors (lens, light exposure and reflections) than by material properties and defects. The shaded or less-illuminated areas (lower direct solar exposure) exhibit lower temperature values than expected, suggesting that the tree has deteriorated [2–4]. The IRT application requires a strong thermal contrast between the object to observe the environment and the objects that surround it. There must be a significant difference between the radiative power of the environment and the object analysed [9].

The thermograms and photographs presented are illustrative of relevant aspects to be taken into account in the observation and thermographic analysis of tree salubrity. It can be seen that on the thermogram of Figure 6, Sp1 and Sp5 are exposed to direct sunlight, and they are at a higher temperature than the atmospheric temperature, whereas the spots in the shadow are at a lower temperature than the atmospheric temperature. On the thermogram of Figure 7, all temperatures are lower than the atmospheric temperature. The colour patterns of thermograms of Figures 6 and 7 appear different. The various pattern colours correspond to different temperatures in the thermogram as shown in Table 4. The temperature differences by themselves do not indicate deterioration, as they can result from sunlight exposure and shadows. Figure 7 shows a general homogeneous temperature distribution, in which the different shades of Sp3 in Figure 7 is on a tree stump. Even its diameter is larger than the other trunks and it is at the lowest temperature and considered lifeless, and therefore not functional. It is probably because it does not circulate sap and its water content is much lower than its surroundings. In the upper part of the tree, the branches are of smaller diameter, as in Sp5. The branches heat and cool faster.

Therefore, at 2 p.m. (under the action of sun exposure) Sp3 is at a higher temperature than Sp4, and at 9:45 p.m. (night), the opposite is registered.

Under closer inspection, there are subtle differences of the temperature in the thermograms, which result from the bark pattern, pruning wounds, and epiphytic vegetation such as mosses and lichens. They appear as spots of slightly lower temperature than the adjacent ones. In any of these cases, these small thermal contrasts are not deterioration signals. Another case that leads to some doubts is the self-defence process (CODIT) that trees develop.

It is important to point out that thermographic inspection requires photographs because they are visual inspection tools (VTA) that facilitate the interpretation of thermograms. By improving the interpretation of the results, it is often possible to avoid invasive methods [8].

Finally, it is relevant to highlight the main merits and weaknesses of the IRT technique. IRT has an enormous capacity to analyse the trees as a whole and differentiate functional tissue from dysfunctional tissue. This is crucial for the inspection of the vitality and health status, representing a fast, economical, nondestructive and environmentally friendly monitoring tool. However, as in other non-invasive methods, the main limitation is that IRT does not identify specifically the damage detected, i.e., does not identify the pathology, nor its causative agent. Nor can it also give precise indications of the magnitude of the damage. More studies are needed to optimise the technology and training, in order to make the system even more efficient and reliable.

5. CONCLUSIONS

In this study, the general principles of the IRT methodology for health status of trees was presented. A qualitative IRT approach of two sample trees, namely species *Quercus pyrenaica* Willd and *Olea europaea* L., was used. Several details were highlighted to describe the IRT analysis applied to trees. The results show the IRT is a non-invasive, sustainable and expedited technique with high potential for tree inspection. It allows for the early diagnosis of damage, even those that are not yet visually noticeable, which is relevant to advanced tree maintenance. As in any other technique, its correct application requires a deep and multidisciplinary knowledge of the phenomena and a high familiarisation with the technique. The study intends, therefore, that thermography and other related methodologies for tree diagnosis result in interventions that privilege sustainability to benefit the economy and nature.

6. RECOMMENDATIONS

IRT is a well-established technique in many fields. However, it is relatively new in agriculture where it remains residually implemented. Most farmers are ready to accept technology if it is profitable, less complex and makes their life easier [26]. For this to happen, more detailed studies to establish a solid application basis, which can guide practitioners needs to be created. In fact, reliable guidelines are not available to describe the acceptable protocol and parameters tailored to adopt for a more straightforward approach for many tree health problems. Several challenges need to be addressed: The complexity of the technique; the new and atypical aspects of the problem; several knowledge gaps related to the technical issues and applicability specificities. Thus, more research is required to ensure its potential and applicability on a large scale. The latest high resolution and sensitivity thermal cameras can also contribute to overcoming these challenges.

References

1. ICNF. Instituto da Conservação da Natureza e das Florestas, Árvores Monumentais de Portugal Árvores Monumentais de Portugal. Available online: http://www2.icnf.pt/portal/florestas/aip/aip-monum-pt (accessed on 14 March 2019).
2. Catena, G. A new application of thermography. *Atti della Fondazione Giorgio Ronchi* **1990**, 947–952.
3. Catena, A. Thermography Reveals Hidden Tree Decay. *Arboric. J.* **2003**, 27, 27–42. [CrossRef]
4. Catena, A.; Catena, G. Overview of Thermal Imaging For Tree Assessment. *Arboric. J.* **2008**, 30, 259–270. [CrossRef]
5. Bellett-Travers, M.; Morris, S. The Relationship between Surface Temperature and Radial Wood Thickness of Twelve Trees Harvested in Nottinghamshire. **Arboric. J. 2010**, 33, *15*–26. [CrossRef]
6. Shigo, A.L.; Marx, H.G. *Compartmentalization of Decay in Trees*; USDA Forest Service: Washington, DC, USA, 1977.
7. Goh, C.L.; Abdul Rahim, R.; Fazalul Rahiman, M.H.; Mohamad Talib, M.T.; Tee, Z.C. Sensing wood decay in standing trees: A review. *Sens. Actuators A Phys.* **2018**, *269*, 276–282. [CrossRef]
8. Mattheck, C.; Breloer, H. Field guide for visual tree assessment (Vta). *Arboric. J.* **1994**, *18*, 1–23. [CrossRef]
9. Holst, G.C. *Common Sense Approach to Thermal Imaging*; JCD Publishing: Winter Park, FL, USA; SPIE Optical Engineering Press: Bellingham, WA, USA, 2000; ISBN 978-0-9640000-7-0.

10. Pitarma, R.; Crisóstomo, J.; Jorge, L. Analysis of Materials Emissivity Based on Image Software. In *New Advances in Information Systems and Technologies;* Rocha, Á., Correia, A.M., Adeli, H., Reis, L.P., Mendonça Teixeira, M., Eds.; Springer International Publishing: Cham, Switzerland, 2016; Volume 444, pp. 749–757. ISBN 978-3-319-31231-6.

11. Crisóstomo, J.; Pitarma, R. The Importance of Emissivity on Monitoring and Conservation of Wooden Structures Using Infrared Thermography. In *Advances in Structural Health Monitoring [Working Title];* Hassan, M., Ed.; IntechOpen: London, UK, 2019.

12. FLIR. Assessing Tree Health with Infrared, Application Story. Available online: http://www.flirmedia.com/MMC/THG/Brochures/T559239/T559239_EN.pdf (accessed on 28 April 2019).

13. Leong, E.-C.; Burcham, D.C.; Fong, Y.-K. A purposeful classification of tree decay detection tools. *Arboric. J.* **2012**, 34, 91–115. [CrossRef]

14. Johnstone, D.; Moore, G.; Tausz, M.; Nicolas, M. The measurement of wood decay in landscape trees. *Arboric. Urban For.* **2010**, 36, 121–127.

15. Rudnicki, M.; Wang, X.; Ross, R.J.; Allison, R.; Perzynski, K. *Measuring Wood Quality in Standing Trees: A Review*; U.S. Department of Agriculture, Forest Service, Forest Products Laboratory: Madison, WI, USA, 2017; p. 13.

16. Burcham, D.C.; Leong, E.-C.; Fong, Y.-K. Passive infrared camera measurements demonstrate modest effect of mechanically induced internal voids on Dracaena fragrans stem temperature. *Urban For. Urban Green.* **2012**, 11, 169–178. [CrossRef]

17. Hoffmann, N.; Schröder, T.; Schlüter, F.; Meinlschmidt, P. Potenzial von Infrarotthermographie zur Detektion von Insektenstadien und -schäden in Jungbäumen. *J. für Kulturpflanzen* **2013**, *65*, 2013.

18. Burcham, D.C.; Leong, E.-C.; Fong, Y.-K.; Tan, P.Y. An Evaluation of Internal Defects and Their Effect on Trunk Surface Temperature in *Casuarina equisetifolia* L. (Casuarinaceae). *Arboric. Urban For.* **2012**, *38*, 277–286.

19. Vidal, D.; Pitarma, R. Infrared Thermography Applied to Tree Health Assessment: A Review. *Agriculture* **2019**, 9, 156. [CrossRef]

20. FLIR. *FLIR Systems T10xx Series User's Guide*; FLIR Systems, Inc.: Wilsonville, OR, USA, 2016.

21. Ferreira, M.; Crisóstomo, J.; Pitarma, R. Infrared Thermography Technology to Support Science Teaching— Meaningful Learning about Trees with University Students. In Proceedings of the 13th International Technology, Education and Development Conference (INTED2019), Valencia, Spain, 11–13 March 2019.

22. Meola, Carosena Origin and Theory of Infrared Thermography. In *Infrared Thermography Recent Advances and Future Trends*; Meola, C. (Ed.) Bentham Science Publishers: New York, NY, USA, 2012; pp. 3–28. ISBN 978-1-60805-143-4.

23. FLIR. *FLIR Tools+ User's Guide*; FLIR Systems, Inc.: Wilsonville, OR, USA, 2016.

24. FLIR. *FLIR ResearchIR 4 User's Guide*; FLIR Systems, Inc.: Wilsonville, OR, USA, 2016.

25. Humphries, C.J.; Press, J.R.; Sutton, D.A. *Árvores de Portugal e Europa*; Fapas: Porto, Portugal, 2005; ISBN 978-972-95951-2-7.

26. Sharma, S.; Kaushik, A. Views of Irish Farmers on Smart Farming Technologies: An Observational Study. *AgriEngineering* **2019**, 1, 164–187.

Chapter 10

EARLY DETECTION OF *ZYMOSEPTORIA TRITICI* IN WINTER WHEAT BY INFRARED THERMOGRAPHY

1. INTRODUCTION: AN OVERVIEW

Wheat is a staple food for about two billion people in the world [1]. Yield losses in agricultural production are caused by several biotic and abiotic stresses. Abiotic stresses such as frost, salinity, heat, and drought contribute to about more than 50% yield loss and are identified as the primary cause of yield loss worldwide [2–4]. Biotic factors such as pests and diseases also cause substantial damage to the crops. For example, in wheat, yield loss caused by diseases varies between 14 and 27%, depending on the different diseases and region [5]. High yield losses are mostly observed in susceptible genotypes, which usually display high disease severity [6]. The most common fungal diseases in temperate wheat growing regions are leaf rust (*Puccinia triticina*), stripe or yellow rust (*P. striiformis*), Fusarium head blight (*Fusarium spp.*), powdery mildew (*Blumeria graminis*), and Septoria tritici blotch (*Zymoseptoria tritici*) [7].

The symptoms of fungal infection usually appear after a certain period of time depending on weather conditions. Early detection and diagnosis of plant pathogens can provide adequate information to predict the onset of disease and adequate measures can be taken to protect the crop before the disease is widespread [8]. There is a need for an advanced technique for rapid, accurate, and reliable detection of plant diseases especially at the time when symptoms are not yet visible on the crop [9]. Though studies have been conducted to apply imaging technologies such as fluorescence imaging [10], multispectral or hyperspectral imaging [11,12], and nuclear magnetic resonance (NMR) spectroscopy [13] to detect fungal diseases. However, these methods work only on a certain wavelength, which are plant and disease specific [9]. For example, Bauriegel et al. [14] detected head blight in wheat using hyper-spectral imaging in the wavelength range of 400–1000 nm. In contrast, infrared thermography works in a wide range of wavelength and can be applied on large number of plant types and varieties. Application of infrared thermography to determine crop water status for irrigation scheduling has been established [15,16]. Crop water stress index (CWSI) is widely used to quantify the plant water stress based on the principle that, under water stress, plants close their stomata and this leads to the increase in leaf temperature [17,18]. The same principle is applied in this study, i.e., the plant transpiration and photosynthetic activity is influenced by Z. tritici infection and this results in the change in canopy temperature which can be detected by thermography. The primary host penetration of Z. tritici occurs by penetrating the stomata 24 to 48 hours after contact with the leaf surface. About 12 days later a rapid change from the biotrophic growth stage to necrotrophic growth takes place with appearance of lesions on the leaves [19]. In cooler climates, like Germany, the occurrence of the first symptoms can be as late as 21 days after infection. This long latency period makes it difficult for the farmer to decide the correct fungicide application date. The chlorotic and necrotic symptoms of Z. tritici can lead to a decrease in leaf photosynthesis [20].

Recently, a number of studies have been conducted to determine the suitability of thermal imaging to detect biotic stress, both in the field and greenhouse: for example, the effect of downy mildew on transpiration of cucumber [21] and the effects of *Z. tritici* on wheat leaf gas exchange [20]. However, further studies need to be carried out in order to assess the potential of using thermography for other crops and for different locations with varying environmental conditions. The objective of this study was to detect fungal colonization at canopy level in field conditions at an early stage. The sensitivity of the methodology and the technology will be tested on different wheat genotypes. Finally, the earliest possible time of *Z. tritici* attack will be determined.

2. MATERIAL AND METHODS

2.1 Experimental Setup and Wheat Varieties

Twenty-five wheat varieties were planted in the field at Stuttgart-Hohenheim in a split-plot design with three replications. The main plots consisted of the two treatments (non-inoculated, artificially inoculated by *Z. tritici*) that were arranged in a complete randomised block design, the subplots of the 25 wheat varieties were randomized according to a 5 × 5 lattice design. Each plot was sown by 60 seeds in two rows 1.2 m long with a distance of 0.21 m between rows resulting in an area of 0.5 m2 per variety. Each main plot was separated by a strip of tall-growing winter triticale ("Cando") to prevent the spreading of the pathogen by spray drift during inoculation or secondary spore production. Likewise, a strip of winter rye around the experimental field as a border was planted.

The seeds were sown on 6 October 2011 by using a small tractor with a planter. No irrigation was applied throughout the growing period, as precipitation was sufficient to meet the crop water requirement. The selection of wheat varieties was done by the State Plant Breeding Institute, University of Hohenheim. In Table 1 the varieties were marked according their resistance against Z. *tritici* as classified by the German Federal Plant Variety Office [22].

Table 1. Tested wheat varieties and their susceptibility to Z. tritici as classified by the German Federal Plant Variety Office (Bundessortenamt, 2012); 1 = low susceptibility, 9 = high susceptibility, n.i. = no information.

Variety	Country of Origin	Susceptibility
Akratos	G	5
Apache	F	n.i.
Arina	CH	n.i.
Batis	G	4
Biscay	G	7
Bussard	G	7
Cubus	G	6
Dream	G	n.i.
Egoist	G	4
F201-R	ROM	n.i.
Florett	G	n.i.
History	G	n.i.

Impression	G	4
Julius	G	3
MES130	C	n.i.
Meteor	G	4
Naturastar	G	6
Nelson	G	3
Pamier	G	3
Rubens	G	n.i.
Sailor	G	5
Skalmeje	G	4
Solitär	G	n.i.
Toras	G	4
Tuareg	G	5

C = China, CH = Switzerland, F = France, G = Germany, ROM = Romania.

2.2 Inoculation with Z. *tritici*

Fungal spores were collected from infected wheat leaves, which were shaken before inoculation to remove the old spores. To prepare a medium for inoculation, Z. *tritici* was cultivated on culture medium made of 4 g yeast extract, 4 g malt extract (powdered), 4 g of glucose, 15 g of agar-agar, and 1000 mL distilled water. After autoclaving at 120 ○C for 20 min the agar was poured into petri dishes and cooled down. For propagation, Z. *tritici* was placed on agar and exposed to white- and UV-light (16 h daytime at 18 ○C and 8 h night time at 12 ○C) for 3–5 days of incubation. For the mass reproduction of Z. *tritici* 150 mL fluid culture, made of 4 g yeast extract, 4 g malt extract, 4 g glucose, and 1000 mL distilled water, was added into a 300 mL Erlenmeyer flask and closed with an aluminium foil at 120 ○C for 30 min. When the medium had cooled after autoclaving, Z. *tritici* was taken by an inoculation loop from the plate on which the spores were well developed and brought into the Erlenmeyer flask. The flask was placed on a shaker for 5 days under white- and UV-light in the same condition as mentioned above. The spores were counted and adjusted to a concentration of 5×10^6 spores/mL and later used for inoculation. Finally, inoculation of Z. *tritici* was done for all varieties on 21 May 2012 in the field. Spores were than sprayed on the fully developed flag leaves of the wheat plants by using a knapsack sprayer with a constant air pressure.

2.3 Acquiring Thermal Images

An infrared camera (VarioCAM®, InfraTec GmbH, Dresden, Germany) was used for thermal imagining. Images were acquired from 17 May until 28 June 2012 between

13:00 and 14:00 to minimize the influence of changing solar angle. Images were taken only on the sunny days when transpiration rate of the crop was high. Each thermal image comprised two wheat varieties as shown in Figure 1A,B.

The thermal images were analysed using thermography software IRBIS (InfraTec GmbH, Dresden, Germany) and the chosen value for emissivity was 0.95. To differentiate between canopy and soil surface, an upper temperature threshold was defined and pixels with higher temperature were excluded by image processing, Figure 1C. A polygon was manually drawn along the canopy area for each variety representing the crop temperature, Figure 1D. The average crop temperature (ACT) was calculated from the pixels within the canopy polygon and calculated as

$$ACT = \frac{\sum T_{Pixel}}{n_{Pixel}}$$

The maximum temperature difference (MTD) within the canopy polygon for each variety was calculated as

$$MTD = T_{Pixel,\,max} - T_{Pixel,\,min}$$

The canopy temperature difference (CTD) was defined as

$$CTD = T_{air} - ACT$$

where, air temperature T_{air} was measured by the weather station installed at the station.

Figure 1. Canopy temperature measurement, (A) visual image, (B) thermal image, (C) histogram with upper temperature threshold, (D) polygons for calculating crop temperature.

2.4 Visual Scoring

The visual scoring was done by an experienced staff on plot basis. *Septoria tritici* blotch was scored according to the severity of symptoms observed on the flag leaf and ranged between 0% and 100%. Visual scoring was done on 13 June (23 DAI), 18

June (28 DAI), 21 June (31 DAI), 25 June (35 DAI), 28 June (38 DAI), and a last time at 3 July (43 DAI).

2.5 Data Analysis

SPSS version 16.0 for Windows (SPSS Inc., Chicago, IL, USA) was used for analysis of variance (ANOVA) to evaluate the effect of Z. *tritici* infection on ACT, MTD, and CTD. A linear regression analysis was performed between the disease level and ACT, MTD, and CTD for the last day of the experiment (DAI 38) across all varieties. Furthermore, a t-test for mean comparison was conducted individually for each variety to determine the date when the first significant difference of ACT, MTD, and CTD to control appeared at a least significant level of 0.05.

3. RESULTS

3.1 Disease Level and Temperature Effects across 25 Wheat Varieties

Figure 2 shows summarized results across all varieties at the last day of the experiment (38 DAI). The percentage of Z. tritici disease levels on inoculated plants was higher than the control treatment. At the end of the experiment, all the wheat varieties were Z. tritici infested irrespective of their susceptibility level to the disease. Variety "Solitär" proved to be the most Z. *tritici* resistant showing only 18% of disease level whereas, variety "MES130" showed the highest disease level of 93%. ACT of the Z. tritici inoculated group showed a 2.3°C higher value than the control group, which was highly significant ($p \leq 0.001$). MTD of Z. *tritici* treatment was 1.4°C higher than that of the control group ($p \leq 0.001$). CTD from Z. tritici treatment was below zero for all the varieties except of "Solitär" and "Toras", showing values of 0.27°C and 0.22°C. In contrast, CTD for the control treatment was always above zero. The difference of means between treatment and control was significant ($p \leq 0.001$).

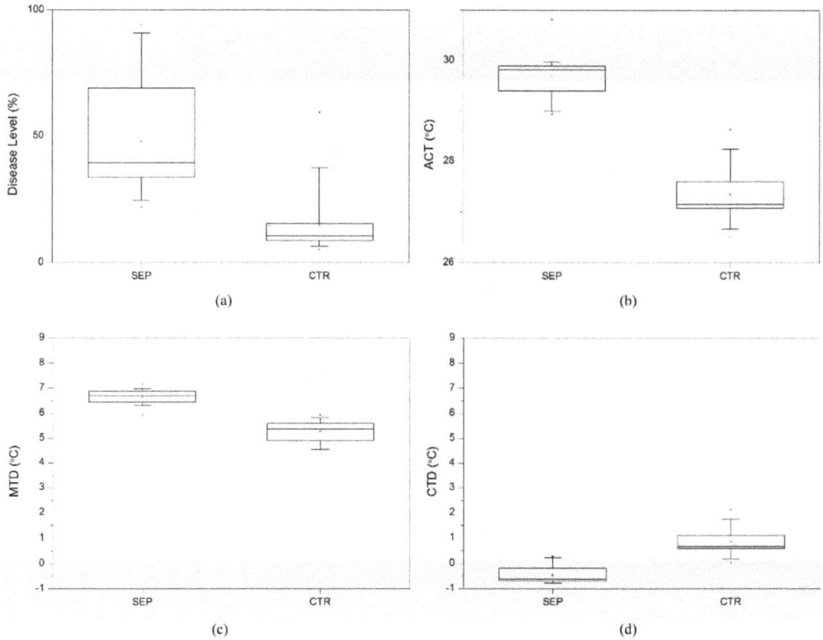

Figure 2. Crop parameters at the end of the experiment (38 DAI), summarized across 25 wheat varieties: disease level, (b) average canopy temperature (ACT), (c) maximum temperature difference (MTD), (d) canopy temperature difference (CTD); SEP = Z. tritici treatment and CTR = control.

3.2 Temperature Effects Related to Disease Level for 25 Wheat Varieties

ACT, CTD, and MTD on the last day of the experiment were plotted against disease level of the varieties as shown in Figure 3. It can be observed that ACT and MTD increased as the disease level increased, whereas CTD decreased with the disease severity. A stepwise regression analysis showed that CTD and MTD are the only parameters which can be used to explain the diseases level in all 25 wheat varieties. The significance level of MTD was however higher ($p \leq 0.01$) compared to that of CTD ($p \leq 0.05$). The estimation of the disease level of Z. *tritici* based on MTD and CTD is expressed by following equation (Adj.R2 = 0.7):

$$P_{\text{infested area}}\ (\%) = -183.426 + 35.485 \times \text{MTD} - 17.192 \times \text{CTD}$$

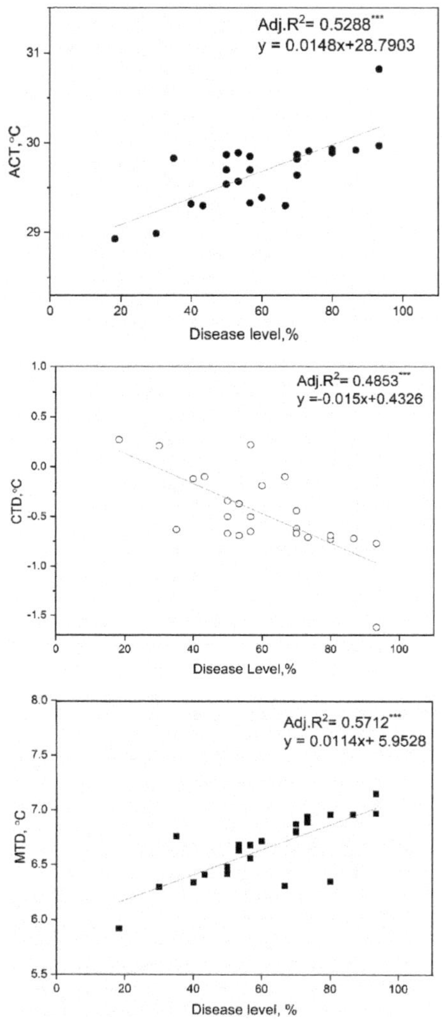

Figure 3. Regression analyses between Average Canopy Temperature (ACT), Canopy Temperature Depression (CTD), and Maximum Temperature Depression (MTD) and disease level at the end of the experiment (DAI 38) for 25 wheat varieties.

3.3 Temperature Effect on Wheat Varieties

As shown in Table 2, ANOVA test was conducted to compare the difference between the control and Z. *tritici* treatments in terms of CTD and MTD. It was observed that throughout the experiment the Z. tritici treatment had lower CTD than the control group and 12 varieties out of 25 showed significant difference at 0.05 levels. Before inoculation MTD of the control treatment was lower than the Z. *tritici* treatment however, these differences were not statistically significant. Later, after inoculation from 2 DAI till 28 DAI significant differences were observed in most of the varieties. An early significant difference in CTD and MTD were observed, this shows that these parameters can describe the health of the plant long before the naked eye observed

Table 2. ANOVA test between non-inoculated and Z. tritici treatment on the last day of the field experiment (38 DAI), difference in canopy temperature difference (CTD), maximum temperature difference (MTD) and the maximum difference from control treatment (DAI = days after inoculation).

Variety	First Visual Symptoms	ANOVA Test		Maximum Difference from Control				First Significant Difference from Control			
	DAI	CTD	MTD	ΔCTD_{max}	DAI	ΔMTD_{max}	DAI	ΔCTD	DAI	ΔMTD	DAI
Akratos	23	-	**	2.74	7	5.42	38	-	-	1.06	3
Apache	23	**	-	3.23	9	1.06	5	2.06	4	-	-
Arina	28	**	***	3.28	5	2.33	28	2.87	7	1.92	7
Batis	23	**	***	4.6	7	2.12	28	4.60	7	0.91	5
Biscay	28	**	**	4.09	9	1.64	38	3.82	8	0.51	4
Bussard	23	**	***	4.31	28	2.35	38	2.61	7	0.90	5
Cubus	23	**	**	3.74	12	1.35	38	1.91	5	0.78	4
Dream	23	-	***	4.88	3	2.05	29	-	-	0.81	4
Egoist	23	***	***	3.63	5	2.21	28	2.98	4	0.45	3
F201-R	23	***	*	4.25	7	1.51	38	2.34	4	0.29	3
Florett	23	**	-	3.20	9	1.06	5	1.70	4	-	-
History	23	**	*	3.45	9	1.69	38	1.75	5	1.04	5
Impression	23	**	-	3.50	8	0.98	26	1.52	3	-	-
Julius	28	-	**	1.69	9	1.43	8	-	-	1.09	4
MES130	23	***	**	3.45	38	2.12	38	1.83	4	0.86	7
Meteor	23	**	**	3.41	3	1.91	38	2.24	1	1.04	5
Naturastar	23	**	***	4.03	7	2.72	3	2.22	4	1.65	4
Nelson	23	***	***	4.54	9	2.57	12	3.51	4	1.17	4
Pamier	23	**	**	3.35	8	1.86	5	2.27	4	1.86	5
Rubens	23	*	*	3.06	29	1.7	29	1.34	1	0.84	3
Sailor	23	***	**	3.78	9	1.15	9	3.77	7	0.95	5
Skalmeje	28	-	**	2.86	26	1.49	1	-	-	1.23	3
Solitär	28	*	-	2.67	26	1.28	38	2.03	4	-	-
Toras	23	*	**	3.24	3	1.91	12	2.50	8	0.90	4
Tuareg	23	-	**	2.76	38	2.01	7	-	-	1.87	5

*- Not significant, * Significant at 0.1 levels, ** Significant at 0.05 levels, *** Significant at 0.01 levels.*

3.4 Temporal Development of Disease Level and Temperature Effects

The development of CTD, MTD, and disease level is displayed in Figure 4. Two varieties were selected, viz. the one with the highest and the one with the lowest difference from the control during the experimental study. The variety "Dream" had the highest ΔCTD value with a twice-higher disease level (20%) at the end of the experiment than that of Julius (10%), which had the lowest value in ΔCTD. The CTD of the control treatment was clearly higher than that of the Z. *tritici* treatment throughout the experimental study of both varieties, i.e., "Dream" and "Julius".

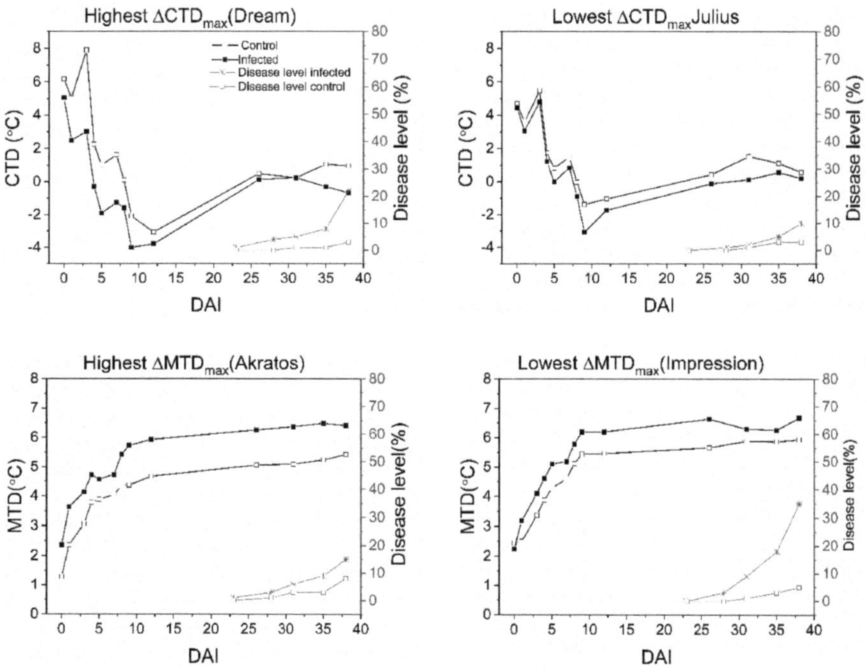

Figure 4. Effects of Z. tritici inoculation on Canopy Temperature Depression (CTD), Maximum Temperature Difference (MTD), Crop Water Stress Index (CWSI) and disease level.

The variety "Impression" had the lowest value of ΔMTDmax and has nearly more than doubled the disease level (35%) of variety "Akratos" (15%) which was the highest in ΔMTDmax. Even though the variety "Akratos" has a similar resistance level as the variety "Impression" according to Bundessortenamt (2012) in Table 1, in Z. *tritici* treatment, the MTD parameter of varieties "Akratos" and "Impression" had a similar trend up to 4 DAI. However, from 5 DAI onwards, the MTD of variety "Impression" was always higher than that of variety Akratos until the end of the experiment. Though it can be seen that the difference of MTD between the varieties was not very high, it did range between 0.27 ∘C and 0.53 ∘C. In addition, the MTD parameter of the Z. *tritici* was always higher than that of the control treatment in both varieties throughout the experiment.

4. DISCUSSION

In this paper, Infrared thermography as a useful tool for early fungal disease detection in winter wheat was studied. Significant differences in ACT, CTD, and MTD between Z. *tritici* and control treatment were observed. It was interesting to find out that a significant difference in CTD, MTD was observed within a week from inoculation (Table 1) which is far earlier than the disease symptoms appeared on the plant. This is due to the fact that Z. *tritici* can germinate within 12 h and penetrate through stomata within 24–48 h after coming in contact depending on the climatic conditions [19]. Robert et al. [20] observed the chlorotic symptoms of Z. tritici on the flag leaves from 10 DAI and on the second leaves on 15 DAI. Plants resist the invasion of these pathogens by adopting various mechanisms such as reinforcing the cell wall, and producing compounds and inhibitors as a resistance reaction [23]. These processes could affect some of the physiological processes and could cause stress on the plant [24]. The chlorotic and necrotic symptoms of Z. *tritici* leads to a reduction in leaf photosynthesis [20]. However, how early the significant difference in CTD and MTD appeared was not related to the susceptibility level of wheat varieties. It agrees with the finding of Kema et al. [25] who found out that the spore germination is not influenced by the susceptibility of wheat varieties.

A decrease in CTD of Z. *tritici* treatment occurred as the disease level increased (Figure 3) which is caused by the reduction in the photosynthetic activity. This is in line with the results of Rosyara et al. [26] who mentioned that the stress from spot blotch caused by *Cochliobolus sativus* lead to a decrease in CTD because of the reduction in photosynthetic activity in infected plants whose green area is less than the healthy leaves. In addition, the invasion of these pathogens could negatively affect transpiration [27,28].

Reduction in the transpiration rate will result in higher temperature, and this could account for the reason why most of the wheat varieties showed significant differences in CTD and MTD towards the end of the experiment. Similarly, Shtienberg [28] mentioned that the colonisation and the level of damage caused by the fungi results in the reduction of photosynthesis and transpiration, and reduced transpiration would lead to a rise in the canopy temperature. The highest correlation of $R^2 = 0.6$ (Figure 3) was found between disease severity and MTD. This implies that the parameter MTD can be used as an indicator of early detection of Z. tritici in winter wheat. Also Oerke et al. [29] found out a strong correlation ($R^2 = 0.8$) between MTD and diseased severity of cucumber with inoculation of Pseudoperonospora cubensi. In our study the result of stepwise regression showed that only MTD and CTD parameters are related with the disease level with $R^2 = 0.7$ (Equation (7)). This is better than only considering one parameter for the prediction of disease level of Z. *tritici*. However, the reliability of CTD and MTD for evaluating the disease severity depends on the environmental conditions. CTD is related to ambient VPD and net radiation according to the energy balance model [30]. MTD of the leaf is also influenced by the environmental conditions such as ambient temperature and relative humidity which have an effect on the transpiration and hence on the leaf surface temperature distribution. In addition, as the disease severity increased, the CTD and MTD parameters of the healthy (control) and Z. tritici treatment became more prominent (Figure 4). This can be explained by the relationship between CTD and transpiration rate [31]. As the transpiration rate of the infected treatment decreases, the canopy temperature increases.

5. CONCLUSIONS

In this study, all the wheat varieties with Z. *tritici* showed significant differences in ACT, CTD, and MTD compared to the healthy plants (control) throughout the study period. However, only CTD and MTD parameters were significantly related to the disease level and therefore, can be used to predict the onset of fungal disease. The results showed that in some varieties, the earliest disease symptoms according to the MTD parameter could be determined as early as 3 DAI ($p \leq 0.01$). Similarly, according to the CTD parameter, the earliest disease symptom was found as early as 4 DAI ($p \leq 0.01$).

However, in the same varieties, the first visual symptoms appeared 23 DAI. Hence, it can be concluded that thermography can be used as high throughput to accelerate monitoring of fungal disease in the field. When mounted on a drone, breeders could highly benefit from thermography for selecting disease-resistant varieties from thousands of progenies in a high-throughput manner.

References

1. Siddiqui, K. Green biotechnology at the crossroads of nanobiotechnology, globalization, poverty alleviation and food sovereignty. *Indian J. Crop. Sci.* **2007**, 2, 1–5.

2. Dhlamini, Z.; Spillane, C.; Moss, J.P.; Ruane, J.; Urquia, N.; Sonnino, *A. Status of Research and Application of Crop Biotechnologies in Developing Countries: Preliminary Assessment;* Food and Agriculture Organization of the United Nations: Rome, Italy, 2005.

3. Jarvis, D.; Mar, I.; Sears, L. (Eds.) *Enhancing the Use of Crop Genetic Diversity to Manage Abiotic Stress in Agricultural Production Systems;* IPGRI: Rome, Italy, 2006.

4. Wang, W.; Vinocur, B.; Altman, A. Plant responses to drought, salinity and extreme temperatures: Towards genetic engineering for stress tolerance. *Planta* **2003**, *218*, 1–14. [CrossRef] [PubMed]

5. Kosina, P.; Reynolds, M.; Dixon, J.; Joshi, A. Stakeholder perception of wheat production constraints, capacity building needs, and research partnerships in developing countries. *Euphytica* **2007**, *157*, 475–483. [CrossRef]

6. Ahmad, S.; Afzal, M.; Noorka, I. Prediction of yield losses in wheat (*Triticum aestivum L.*) in relation to epidemiological factors in Faisalabad. *Pak. J. Bot.* **2010**, 42, 401–407.

7. Duveiller, E.; Singh, P.R.; Nicol, J.M. The challenges of maintaining wheat productivity: Pests, diseases, and potential epidemics. *Euphytica* **2007**, *157*, 417–430. [CrossRef]

8. Lucas, R.; Ubeda, A.; Paya, M.; Alves, M.; del Olmo, E.; Lopez, J.L.; San Feliciano, A. Synthesis and enzyme inhibitory activities of a series of lipidic diamine and amino alcohol derivatives on cytosolic and secretory phospholipases A2. *Bioorg. Med. Chem. Lett.* **2000**, 10, 285–288. [CrossRef]

9. Sankaran, S.; Mishra, A.; Ehsani, R.; Davis, C. A review of advanced techniques for detecting plant diseases. *Comput. Electron. Agric.* **2010**, 72, 1–13. [CrossRef]

10. Moshou, D.; Bravo, C.; Oberti, R.; West, J.; Ramon, H.; Vougioukas, S.; Bochtis, D. Intelligent multi-sensor system for the detection and treatment of fungal diseases in arable crops. *Biosyst. Eng.* **2011**, 184, 311–321. [CrossRef]

11. Huang, W.; Lamb, D.W.; Niu, Z.; Zhang, Y.; Liu, L.; Wang, J. Identification of yellow rust in wheat using in-situ spectral reflectance measurements and airborne hyperspectral imaging. *Precis. Agric.* **2007**, 8, 187–197. [CrossRef]

12. Mehl, P. Development of hyperspectral imaging technique for the detection of apple surface defects and contaminations. *J. Food Eng.* **2004**, *61*, 67–81. [CrossRef]

13. Choi, Y.; Angenot, L.; Harnischfeger, G. Analysis of *Strychnos nux-vomica, Strychnos icaja and Strychnos ignatii* extracts by 1H nuclear magnetic resonance spectrometry and multivariate analysis techniques. *Phytochemistry* **2004**, *65*, 1993–2001.

14. Bauriegel, E.; Giebel, A.; Geyer, M.; Schmidt, U.; Herppich, W.B. Early detection of fusarium infection in wheat using hyper-spectral imaging. *Comput. Electron. Agric.* **2011**, *75*, 304–312. [CrossRef]

15. Padhi, J.; Misra, R.K.; Payero, J. Prospects of using infrared thermography for irrigation scheduling of wheat crop. In Proceedings of the Australian Irrigation Conference and Exhibition 2010: One Water Many Futures, Sydney, Australia, 8–10 June 2010.

16. Zia, S.; Spohrer, K.; Merkt, N.; Wenyong, D.; He, X.; Müller, J. Non-invasive water status detection in grapevine (*Vitis vinifera L.*) by thermography. *Int. J. Agric. Biol. Eng.* **2009**, 2, 46–54.

17. Jones, H.G. Use of infrared thermometry for estimation of stomatal conductance as a possible aid to irrigation scheduling. *Agric. For. Meterorol.* **1999**, 95, 139–149. [CrossRef]

18. Jones, H.G.; Stoll, M.; Santos, T.; de Sousa, C.; Chaves, M.M.; Grant, O.M. Use of infrared thermography for monitoring stomatal closure in the field: Application to grapevine. *J. Exp. Bot.* **2002**, 53, 2249–2260. [CrossRef] [PubMed]

19. Eyal, Z.; Sharen, A.L.; Prescott, J.M. *The Septoria Diseases of Wheat, Concepts and Methods of Disease Management*; CIMMYT: Texcoco, Mexico, 1987.

20. Robert, C.; Bancal, M.-O.; Lannou, C.; Ney, B. Quantification of the effects of *Septoria tritici* blotch on wheat leaf gas exchange with respect to lesion age, leaf number, and leaf nitrogen status. *J. Exp. Bot.* **2006**, 57, 225–234. [CrossRef]

21. Lindenthal, M.; Steiner, U.; Dehne, H.W.; Oerke, E.C. Effect of downy mildew development on transpiration of cucumber leaves visualized by digital infrared thermography. *Phytopathology* **2005**, 95, 233–240. [CrossRef]

22. *Beschreibende Sortenliste 2012-Getreide, Mais, Öl- und Faserpflanzen, Leguminosen, Rüben, Zwischenfrüchte*; Bundessortenamt: Hannover, Deutschland, 2012.

23. Ding, L.; Xu, H.; Yi, H.; Yang, L.; Kong, Z.; Zhang, L.; Xue, S.; Jia, H.; Ma, Z. Resistance to hemi-biotrophic *F. graminearum* infection is associated with coordinated and ordered expression of diverse defense signaling pathways. *PLoS ONE* **2011**, 6, 1–17. [CrossRef]

24. Kuc, J. Phytoalexins, stress metabolism, and disease resistance in plants. *Ann. Rev. Phytopathol.* **1995**, 33, 275–297. [CrossRef]

25. Kema, G.H.J.; Sayoud, R.; Annone, J.G.; Van Silfhout, C.H. Genetic variation for virulence and resistance in the wheat-*Mycosphaerella graminicola* pathosystem. II: Analysis of interactions between pathogen isolates and host cultivars. *Phytopathology* **1996**, 8, 213–220. [CrossRef]

26. Rosyara, R.U.; Subedi, S.; Duveiller, E.; Sharma, R.C. The effect of spot blotch and heat stress on variation of canopy temperature depression, chlorophyll fluorescence and chlorophyll content of hexaploid wheat genotypes. *Euphytica* **2010**, *174*, 377–390. [CrossRef]

27. Rabbinge, R.; Jorritsma, I.T.M.; Schans, J. Damage components of powdery mildew in winter wheat. *Neth. J. Plant. Pathol.* **1985**, 91, 235–247. [CrossRef]

28. Shtienberg, D. Effect of Folia Diseases on Gas Exchange Processes: A comparative Study. *Phytopathology* **1992**, *82*, 760–765. [CrossRef]

29. Oerke, E.C.; Steiner, U.; Dehne, H.W.; Lindenthal, M. Thermal imaging of cucumber leaves affected by downy mildew and environmental conditions. *J. Exp. Bot.* **2006**, 57, 2121–2132. [CrossRef] [PubMed]

30. Jackson, R.G.; Idso, S.B.; Reginato, R.J.; Pinter, P.J. Canopy temperature as a crop water stress indicator. *Water Resour. Res.* **1981**, 17, 1133–1138. [CrossRef]

31. Balota, M.; Payne, A.W.; Evett, R.S.; Lazar, D.M. Canopy Temperature Depression Sampling to Assess Grain Yield and Genotypic Differentiation in Winter Wheat. *Crop Sci.* **2007**, *47*, 1518–1529. [CrossRef]

Chapter 11

UTILISATION OF GROUND AND AIRBORNE OPTICAL SENSORS FOR NITROGEN LEVEL IDENTIFICATION AND YIELD PREDICTION IN WHEAT

1. INTRODUCTION: AN OVERVIEW

Agricultural systems using Precision Farming (PF) technologies have already been introduced in the market. The range varies from entry level guidance to data acquisition systems integrated into the farm management software. Most of these systems gather tractor-implement information, or perform tailor made applications [1]. The more intensive the crop production system is, the more advanced the technology adaptation on farms is [2]. This serves the goal of higher yields and better crop quality, with the support of sensor systems. The increasing number of available sensors, along with the high diversity of sensor technologies, e.g., imaging sensors, multi- and hyperspectral optical sensors, fluorometers, etc., has increased the possibility for integrating these sensor systems into the daily farm operation. Each sensor has advantages and disadvantages, and can provide important information concerning the field status [2–4]. Yet each sensor type has limitations to overcome. By merging the data of different sensors and sensor types, their limitations can be reduced, since data can be complementary or more informative [5]. In that sense, data fusion approaches are necessary, achieving better results by merging numerous sensor data deriving from the field and comparing them with ground truth data like yield or biomass.

Hall and Llinas [6] defined data fusion as "the integration of information from multiple sources to produce specific and comprehensive unified data about an entity". Brooks and Iyengar [5] classified four categories for sensor data fusion: (1) redundant; (2) complementary; (3) coordinated; or (4) independent fusion. Dasarathy [7] defined three levels: (I) raw data fusion; (II) feature fusion with feature extraction; and (III) decision fusion, which includes inter alia weighted decision methods [8]. Many different terms are used in literature to describe and discuss "fusion" concerning data. Dasarathy [9] also decided to use "information fusion" instead, as the overall term. In all cases, fusion of the sensor information can improve our knowledge of the field conditions [6].

For agricultural applications many sensors have been proposed. Several research studies based on spectral data are available, e.g., using data mining techniques with a genetic algorithm for nitrogen (N) status and grain yield estimation [10], or acquiring multispectral aerial images for the detection of wheat crop and weeds [11]. They are often based on measurements with one single sensor. There is a lack of information, of how informative different sensors and combination of sensors are, in the variability presented at the field level. Peteinatos et al. [12] measured stress levels in outdoor wheat pots with three optical sensors. Yet there is work to be done, connecting ground data with aerial data, even more in real field conditions. Using mobile platforms for data acquisition offers the possibility of system automation with fusion approaches. The advantage of ground platforms is their ability of carrying higher loads and more equipment than it would be possible with Unmanned Aerial Vehicles (UAV) [13].

In the current paper, the investigated research fields were planted with winter wheat utilising different N levels. These fields were examined with a fluorescence sensor and spectrometers, one spectrometer installed on an UAV, the other two spectrometers and the fluorescence sensor on a ground platform. The aim of this research was to test research sensors on field trails close to normal, practical farming conditions. This publication will discuss redundant and complementary fusion approaches, on a raw data and feature fusion level. It investigates the questions; (i) how the used research sensors perform in a large field; (ii) which of the calculated features are statistically significant for assessments of wheat yield, biomass and the available N for the plant; and (iii) how information fusion can support farming decisions.

3. MATERIAL AND METHODS

3.1 Experimental Site

The investigations have been made at the facility Ihinger Hof in Renningen (Germany), an institution of the University of Hohenheim, Stuttgart in South-West Germany. The location of Ihinger Hof (N 48°44!41!!, E 8°55!26!!) has a mean annual precipitation of 690 mm (710 mm in 2011 and 727 mm in 2012), and an average annual temperature of 7.9 °C. The measurements about four site-years have been carried out with winter wheat (*Triticum aestivum* L., cv. Toras and Schamane) on the experimental fields "Inneres Täle" in 2011 respectively site-year (1), "Riech" in 2011 and 2012 (2) + (3), as well as "Lammwirt" in 2012 (4). The term "site-year" is a combination of two factors site and year, according to Beres et al. [14], where site relates to an individual field of a farm.

On site-year 1 and 4, the N levels ranged from 60 to 180 kg N·ha^{-1} in five distinct levels. Additionally a dosage of 170 kg N·ha^{-1}, that is usually applied on this farm, were used as the conventional application level. Figure 1 represents the N levels of site-year 1; site-year 4 had the same levels. Nitrogen was distributed in three fertiliser applications in the early growing periods (Zadoks' Scale (Z) 27–Z 47) [15] with a pneumatic fertiliser spreader and a tractor with an automatic steering system and GPS Real-Time Kinematic (RTK) precision (approx. ±2.5 cm). The first N application of 60 kg N·ha^{-1} has been distributed equally over the whole field. The second application had 0–80 kg N·ha^{-1} based on the treatment; and the third application has been carried out with 0–40 kg·N ha^{-1} to reach the planned total amount of N for the respective N level.

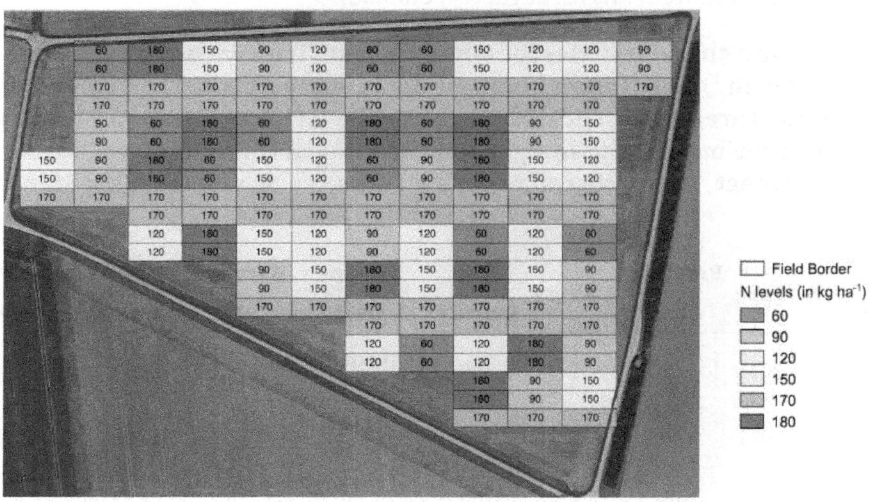

Figure 1. The colored plots reflect the different N levels in kg N·ha-1 for site-year 1, as shown in the legend. Each plot has a size of 36 m × 12 m (L × W).

For site-year 2 + 3, the fertilisation was applied in repeating rows over the whole field: (1) control; (2) APOLLO model output [16]; and (3) Yara N-sensor control. The field design in this case was different compared to site-year 1 and 4, however it provided the required randomisation for the data analysis, with N levels from 60–170 kg N·ha^{-1} in eight distinct levels (see Figure 2).

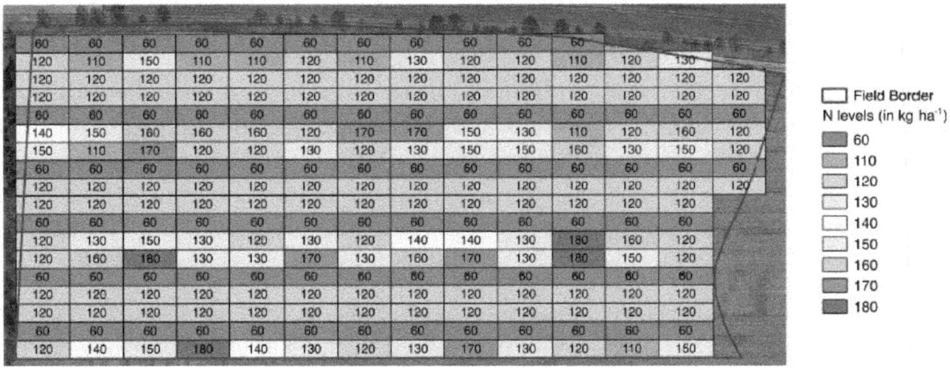

Figure 2. The colored plots reflect the different N levels for site-years 2 and 3, as shown in the legend. Each plot has a size of 36 m × 12 m (L × W).

Table 1 gives an overview of the site characteristics for the research fields. The research fields at the location Ihinger Hof have a high, natural field variability with soil types reaching from pure clay to silty loam. All fields of the location Ihinger Hof were investigated in the year 2009 on their electrical soil conductivity with an EM38 sensor (Geonics Limited, Mississauga, ON, Canada).

Table 1. Site characteristics for winter wheat (*Triticum aestivum* L.): plant density (No.·m^{-2}), seeding and harvest dates, and electrical soil conductivity (mS), of the three research fields Inneres Täle (IT), Riech (RI) and Lammwirt (LW) at experimental site Ihinger Hof, Renningen (Germany). C = Corn, WW = Winter wheat, OSR = Oilseed rape, T = Toras, S = Schamane, SD = Standard deviation.

Site-Year	Site	Year	Previous Crop	Variety	Plant Density	Seeding Date	Harvesting Date	Soil Conductivity Min	Mean	Max	SD
1	IT	2011	C	T	340	27 November 2010	11 August 2011	19.68	38.19	56.98	4.93
2	RI	2011	WW	S	300	14 October 2010	4 August 2011	14.48	31.82	69.53	6.01
3	RI	2012	WW	S	300	17 October 2011	31 July 2012	14.48	31.82	69.53	6.01
4	LW	2012	OSR	T	300	14 October 2011	1 August 2012	52.49	64.48	85.57	6.38

On all four site-years, biomass (BM) samples have been collected over the whole field at three growing stages: stem elongation (approx. Z 35), flowering (approx. Z 61), and before harvest (approx. Z 93). To determine the N content in the soil (N_{min}-method), the samples were analysed on three soil depths: (1) 0–30 cm; (2) 30–60 cm; and (3) 60–90 cm. This took place at the end of tillering (Z 29) and after the harvest. The BM samples have been analysed for grains per ear, the number of tillers, the protein content and the BM weight. The wheat fields have been harvested with a standard New Holland combine harvester, equipped with a header of 6 m cutting width and a GPS receiver with RTK precision to geo-reference the yield data. The laboratory analysis and the yield logging are considered in the current manuscript as the ground truth data, with which the sensor data will be compared. The available N for the plant (N_{avail}), used in this manuscript, is defined as the sum of N_{min} and applied N until the respective sensor measurement date. N_{avail} is a simplified form to express the N supply for the plants in field, as atmospheric entries and mineralisation may provide additional N during the growing season. In spring, soil samples over the whole field have been taken and after harvesting. Table 2 gives an overview of the measurement dates for the ground and UAV mounted sensors for site-year 1. A similar frequency of the field sampling applies to the rest of the site-years.

Table 2. Exemplary for the other site-years, the overview shows the dates for site-year 1 (2011) regarding ground and aerial sampling in the different growing stages (Z). A = aerial spectrometer, G = ground spectrometer.

Z	Spectrometer	Fluorescence Sensor
30	G: 28 April 2011	28 April 2011
37	A + G: 20 May 2011	20 May 2011
75	G: 16 June 2011	16 June 2011
77	G: 28 June 2011	28 June 2011
85	A: 4 July 2011, G: 6 July 2011	6 July 2011

3.2 Measurement Set-Up

The sensor measurements derive from data of three sensors, two spectrometer devices, FieldSpec Handheld (FS—Analytical Spectral Devices, Boulder, CO, USA), HandySpec Field® (HS—tec5 AG, Oberursel, Germany), and the fluorescence sensor Multiplex® Research (MP—Force-A, Orsay, France). The ground sensors (Table 3) were mounted on a rebuilt self propelled Hege 76 multi-equipment carrier (Wintersteiger AG, Ried, Austria), the so called Hohenheim multi-sensor platform "Sensicle"; for more information and image see Keller et al. [3] and Zecha et al. [4]. The sensors mounted to the Sensicle have been adjusted at every measurement date at a specific height for each sensor relative to the canopy (Table 3). The spectrometers are passive sensors, highly dependent on the sun illumination. On the other hand, the MP fluorometer is insensitive to the ambient light conditions due to its light-emitting diodes (LEDs) used for signal excitation. More information about the sensors can be found in Table 3. The Monolithic Miniature-Spectrometer (MMS) 1 NIR enhanced (Carl Zeiss Jena GmbH, Jena, Germany & tec5 AG, Oberursel, Germany) has been selected due to the compact dimension, the low weight of only 500 g, and the high spectral resolution [17]. It has similar technical properties like the HS sensor mounted on the Sensicle ground platform (Table 3). It was mounted in the centre of a fixed-wing UAV pointing with the detector to the ground and set to a flight altitude of 100 m above ground; for more information and images see Link et al. [17].

Table 3. Used sensor devices and sensor details. BGF = Blue-Green Fluorescence; RF = Red Fluorescence; FRF = Far-Red Fluorescence.

Type	Manufacturer	Sensor Model	Wavelength Range	Spectral Resolution	Footprint	Classification
Spectrometry	Analytical Spectral Devices	FieldSpec Handheld	325–1075 nm	1 nm	2.74 m^2	Passive, Ground
	tec5 AG	HandySpec Field®	360–1000 nm	10 nm	0.44 m^2	Passive; Ground
	Carl Zeiss Jena GmbH & tec5 AG	MMS1 NIR enhanced	310–1110 nm	3.3 nm	50.27 m^2	Passive, Aerial
Fluorescence	Force-A	Multiplex® Research	BGF, RF and FRF	-	0.005 m^2	Active, Ground

3.2 Information Fusion and Statistical Data Analysis

The ground sensor software for triggering the measurements has been developed by the respective sensor hardware companies. The data logging software for the aerial spectrometer has been developed in C++ for Windows mobile 5 on a Personal Digital Assistant (PDA) [17]. The sensor raw data have been processed and converted to features (indices and ratios) that correlate with field information and biological parameters. This has been done using Unix-Shell and awk scripts on Ubuntu 12.04 Long Term Support, in combination with the statistical software R [18]. For the spectral data, several indices were derived, allowing a comparison with other sensor data. Common plant characteristics like the chlorophyll content, are commonly used to determine the presence of stress or correlate with the field biomass [19,20]. In the current measurements, the following indices were calculated:

Red-Edge Inflection Point [21]

$$REIP = 700 + 40 \times \frac{(R_{670} + R_{780})/2 - R_{700}}{R_{740} - R_{700}} \quad (1)$$

Normalised Difference Vegetation Index [22]

$$NDVI = \frac{(R_{780} - R_{680})}{(R_{780} + R_{680})} \quad (2)$$

CropSpec [23]

$$CropSpec = (\frac{R_{808}}{R_{735}} - 1) \times 100 \quad (3)$$

Hyperspectral Vegetation Index e.g., [24]

$$HVI = \frac{R_{750}}{R_{700}} \quad (4)$$

Optimised Soil-Adjusted Vegetation Index [25,26]—factor L varies between 0 and 1

$$OSAVI/SAVI = \frac{(R_{800} - R_{670})}{(R_{800} + R_{670} + L)} \times (1 + L) \quad (5)$$

factor) was used, 0.16 for OSAVI and 0.20 for SAVI. Concerning the Multiplex® Research fluorescence sensor, the following

signals and ratios were used, as described in Cerovic et al. [27] and Ghozlen et al. [28]. The index denotes the fluorescence type while the subindex denotes the wavelength excitation of the LEDs:

$BGF_{UV/G}$ = Yellow Fluorescence
$RF_{UV/G}$ = Red Fluorescence
$FRF_{R/G}$ = Far-Red Fluorescence

Anthocyanins

$$ANTH = \log(\frac{FRF_R}{FRF_G}) \quad (6)$$

Flavonols

$$FLAV = \log(FER_{RUV}) \quad (7)$$

Fluorescence Excitation Ratio Anthocyanin Relative Index

$$FERARI = \log(\frac{5000}{FRF_R}) \quad (8)$$

Simple Fluorescence or Chlorophyll Ratio

$$SFR_{R/G} = \frac{FRF_{R/G}}{RF_{R/G}} \quad (9)$$

The geographic information system (GIS) Quantum GIS [29] has been used for data visualisation and for merging the geo-referenced features in form of indices, signals and ratios with the field design. Linear regression, analysis of variance (ANOVA) and branch-and-bound algorithm have been employed to the sensor data features with the aid of R. After post-processing the data, all features (independent variables—IDV) have been intensively analysed and correlated against the ground truth data (dependent variables—DV).

4. RESULTS

4.1 Field Conditions

The average yield amounts of all site-years per N level are presented in Table 4.

Table 4. Average grain yield (t·ha-1) for winter wheat (Triticum aestivum L.) with 14% grain moisture content at the different N levels (kg·ha-1) of the four site-years.

Yield for N Levels

Site-Year	Yield for N Levels					
	60	90	120	150	170	180
1	7.2	6.7	6.7	7.1	7.3	7.3
2	4.9	-	4.9	5.4	5.3	5.9
3	6.5	-	6.3	6.6	6.9	6.6
4	6.2	6.5	7.5	7.8	7.9	7.9

For the six N levels of site-year 1, yield showed an increasing amount with more N, except for the N level of 60 kg·ha^{-1} (Table 4). The 60 kg·ha^{-1} plots had a similar yield than the plots at a higher N level. On site-year 4, the average yield increased with the N levels until the level of 150 kg·ha^{-1}. For the treatments of 150–180 kg·ha^{-1}, the yield remained on a similar amount. For site-years 2 and 3, these fields have been fertilised with a different strategy, so the average yield amounts per N level are not directly comparable to site-years 1 and 4. As shown in Table 1, all fields used in this research have a high deviation for the electrical soil conductivity with values ranging from 14.48 to 85.57 Milli-Siemens (mS). The field belonging to site-year 2 and 3 has the highest range of all investigated fields.

4. 2 Regression Analysis

The feature extraction of the sensor raw data, presented as wavelength indices and fluorescence ratios, have been taken as independent variables (IDV) for the linear regression. As dependent variables (DV) for the following results have been chosen: (1) wheat yield; (2) BM weight; (3) leaf area index (LAI); and (4) available N (N_{avail}). The data analysis was carried out separately for each measurement date, to better observe the changes in correlation over time.

The linear regression results of the aerial sensor MMS1 data with the IDV's of site-years 2 and 3 are shown in Table 5. These results were not significant for DV's (1) and (3) of site-years 2 and 3, whereas DV LAI showed low correlations for site-year 2. The correlations with the DV's BM and LAI could not be measured for site-year 3. The number of valid UAV data fitting to the design layout of site-years 1 and 4 was too low for a significant data analysis.

Table 5. Linear regression analysis of UAV MMS1 sensor indices for site-years 2 and 3. Z = Zadoks' Scale, DV = Dependent Variable, IDV = Independent Variable, RMSE = Root mean square error, BM = Biomass, LAI = Leaf Area Index, HVI = Hyperspectral Vegetation Index, OSAVI = Optimized Soil-Adjusted Vegetation Index, REIP = Red-Edge Inflection Point, PVR = Plant Vigor Ratio, TCARI = Transformed Chlorophyll Absorption Reflectance Index.

Season	Z	DV	IDV	Adj. r^2	RMSE	p-Value
2011	73	LAI	HVI	0.15	0.31	0.0144
	73		NDVI	0.24	0.30	0.0020
	73		OSAVI	0.23	0.30	0.0024
	73		REIP	0.18	0.31	0.0077
	66	BM Weight	HVI	0.13	1521.64	0.0000
	73		HVI	0.21	852.72	0.0039
	34		NDVI	0.22	794.57	0.0246
	73		NDVI	0.23	838.76	0.0022
	34		OSAVI	0.22	793.42	0.0240
	73		OSAVI	0.23	842.12	0.0025
	34		PVR	0.19	812.01	0.0374
	73		REIP	0.17	875.04	0.0095
2012	61	LAI	REIP	0.09	0.47	0.0153
	85		NDVIg	0.08	0.49	0.0286
	85		HNDVI	0.08	0.49	0.0304
	85		OSAVI	0.07	0.49	0.0315
	85		NDVI	0.07	0.49	0.0332
	61	BM Weight	TCARI	0.08	117.17	0.0229

Tables 6 and 7 present the correlation results of site-year 4 for the ground sensors MP and FS, only for Adj. r^2 values > 0.46. The FERARI and SFR ratios are significant with yield and BM weight for end of heading and flowering growing stages onwards; the Yellow Fluorescence (BGF) correlates already at the beginning of stem elongation. For N_{avail}, the SFR ratio and the RF signal show significant results (Table 6). The calculated indices HVI, NDVI, OSAVI of the FS sensor show correlations with yield over several measurements of the growing season. The CropSpec and REIP indices highly correlate with N_{avail} for the end of heading stage and further on, HVI and NDVI on the other hand have a lower correlation (Table 7).

Table 6. Linear regression analysis of signals and ratios from Multiplex® Research fluorescence sensor for site-year 4. Z = Zadoks' Scale, DV = Dependent Variable, IDV = Independent Variable, RMSE = Root mean square error, BM = Biomass, FERARI = Fluorescence Excitation Ratio Anthocyanin Relative Index, SFR = Simple Fluorescence Ratio, BGF = Yellow Fluorescence. Significance level: p-value < 0.001.

Z	DV	IDV	Adj. r^2	RMSE
66	Yield	FERARI	0.48	0.75
85		FERARI	0.49	0.74
59		SFRG	0.53	0.71
85		SFRG	0.61	0.65
59		SFRR	0.56	0.69
85		SFRR	0.63	0.64
59	N_{avail}	SFRG	0.67	16.52
66		RFUV	0.63	26.77
31	BM Weight	BGFG	0.46	97.66
59		BGFG	0.61	83.32
85		BGFG	0.73	69.21
91		BGFG	0.74	68.11
85		FERARI	0.83	55.25
85		FLAV	0.62	81.98
85		FRFR	0.86	50.40
85		RFG	0.83	54.91
85		SFRR	0.85	52.14

Table 7. Linear regression analysis of indices from spectrometer FieldSpec HandHeld for site-year 4. Z = Zadoks' Scale, DV = Dependent Variable, IDV = Independent Variable, RMSE = Root mean square error, BM = Biomass. Significance level: p-value < 0.001.

Z	DV	IDV	Adj. r^2	RMSE
59	Yield	HVI	0.56	0.68
66		HVI	0.54	0.70
85		HVI	0.59	0.65
66		NDVI	0.56	0.70
85		NDVI	0.63	0.62
59		OSAVI	0.55	0.69
85		OSAVI	0.67	0.58
85		REIP	0.54	0.69
85		CropSpec	0.53	0.70
59		CropSpec	0.79	13.12
66		CropSpec	0.68	24.57
85		CropSpec	0.78	20.56

59		HVI	0.69	15.98
66		HVI	0.67	25.08
85	N_{avail}	HVI	0.68	24.78
66		NDVI	0.62	27.15
85		NDVI	0.62	26.91
59		REIP	0.86	10.76
66		REIP	0.76	21.40
85		REIP	0.83	18.04
31	BM Weight	REIP	0.78	67.07
85		SAVI	0.75	67.22

4.3 Data Validation

Basis of the data validation were the results of Zecha et al. [4], in which four mixed correlation models were presented, based on measurements with the same spectral and fluorescence sensors. From this research, Model 4 is proposed by the authors for cross-validation with the sensor data of site-years 1–3. The parameters A_x, B_x, C_x and D_x denominate the modelled coefficients of the linear regression—Model 4 as shown in [4]:

$$y = A_4 + B_4 \times \text{CropSpec} + C_4 \times \text{HVI} + D_4 \times RF_{UV} \quad (10)$$

By cross-validating the above model with the data from site-years 1–3, the correlations were low. Using a Find-Best-Model-Algorithm in R for the data sets of all site-years, the following yield model has been discovered:

$$\text{Yield}_{predicted} = A_x + B_x \times \text{FERARI} + C_x \times \text{HVI} + D_x \times \text{RFUV} \quad (11)$$

Table 8 highlights the corresponding correlations for Model 4 of site-year 4 and for the new model $\text{Yield}_{predicted}$, employed to the data sets of all site-years.

Table 8. Adj. r2 values of Model 4 and model Yield predicted for the corresponding FS indices and MP signal, grouped by months and site-years. n.a. = not available, n.s. = not significant. p-values for Model 4 < 0.05. p-values for Yieldpredicted < 0.001.

Model	Site-Year	April	May	June	July
Model 4	4	0.52	0.52 + 0.79	0.75 + 0.77	0.20
	1	0.18	0.24	0.40 + 0.47	0.47
	2	0.17	n.a.	0.32	0.37
$\text{Yield}_{predicted}$	3	n.s.	0.31	0.40	0.39
	4	0.26 + 0.40	0.50 + 0.71	0.64 + 0.74	0.66

5. DISCUSSION

This study describes the performance of the used optical sensors, and their ability of wheat yield, biomass and N_{avail} assessment. Based on the yield amounts, the crop development had a steady growth for all site-years, despite of an irregular high yield amount of site-year 1 at the field plots with an N level of 60 kg·ha^{-1}. The reason for this irregularity may be caused by the previous season in 2010. There, corn was planted which can have positive effects on the organic humus content of the field, e.g., Singh Brar et al. [30]. For site-years 1 and 4, the yield at N levels between 150 an 180 kg N·ha^{-1} had no increasing effect on the grain quantity or quality [4]. A lower N level can be recommended for the fertiliser management of these fields for the cultivation of wheat. The total average yield of site-year 3 was 26.7% higher than on site-year 2, which is an indication of more BM in the field, that is able to produce more grain.

The UAV MMS1 spectrometer has similar technical properties like the HS spectrometer, however, the results of both sensors are on a different prediction level for the IDV's in the presented research design. The analysis with the chosen DV's yield and N_{avail} for the MMS1 spectrometer data did not show any correlations. For site-year 2, there are low correlations for BM Weight and LAI; they were not repeatable for site-year 3 (Table 5). Reasons for the low or non existent correlations, based on the findings of Link et al. [17], are (1) a limited path accuracy with the consequence of outlaying data points not fitting to the research field design; (2) height inaccuracy of the UAV; (3) a short flight time of 15 min which required several flight missions to cover the entire research field; (4) that data post processing relies on accurate data from the autopilot system for pitch and roll correction of each data point, and on the control measurement of the MMS1 sensor at the start of the UAV. As the sensor in this setup only could be configured for continuous measurements, a lot of the logged data were of no use as they included the necessary flight turns and the surface measurements on the flight to the research field, Changing light conditions during the following flight mission affected the measurement precision in each design plot; and (5) the sensor footprint of 50.27 m^2 with an overlapping factor of 0.33 [17], covering a larger area at each measurement than the ground sensors were able to acquire (Table 3). As a consequence, the MMS1 data had a higher averaged value than the ground sensor data, which results in a lower resolution and a lower detection accuracy. However, this may be sufficient depending on other investigation purposes, ensuring a stable flight altitude and an integrated fusion approach on a raw data or feature fusion level. Other aerial platform approaches, like an electric multicopter, may lead to better results due to its better flight stability and easier point to point navigation behaviour. Geipel et al. [31] took the same MMS1 spectrometer like in the presented manuscript and mounted it to a hexacopter. With the same ground-truth information via sampling the above-ground BM they were able to measure higher correlations with BM and grain yield, taking into account a data acquisition system for all involved sensors [32].

The MP fluorometer was able to detect significant correlations with grain yield (Adj. r^2 of 0.48–0.63), notably in the ratios SFR with green and red excitation as well as in FERARI. They are linked to the chlorophyll content of the crop [33,34]. The correlations with the available N are high and reach Adj. r^2 values of 0.63–0.67 at a later growing stage (Z 59 and Z 66) with the RF signal and the SFR ratio. The highest correlations are with BM related properties. The correlations with the BM weight range from an Adj. r^2 of 0.46 at the early growing stage (Z 31), up to an Adj. r^2 of 0.86 at ripening (Z 85) and senescence (Z 91) stages. Fluorescence sensors for agricultural usage on tractors or other mobile platforms are barely in use. Their required contact with the crop canopy is one of the reasons why most of the used agricultural sensors are based on spectral characteristics [35]. However, due to the active LED emission source of the MP sensor, it provides a profound, reliable and repeatable technology especially for measurements on the field with changing illumination. Hyperspectral line scanners do not require close contact with the crop canopy and use sun induced fluorescence, however their field application is still in discussion and used on a research level [36].

Spectral sensors are already well adopted at large modern farms, and are able to fuse the measured data with previously gathered data sets via a map overlay approach [2,37,38]. Also in scientific research spectral sensors have a high acceptance, as more than 90% of the spectral information on crop canopy is contained in the red and near infrared (NIR) spectral bands [39,40]. For the FS indices HVI, NDVI, OSAVI, REIP and CropSpec, the correlations with yield increased, starting at heading stage (Z 51) to a high level of an Adj. $r^2 = 0.67$ at ripening stage (Z 85). Especially the indices CropSpec and REIP correlate very high with N_{avail} and provide an Adj. r^2 up to 0.86. For the BM characteristics, REIP, SAVI and CropSpec have high Adj. r^2 values > 0.63 already from stem elongation stage onwards (Z 30). The r^2 values of the HandySpec sensor data analysis was at a lower level than the ones from the FieldSpec sensor. They conclude in a maximum correlation of an Adj. $r^2 \leq 0.64$ at a significance level <0.05, with the presented DV's and IDV's.

For research, the high correlations of the MP fluorometer and the FS sensor can be merged on a feature fusion level. This has been done by Zecha et al. [4] and in the presented manuscript with a data post-processing method. The developed Model 4 from site-year 4 has been applied to the data sets of site-years 1–3. Model 4 did not correlate on a significant level with the gathered sensor data in these three site-years. However, a similar combination of indices and ratios (model $Yield_{predicted}$) resulted in significant correlations for all four site-years, by changing only one parameter (FERARI with CropSpec). By this change, the Adj. r^2 was between 0.32 and 0.74 two months before harvest for all site-years. The data analysis reveals that the more mixed indices and ratios are in a model, the higher and more robust the Adj. r2 values became, like RFUV and HVI, combined with index CropSpec or ratio FERARI, in the investigated linear models.

This model has a potential to continue working. Three out of the four parameters are exactly the same, providing results for the other three site-years. On the other hand, the ability of the presented model, predicting wheat yield by using unknown or different data, has not yet been validated, e.g., with machine learning methods proposed by Peña et al. [41] or as comparison with the linear models of Mortensen et al. [42] estimating above-ground biomass and N-uptake through aerial images. Future work needs to be done to train and test the real capabilities of this model, and to prove if it works.

6. CONCLUSIONS

i. The used aerial data collection system, as a combination of a fixed wing UAV and the MMS1 spectrometer, cannot be recommended for multispectral data acquisition like it has been done in the presented setup. A limited path accuracy, a short flight time of approx. 15 min including take-off, flight turns and landing, the MMS1 sensor setup in continuous measurement mode, independent sensor data logging and the related huge post-processing efforts, and the footprint along with an overlap of 30% make it unfavorable for a qualitative data analysis and feature correlation with ground truth data. For the aerial data acquisition, the authors recommend an integrated data acquisition system with all sensors connected via a sensor data infrastructure.

ii. Two ground sensors mounted to the Sensicle platform, the fluorometer Multiplex® Research (MP) and the FieldSpec HandHeld (FS) spectrometer, had high correlations with wheat yield, available nitrogen and the sampled biomass characteristics from the field plots. The HandySpec Field® (HS) spectrometer had lower significant correlations in all site-year than the FS sensor. The usage of the three ground sensors in continuous measurement mode is most reliable for the fluorometer MP. With an internal GPS sensor and an active LED source, measurement starts with one click and data storage on a SD card. The FS and HS spectrometer require an additional device for measurement triggering, and do rely on an external GPS receiver. The raw data post-processing cannot be handled without scripts, converting the raw data in features like indices, while calculating them with the white reference measurements, taken at the start of each continuous measurement series. The ability of the presented model, predicting wheat yield by using unknown or future data, has not yet been validated. Recommending the developed model for a general performance, further model training and model testing need to take place.

iii. An enhanced algorithm during the raw data calculation of the spectrometer, taking into account the ambient solar radiation during each continuous measurement mission, may improve the correlations and make the developed model more robust to apply it in earlier growing stages with high correlations. Advanced algorithms considering the factors (1) ambient solar radiation; (2) electrical soil conductivity; (3) aerial images with feature extraction; or (4) soil scoring may result in better yield predictions by providing the right decision for each spot in a field. In combination with the existent map overlay approaches of today's spectral sensor systems, these complete and weighted decision can save field inputs and ensure the perfect crop development to reach the maximum yield for the specific field. Once, the field data collection and analysis process can be accomplished with sensors and software in an convenient way also for a farmer, the adoption of sensor technology in agriculture will increase.

References

1. Auernhammer, H. Precision farming—The environmental challenge. *Comput. Electron. Agric.* **2001**, 30, 31–43. [CrossRef]
2. Adamchuk, V.I.; Rossel, R.A.V.; Sudduth, K.A.; Lammers, P.S. Sensor Fusion for Precision Agriculture. In *Sensor Fusion—Foundation and Applications*; Thomas, C., Ed.; InTech: Rijeka, Croatia, 2011; Chapter 2. [CrossRef]
3. Keller, M.; Zecha, C.; Weis, M.; Link, J.; Gerhards, R.; Claupein, W. Competence centre SenGIS-exploring methods for georeferenced multi-sensor data acquisition, storage, handling and analysis. In Proceedings of the 8th European Conference on Precision Agriculture (ECPA), Prague, Czech Republic, 11–14 July 2011; Czech Centre for Science and Society: Ampthill, UK; Prague, Czech Republic, 2011; pp. 491–500.
4. Zecha, C.W.; Link, J.; Claupein, W. Fluorescence and reflectance sensor comparison in winter wheat. *Agriculture* **2017**, 7, 78. [CrossRef]
5. Brooks, R.R.; Iyengar, S. *Multi-Sensor Fusion: Fundamentals and Applications with Software*; Prentice-Hall, Inc.: Upper Saddle River, NJ, USA, 1998.
6. Hall, D.L.; Llinas, J. A challenge for the data fusion community I: Research imperatives for improved processing. In Proceedings of the 7th National Symposium on Sensor Fusion, Albuquerque, NM, USA, 16–18 March 1994.
7. Dasarathy, B. Sensor fusion potential exploitation-innovative architectures and illustrative applications. *Proc. IEEE* **1997**, *85*, 24–38. [CrossRef]
8. Elmenreich, W. *An Introduction to Sensor Fusion*; Technical Report; Vienna University of Technology, Department of Computer Engineering: Vienna, Austria, 2002.

9. Dasarathy, B.V. Information Fusion—What, where, why, when, and how? *Inf. Fusion* **2001**, 2, 75–76. [CrossRef]

10. Thorp, K.R.; Wang, G.; Bronson, K.F.; Badaruddin, M.; Mon, J. Hyperspectral data mining to identify relevant canopy spectral features for estimating durum wheat growth, nitrogen status, and grain yield. *Comput. Electron. Agric.* **2017**, *136*, 1–12. [CrossRef]

11. Mesas-Carrascosa, F.J.; Torres-Sánchez, J.; Clavero-Rumbao, I.; García-Ferrer, A.; Peña, J.M.; Borra-Serrano, I.; López-Granados, F. Assessing optimal flight parameters for generating accurate multispectral orthomosaicks by UAV to support site-specific crop management. *Remote Sens.* **2015**, 7, 12793–12814. [CrossRef]

12. Peteinatos, G.G.; Korsaeth, A.; Berge, T.W.; Gerhards, R. Using optical sensors to identify water deprivation, nitrogen shortage, weed presence and fungal infection in wheat. *Agriculture* **2016**, 6, 24. [CrossRef]

13. Zecha, C.W.; Link, J.; Claupein, W. Mobile sensor platforms: Categorisation and research applications in precision farming. *J. Sens. Sens. Syst.* **2013**, 2, 51–72. [CrossRef]

14. Beres, B.L.; Turkington, T.K.; Kutcher, H.R.; Irvine, B.; Johnson, E.N.; O'Donovan, J.T.; Harker, K.N.; Holzapfel, C.B.; Mohr, R.; Peng, G.; et al. Winter wheat cropping system response to seed treatments, seed size, and sowing density. *Agron. J.* **2016**, 108, 1101–1111. [CrossRef]

15. Zadoks, J.C.; Chang, T.T.; Konzak, C.F. A decimal code for the growth stages of cereals. *Weed Res.* **1974**, 14, 415–421. [CrossRef]

16. Link, J.; Graeff, S.; Batchelor, W.D.; Claupein, W. Evaluating the economic and environmental impact of environmental compensation payment policy under uniform and variable-rate nitrogen management. *Agric. Syst.* **2006**, 91, 135–153. [CrossRef]

17. Link, J.; Senner, D.; Claupein, W. Developing and evaluating an aerial sensor platform (ASP) to collect multispectral data for deriving management decisions in precision farming. *Comput. Electron. Agric.* **2013**, 94, 20–28. [CrossRef]

18. R Development Core Team. *R: A Language and Environment for Statistical Computing;* R Foundation for Statistical Computing: Vienna, Austria, 2008; ISBN 3-900051-07-0.

19. Thenkabail, P.S.; Lyon, J.G.; Huete, A. (Eds.) *Hyperspectral Remote Sensing of Vegetation*, 1st ed.; Crc Press Inc.: Boca Raton, FL, USA, 2012; doi:10.1201/b11222. [CrossRef]

20. Weis, M.; Andújar, D.; Peteinatos, G.G.; Gerhards, R. Improving the determination of plant characteristics by fusion of four different sensors. In Precision Agriculture '13; Stafford, J.V., Ed.; Wageningen Academic Publishers: Wageningen, The Netherlands, 2013; pp. 63–69.

21. Horler, D.N.H.; Dockray, M.; Barber, J. The red edge of plant leaf reflectance. *Int. J. Remote Sens.* **1983**, 4, 273–288. [CrossRef]

22. Rouse, J.; Haas, R.; Schell, J.; Deering, D. Monitoring vegetation systems in the Great Plains with ERTS. In *Third ERTS Symposium*; NASA SP-351; Freden, S.C., Becker, M.A., Eds.; NASA: Washington, DC, USA, 1973; pp. 309–317.

23. Reusch, S.; Jasper, J.; Link, A. Estimating crop biomass and nitrogen uptake using Cropspec, a newly developed active crop-canopy reflectance sensor. In Proceedings of the 10th International Conference on Positron Annihilation (ICPA), Denver, CO, USA, 18–21 July 2010; p. 381.

24. Thenkabail, P.S.; Smith, R.B.; De Pauw, E. Hyperspectral vegetation indices and their relationships with agricultural crop characteristics. *Remote Sens. Environ.* **2000**, 71, 158–182. [CrossRef]

25. Huete, A. A soil-adjusted vegetation index (SAVI). *Remote Sens. Environ.* **1988**, 25, 295–309. 0034-4257(88)90106-X. [CrossRef]

26. Haboudane, D.; Miller, J.R.; Tremblay, N.; Zarco-Tejada, P.J.; Dextraze, L. Integrated narrow-band vegetation indices for prediction of crop chlorophyll content for application to precision agriculture. *Remote Sens. Environ.* **2002**, 81, 416–426. [CrossRef]

27. Cerovic, Z.; Moise, N.; Agati, G.; Latouche, G.; Ben Ghozlen, N.; Meyer, S. New portable optical sensors for the assessment of winegrape phenolic maturity based on berry fluorescence. *J. Food Compos. Anal.* **2008**, 21, 650–654. [CrossRef]

28. Ghozlen, N.B.; Cerovic, Z.G.; Germain, C.; Toutain, S.; Latouche, G. Non-destructive optical monitoring of grape maturation by proximal sensing. *Sensors* **2010**, 10, 10040–10068, doi10.3390/s101110040. [CrossRef] [PubMed]

29. QGIS Development Team. *QGIS Geographic Information System*; Open Source Geospatial Foundation: Chicago, IL, USA, 2009.

30. Singh Brar, B.; Singh, J.; Singh, G.; Kaur, G. Effects of long term application of inorganic and organic fertilizers on soil organic carbon and physical properties in maize–wheat rotation. *Agronomy* **2015**, 5, 220–238. [CrossRef]

31. Geipel, J.; Link, J.; Wirwahn, J.A.; Claupein, W. A Programmable aerial multispectral camera system for in-season crop biomass and nitrogen content estimation. *Agriculture* **2016**, 6, doi10.3390/agriculture6010004. [CrossRef]

32. Geipel, J.; Jackenkroll, M.; Weis, M.; Claupein, W. A sensor web-enabled infrastructure for Precision Farming. *ISPRS Int. J. Geo-Inf.* **2015**, 4, 385–399. [CrossRef]

33. Lichtenthaler, H.; Buschmann, C.; Rinderle, U.; Schmuck, G. Application of chlorophyll fluorescence in ecophysiology. *Radiat. Environ. Biophys.* **1986**, 25, 297–308. [CrossRef] [PubMed]

34. Buschmann, C. Variability and application of the chlorophyll fluorescence emission ratio red/far-red of leaves. *Photosynth. Res.* **2007**, 92, 261–271. [CrossRef] [PubMed]

35. Tremblay, N.; Wang, Z.; Cerovic, Z.G. Sensing crop nitrogen status with fluorescence indicators. A review. *Agron. Sustain. Dev.* **2011**, 32, 451–464. [CrossRef]

36. Broge, N.; Mortensen, J. Deriving green crop area index and canopy chlorophyll density of winter wheat from spectral reflectance data. *Remote Sens. Environ.* **2002**, 81, 45–57. [CrossRef]

37. Bill, R.; Nash, E.; Grenzdörffer, G. GIS in Agriculture. In *Springer Handbook of Geographic Information;* Danko, D.M., Kresse, W., Eds.; Springer: Berlin/Heidelberg, Germany, 2011; pp. 461–476.

38. Stone, M.L.; Raun, W.R. Sensing Technology for Precision Crop Farming. In *Precision Agriculture Technology for Crop Farming*; Zhang, Q., Ed.; CRC Press: Boca Raton, FL, USA, 2015; Chapter 2, pp. 21–54.

39. Sheffield, C. Selecting band combinations from multispectral data. Photogramm. Eng. Remote Sens. 1985, 51, 681–687.

40. Li, H.; Zhao, C.; Yang, G.; Feng, H. Variations in crop variables within wheat canopies and responses of canopy spectral characteristics and derived vegetation indices to different vertical leaf layers and spikes. Remote Sens. Environ. 2015, 169, 358–374. [CrossRef]

41. Peña, J.M.; Gutiérrez, P.A.; Hervás-Martínez, C.; Six, J.; Plant, R.E.; López-Granados, F. Object-based image classification of summer crops with machine learning methods. *Remote Sens.* **2014**, 6, 5019–5041. [CrossRef]

42. Mortensen, A.K.; Gislum, R.; Larsen, R.; Jørgensen, R.N. Estimation of above-ground dry matter and nitrogen uptake in catch crops using images acquired from an octocopter. In *Precision Agriculture '15*; Wageningen Academic Publishers: Wageningen, The Netherlands, 2015; pp. 227–234.

Chapter 12

HIGH-RESOLUTION MULTISENSOR REMOTE SENSING TO SUPPORT DATE PALM FARM MANAGEMENT

1. INTRODUCTION: AN OVERVIEW

Across the globe, people are suffering from a lack of food availability and food security as populations grow and land becomes unsuitable for farming [1–3]. There is a need for crops with high nutritional value that can withstand the arid and semiarid conditions in countries with limited water resources. The date palm is one such plant [2]. In the Middle East and North Africa alone, 100 million palms are cultivated on one million hectares of cropland [4].

There are several biotic and abiotic factors that affect the management of a healthy date palm crop. Global climate change may pose a threat to plantations of date palm; climate models predict that regions suitable for date palm growth will shrink, especially in the Middle East [5]. Water is a limiting factor for growth, although excess water can also reduce yield. Studies investigating the effects of water on date palm growth have found that insufficient water application slows the growth of the plants [6,7].

Increasing salinity also appears to negatively impact the growth of date palms [8–10]. This is a growing problem in arid regions as groundwater reservoirs are depleted due to excess irrigation.

Plantations of date palm and other palm species are also under threat due to infestations by different pests, the most extensive of which is the red palm weevil, or *Rhynchophorus ferrugineus* (Olivier, 1790). *R. ferrugineus* is spread via the transport of cuttings of palm tree leaves [4,11], and has been detected in 50% of date palm growing countries [12]. The effects of *R. ferrugineus* are severe; in the later stages of infestation the trunk can be damaged to the extent that the tree collapses and will most certainly die. Other pests that affect date palms include the Dubas bug, lesser date palm moth, and many different borer species [13]. The Dubas bug feeds on leaf sap which results in yellowing and wilting of leaves [3]. Lesser date palm moths infest and damage the fruit of the date palm [13]. Borers have similar effects to R. ferrugineus, although they may infest different parts of the tree, such as the middle rib of the frond and the upper section of the tree near the bunches of fruit [13].

At present, a heavy focus is placed on monitoring plantations for R. ferrugineus infestation. Unfortunately, current methods of detection do not allow for effective, reliable identification of infested trees [14]. In fact, the detection of palms in the early stages of infestation has been stated as the main issue in controlling weevil infestation [15]. It is very difficult to recognise whether a tree is infested based on visual inspection as the symptoms vary depending on conditions such as the tree species and initial site of infestation [14,16]. All of the current detection methods, including acoustic detection, x-ray, and sniffer dogs are often ineffective on a large scale because each tree has to be inspected individually [14,16].

Remote sensing has been suggested as a potential detection method, as it allows for large scale monitoring and effective data analysis techniques [16]. Thermal imaging studies have shown it might be possible to detect plants infested with *R. ferrugineus* using aerial thermal photographs [17]. Multi- and hyperspectral images have successfully identified infested trees. WorldView-3 satellite imagery has been used for *R. ferrugineus* detection [18]. Hyperspectral imagery was conducted on oil palms to classify *R. ferrugineus*-infested palms based on vegetation indices [19]. Light Detection and Ranging (LiDAR) has not been applied to the palm assessment case, but it does have potential; for example, LiDAR has been used to classify eucalyptus tree health [20].

Remote sensing has also been applied to assess the effect of other stressors on palm plantations. Cohen et al. (2012) used thermal images to investigate how changes in irrigation affect the health of date palm plantations [6]. The effect of soil salinity on date palm health has been assessed by using Landsat Thematic Mapper (TM) and Enhanced Thematic Mapper Plus (ETM+) images to calculate the Soil Adjusted Vegetation Index (SAVI) for two different years [21].

The studies listed above have all explored how individual stressors affect date palm plantations. However, it is highly likely that a date palm plantation will be subject to multiple stressors at once and identifying the influence of these effects will allow for more comprehensive and efficient plantation management. Remote sensing approaches can provide the tools to conduct this form of management; analysis of the entire plantation is possible using automated techniques in order to differentiate between a combination of stressors.

Therefore, the goal of this study is to explore which remote sensing sensors and derived indicators can detect stressors on palm health at different spatial levels, in order to support a holistic solution for effective date palm management. Plantations are often organised in large rectangular sub-sections or blocks—this will be the first spatial level to be evaluated. The second level of analysis will be conducted on an individual tree level. For this study, several different remote sensing imagery sources have been used, in order to determine whether different techniques are most suitable for highlighting differences in palm status. The available imagery sources are thermal, hyperspectral, and LiDAR, with data collected using airborne platforms. High resolution imagery is achievable with airborne platforms, allowing for analysis of individual tree health.

2. MATERIALS AND METHODS

2.1 Study Area and Data Description

The farm shown in Figure 1 is in the Al-Kharj region in Saudi Arabia, southeast of the capital Riyadh. Saudi Arabia was one of the earliest countries growing date palms to be affected by R. ferrugineus infestation outside of South East Asia, with first infestations occurring since the 1980s [22]. The country has the highest per capita consumption of dates in the world. Flood irrigation is making way for more modern irrigation methods such as drip irrigation in many farms [23].

Of the farm, an area of 168.8 hectares (outlined in blue in Figure 1) has been chosen as the key study area of this analysis. This farm is relatively well managed, with continuous maintenance and targeted irrigation practices in place. As is exemplified in Figure 1b, date palm plantations are commonly divided into regular-sized rectangles of approximately 10 hectares; these sub-sections are hereafter referred to as blocks. Figure 1c illustrates the uniform growth patterns of the date palms; the crowns are distinct and organised in a grid-like structure.

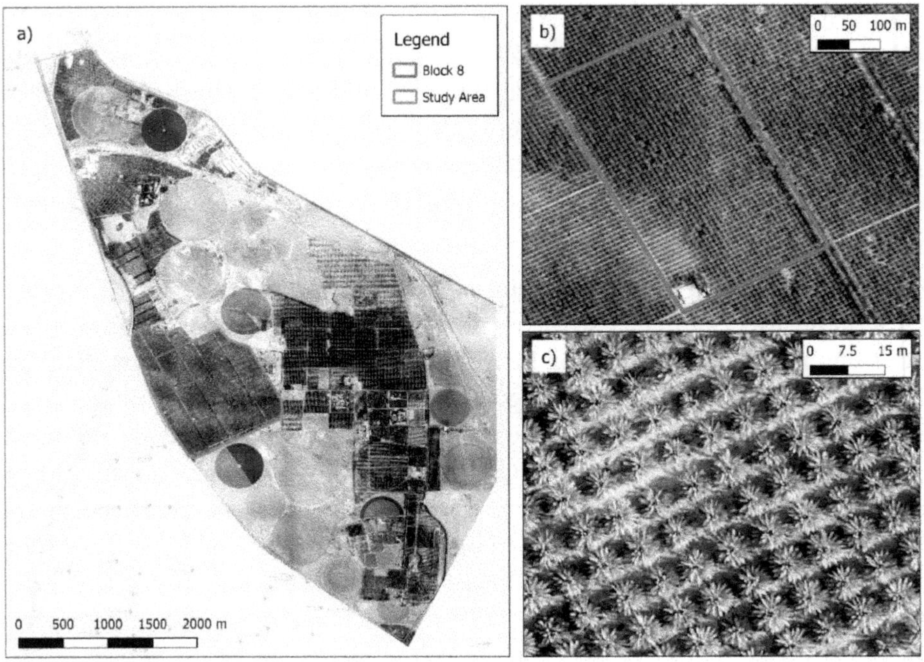

Figure 1. Visual red-green-blue (RGB) images with 15-cm resolution, taken using an aerial platform on 26 May 2016, of (a) the date palm plantation in the study area (outlined in blue), located south-east of Riyadh, the capital of Saudi Arabia; (b) an example of a block within the plantation (outlined in red) and (c) an example of individual palm trees.

For this study, four remote sensing data types were available for analysis: hyperspectral, thermal, LiDAR, and high-resolution visual red-green-blue (RGB) images. Each dataset contains different information that will be useful when assessing date palm health. The data has been collected using a Diamond DA42 MPP GEOSTAR (Diamond Aircraft Industries GmbH, Wiener Neustadt, Austria) aircraft, which is designed to carry many different remote sensing instruments. The flight was conducted on 26 May 2016 between approximately 6:00 and 9:00AM.

Table 1 shows information about the four sensors used for data collection. The spatial resolution refers to the pixel size of the data used in the analysis after preprocessing; the original LiDAR data was provided as a point cloud but has been processed into raster data.

Table 1. Characteristics of the four sensor systems used for data collection.

Data Type	RGB	LiDAR	Thermal	Hyperspectral	
Name of sensor [1]	Phase One iXA-R Camera [2]	RIEGL LMS-Q1560 [3]	VarioCAM®HD head [4]	HySpex VNIR-1800 [5]	Hy-Spex SWIR-384 [5]
Type of sensor	Medium-format camera system	Rotating polygon mirror	Uncooled microbolometer focal-plane array	Pushbroom camera actively cooled and stabilized scientific CMOS detector	Pushbroom camera Mercury cadmium telluride sensor
Spectral range	Visible	Near-infrared	7.5–14 µm	0.4–1 nm, 182 bands	0.93–2.5 nm, 288 bands
Spatial resolution	0.15 m, 0.6 m, 1.8 m	1 m, 2 m	1.8 m	1 m	1 m

[1] Manufacturer information; [2] Phase One Industrial, Frederiksberg, Demark; [3] RIEGL Laser Measurement Systems GmbH, Horn, Austria; [4] InfraTec, Dresden, Germany; [5] Norsk Elektro Optikk AS (NEO), Skedsmokorset, Norway.

]

2.2. Methodology

2.2.1 Preprocessing

Different preprocessing steps were necessary for each of the multiple remote sensing data sources available. These steps will be described briefly in the next few paragraphs. Steps were carried out in various software: R (version 3.3.1, R Foundation for Statistical Computing, Vienna, Austria), QGIS (version 2.1.4 'Essen', Open Source Geospatial Foundation, Chicago, IL, USA), ArcGIS (version 10.4, ESRI, Redlands, CA, USA), LAS tools (version 2016, rapidlasso GmbH, Gilching, Germany), and Agisoft PhotoScan (version 1.2, Agisoft LLC, St. Petersburg, Russia).

The RGB data was provided in JPEG format and was not geographically referenced. The files were given a geographic coordinate system and then projected to WGS 1984 UTM Zone 38N, and also converted to GeoTIFF. This projection was also used for all the other datasets.

The LiDAR dataset was originally received as a point cloud. To generate a tree height raster, Digital Surface Models (DSM) at different resolutions was generated; additionally, the mean, minimum, maximum and standard deviation values of the points within these pixels were used.

During acquisition, the VarioCAM thermal camera was controlled using the IRBIS®Thermography Software (InfraTec, Dresden, Germany). Within this software, the temperature range for acquisition was set to values between 0 and 50 degrees Celsius [24]. This range was broader than the actual temperature range which could be expected but was set by the flight operators to avoid saturation at low or high values. The raw thermal data was provided as individual snapshot images that were then stitched together using the structure-from-motion method [25] as implemented through Agisoft PhotoScan. In some areas where there are a lot of overlapping images, tie-points do not match perfectly, causing some blurring in the image; where this occurs, individual trees are not identifiable. This effect cannot be reduced without compromising the resolution of the entire image. The image was received as a (8 bit) greyscale image with 256 levels. To convert the values into a relative approximation of degrees Celsius, the acquired greyscale image was converted using the multiplication factor of 0.1953125 to fit the temperature range settings during acquisition (0 and 50 degrees Celsius).

The hyperspectral images were received without the inertial measurement unit (IMU) data due to an error in the acquisition procedure. This means that it was not possible to correct for image distortion due to plane movement. The images were also provided in an unknown multiplication factor, and thus the values are unusual. However, an evaluation of this data source is still considered worthwhile as it was acquired at the same time as the other remote sensing types. This meant that extensive steps were required to attain usable imagery.

By using the recorded times and positioning of the RGB imagery, the order in which hyperspectral images were acquired was determined. This meant that a representative image could be constructed and aligned with the study area. The image was then transformed using digitally assigned geometric control points as identified from the RGB images. The QGIS *Georeferencer* tool was used, with transformation type set as Thin Plate Spline which is useful for heavily distorted data, and Nearest Neighbour as a resampling method to maintain the original data values. The pixel data was multiplied by a scaling factor of 0.002 as this ensured they were in the relative range of standard reflectance values. As this is not the standard way to determine actual reflectance values, it is only possible to carry out relative comparisons within the dataset—this dataset cannot be numerically compared with any hyperspectral data collected in later studies.

2.2.2 Identification of Blocks and Trees

As analysis is carried out on two different spatial levels, namely blocks and trees, it is necessary to identify these regions. Block shapefiles were provided for the farm, and these were used for analysis of the thermal and LiDAR-derived height data. As the hyperspectral blocks were geometrically distorted, these were drawn manually.

Several methods were used in order to identify the trees for tree-level analysis. For the non-distorted data, the canopy height map as determined from LiDAR was used to locate individual tree points. This was done using the watershed method, which involves inverting the height rasters, calculating the sinks and converting them to points, and filtering these points at an appropriate buffer width. The software used was LAS tools and ArcMap.

It was also decided that deriving canopy polygons could be interesting as the area of the tree canopy could prove a useful indicator of tree health. The LiDAR data was analysed using an R package called Individual Tree Crown (ITC) Segment (version 0.6, Michele Dalponte, San Michele all'Adige, Italy) [26]. This tool selects local maxima within a height raster and then spreads outwards to select the surrounding pixels that are within a specific height range representing the canopy. A polygon shapefile is generated, for which the height and area are calculated. A 3-m width limit was selected to constrain the growth of the polygon outside of the general tree size. To investigate sensitivity to spatial resolution, the model was run for pixel sizes of both 1 m and 2 m. In this case, the raster's containing the maximum value of the point cloud within each pixel were used for analysis, as these data were available for both 1-m and 2-m resolution.

For the hyperspectral data, a different method for tree identification was used. This is necessary because the distortions in the hyperspectral dataset mean that it is not possible to use the same tree locations for each dataset. It was decided to use vegetation indices, specifically normalised difference vegetation index (NDVI) (see Equation (1)), to identify the trees. Once the NDVI was calculated, based on the values of Equation [1], the *TreeTopFinderTool* in the R package "Forest Tools" (version 0.1.5, Andrew Plowright, BC, Canada) was used to identify local maxima. It is hypothesised that the points where the NDVI is highest represent the top of the tree canopy. A width threshold of four was set, based on the upper limit of the average canopy size, which is 3 to 4 m. Additionally, a minimum NDVI value of 0.3 was set to ensure that maxima in treeless regions were excluded—this value was based on the bimodal distribution of the NDVI values that distinguishes between soil and tree points.

$$NDVI = \frac{NIR_{750} - R_{660}}{NIR_{750} + R_{660}} \tag{1}$$

2.2.3 Block-Level Analysis

The six blocks shown in Figure 2 have been selected for further analysis. These blocks were chosen based on the attributes provided in Table 2.

Two different groups of blocks have been determined based on their characteristics as described in Table 2. The data that was available from the farm manager mainly related to instances of R. ferrugineus, so this was a factor that was considered in block division. A visual assessment of the farm was made based on the homogeneity of canopy area and the sparseness of the block. The first group of trees was chosen with the requirements of having no instances of R. ferrugineus and of having high canopy homogeneity and low sparseness. The second group of blocks had relatively high instances of R. ferrugineus, and the appearance of the blocks revealed many gaps in trees and a wide variation of tree canopy widths. These two indicators were used to differentiate between potentially healthy and unhealthy groups. An additional factor that was considered was the amount of blurring in the thermal image, as discussed in Section 2.2.1; if the blurring in a block was visually estimated as being more than 70%, it was not considered.

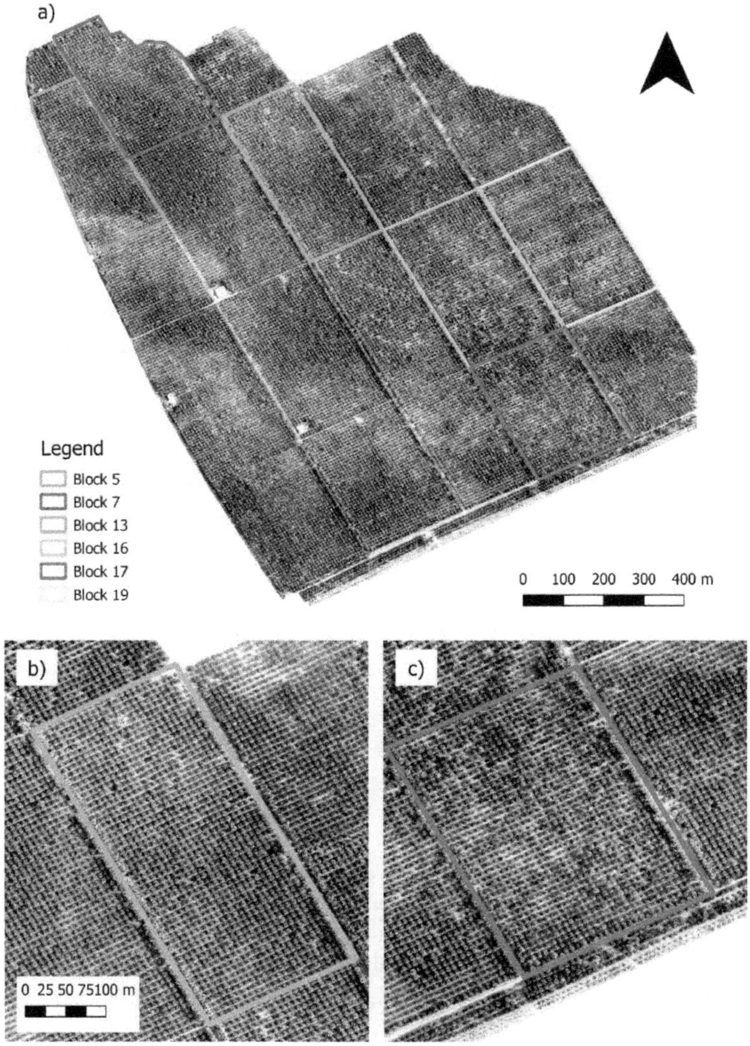

Figure 2. (a) Shows the selected test blocks; Blocks 5, 7, and 13 have been classified as healthy and Blocks 16, 17 and 19 have been classified as unhealthy. (b) and (c) shows an example of blocks classified as healthy (green) and unhealthy (red) respectively.

The assumed healthy test blocks are 5, 7, and 13, and the unhealthy blocks are 16, 17, and 19. Their respective values are shown in Table 2.

Table 2. Criteria used to determine the selection for the six test blocks.

	May 2016	June 2016	RGB Check1	Thermal Image Distortion2
Block No.	No. RPW infested	No. RPW infested	1–5 rating	Percent distorted
5	0	0	1	20
7	0	0	1	30
13	0	0	2	30
16	8	4	5	30
17	10	4	5	30
19	1	9	5	20

1 Indicator of block health status based on visual check for sparseness and heterogeneity, where 1 = homogenous and wide canopies, 5 = many patches and smaller canopies; 2 Visual check to determine how much of the image is affected by blurring due to structure-from-motion processing technique.

To summarise the statistical attributes of these blocks, boxplots were generated for temperature and height of the trees. These illustrate the spread of temperature and height across the test blocks, also in relation to all blocks in the farm. These boxplots illustrate the 25th, 50th and 75th percentiles, extreme values, and outlier values for the canopy pixels. For the temperature values, the derived canopy areas were used as a mask layer for the temperature raster. The height values are also based on the polygon shape. A visual check was conducted to explain the findings of the boxplot analysis.

For each block, a total of 10 'healthy' and 10 'unhealthy' trees were randomly selected after filtering based on their ratio between height and canopy area—based on discussions with the farm manager, it was hypothesised that trees with a high height area ratio (HAR), meaning they have a small canopy in relation to height, would be more likely to be unhealthy. Trees were selected based on whether they were above or below a threshold for this ratio and were also filtered based on height to attempt to select trees of similar height.

The mean temperatures of each group of trees (healthy and unhealthy) have been compared using two-factor analysis-of-variance (ANOVA) analysis. This statistical test is used to determine whether the differences in means of two or more groups at two or more levels are statistically significant; in this case, between the 'healthy' and 'unhealthy' trees, between the blocks, and the interaction effect between them. The post-hoc Tukey honest significant difference (HSD) test is applied following ANOVA—

it is a comparison test that determines which pairs of means are responsible for the statistically significant test result. A significance level of α = 0.05 has been used to determine whether the result is statistically significant or not.

2.2.4 Tree-Level Analysis

Different blocks were used for the individual tree analysis, as several of the blocks were not available in the hyperspectral dataset. Block 11 was selected for analysis for two reasons; firstly, it has the least visible blurring in the thermal dataset, and also it has a wide variety of trees with different canopy sizes.

As described in Section 2.2.2, the tree canopies have been identified based on NDVI. The next step was to select trees that can be used to extract more extensive information. In this case, trees were selected randomly from the hyperspectral dataset, and were matched manually to the trees in the thermal dataset. This was done by starting at the edges of the trees in each block and then counting vertically and the horizontally to identify the correct tree. A shapefile was created, and points were placed on the identified trees. A total of 100 trees was used for this block. As the trees are quite noticeable in each dataset, it is expected that this method was quite accurate at detection, although in some cases the tree may be misidentified.

Fourteen vegetation indices (VI) have been selected for further analysis (See Appendix A Table A1). These VIs can be split into general groups that focus on different regions of in the visible light spectrum, such as green, red, and the red-edge position. For each test tree, the correlation between each vegetation index and temperature and height has been derived, and R squared values have been calculated. A visual check has also been carried out.

3. RESULTS

3.1 Block and Tree Identification Results

The results of the method which identifies the areas of entire tree canopies can be seen in Figure 3. The ITC segment tool was run on two different height rasters (spatial resolutions of 1 m and 2 m) derived from the LiDAR dataset. Table 3 shows the number of tree points identified using the two different height raster maps; these values have been compared with tree numbers provided by the farm manager. For the 1 m resolution data these values exceeded 90% in every block. Fewer trees are detected using the 2 m resolution data: 73–90% were detected when comparing the number of tree polygons to the number of tree points in each block.

Table 3. Number (#) of trees within the test blocks based on the farm managers records compared to results of the Individual Tree Crown (ITC) segment analysis for 1 m and 2 m height rasters.

Block No.	Recorded # of Trees	# of Trees (1 m Resolution)	# of Trees (2 m Resolution)
5	1188	1137	1043
7	1213	1197	1084
13	1417	1277	1131
16	1749	1572	1242
17	1422	1302	1013
19	1691	1617	1440

When comparing Figure 3a,b, it can be seen that in the denser parts of the plantation, using the 1m resolution LiDAR data has a higher detection rate of tree canopies.

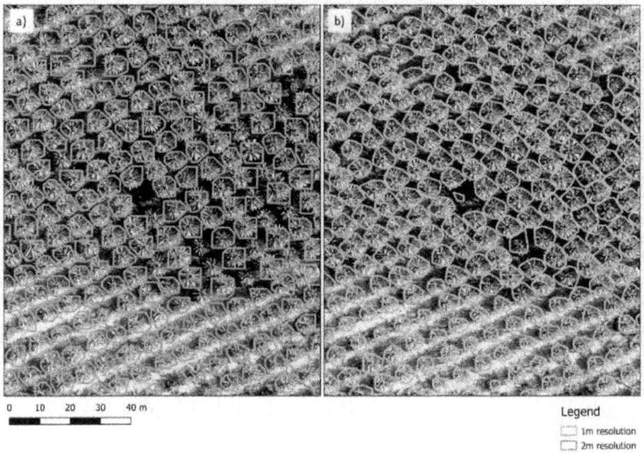

Figure 3. Section of an area of Block 5 in the farm. Canopy polygons derived using the ITC segment tool plotted in relation to the RGB data, for (a) the 2-m spatial resolution height raster (orange), and (b) 1-m spatial resolution (blue).

The second method used the NDVI image that was extracted from the hyperspectral dataset. Two different red bands were used to calculate the NDVI, and the comparisons can be seen in Figure 4. The outcome for both NDVI calculations is similar, although trees are identified more accurately in denser canopies when band 750 is used; for Block 16, for example, the number of trees recorded by the farm manager was 1749, compared to 1605 and 1601 for NDVI 750 and NDVI 800 respectively. For Block 19, the method using NDVI 750 identified 1661 trees, compared to 1656 (NDVI 800), and 1691 (trees recorded by farm manager). This indicates that in this case NDVI 750 is slightly more effective at identifying trees. Misidentification tends to be on the edges where the tree canopies are denser and therefore less distinct.

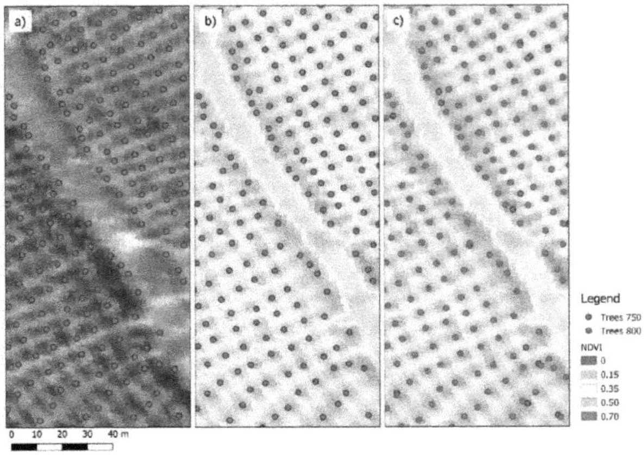

Figure 4. Tree points (red) derived using NDVI with band 750, overlaying (a) the hyperspectral image, and (b) the NDVI-derived using band 750, and (c) tree points derived using band 800 (blue) in comparison NDVI 750 points, overlying NDVI-derived using band 800.

For further analysis, the datasets with the highest detection rates were used. For analysis of the thermal and LiDAR datasets, the canopy polygons derived from the 1 m height rasters were used. The tree points resulting from the NDVI using spectral band 750 were chosen for the hyperspectral data analysis.

3.2 Block-Level Analysis

For block-level analysis, thermal data and height rasters were analysed. Six test blocks were chosen based on their attributes which have been discussed in the methods section. As the same six test blocks are not available for the hyperspectral dataset, this dataset has not been included in block-level analysis.

Boxplots of the thermal and height values for the six selected test blocks are presented in Figures 5 and 6. This illustrates the spread of heights across the test blocks, also in relation to all blocks in the farm. These boxplots illustrate the 25th, 50th and 75th percentiles, extreme values, and outlier values for the canopy pixels.

Figure 5 shows boxplots of the canopy temperature pixels of the test blocks. When treating block 5 as an outlier, it appears that two of the assumed healthy blocks have a mean temperature that is 0.5 ◦C lower than the unhealthy blocks. This is a positive result as the hypothesis was that unhealthy blocks would appear warmer than the healthy blocks. However, the mean temperature of block 5 is 0.5 ◦C higher than the mean temperature of the unhealthy test blocks. This could be because block 5 has been inaccurately classified as healthy. Feedback from the farm manager has indicated that the soil quality in this region is very poor so the trees there do not grow as well.

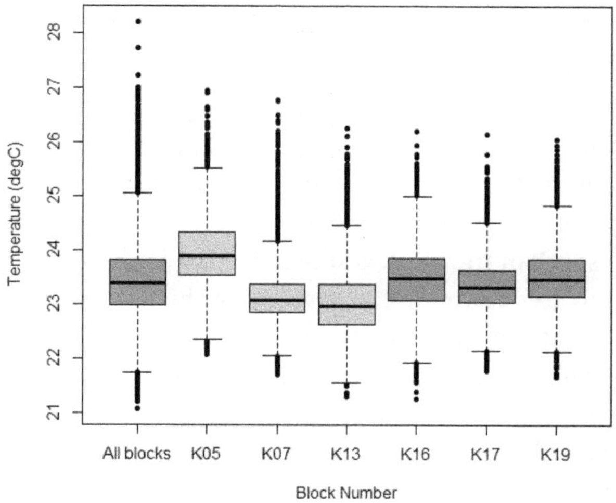

Figure 5. Boxplots of temperature for pixels defined as canopy using a buffer of 3 m and a height threshold of 4 m. The boxplot that is coloured blue shows the temperature for all blocks. The three test blocks in green (5, 7 and 13) are assumed healthy, and the three boxplots coloured red (16, 17 and 19) are unhealthy.

Figure 6 shows the outcome of the height analysis. The blocks classified as healthy appear to have similar heights to all blocks. Interestingly, the unhealthy blocks appear to deviate from the mean heights across all blocks. This is especially true for Block 19, which contains shorter trees than the other blocks. Block 16 appears to contain taller trees than the other blocks, and Block 17 has a wider spread within the upper and lower quartiles.

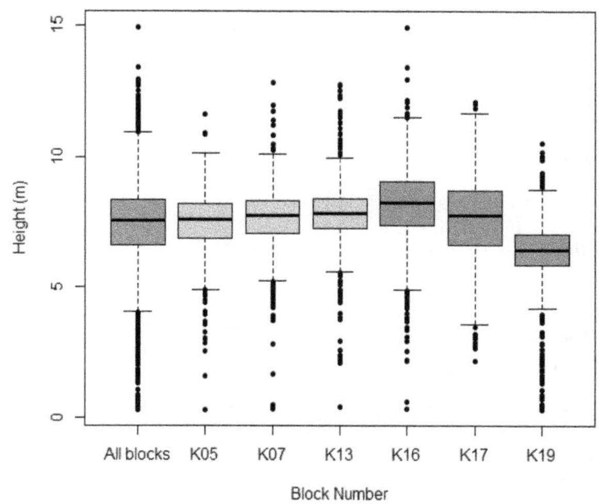

Figure 6. Boxplots for height of the blocks. The blue block shows the values for all blocks in the model farm. The healthy blocks are shown in green and the unhealthy blocks are shown in red.

The results of the boxplot analysis show some interesting trends, that are even more revealing when compared with the spatial plots of the blocks.

When looking at the spatial variation in the canopy pixels of the blocks (Figure 7), we can identify the areas that contribute to the varying values seen in the boxplots (Figure 5). Block 5 is evidently hotter than the other areas, especially in the south of the block, but also across the entire block as well. This is interesting, as the sparseness of the trees in the south of the block is similar to that in the sparse areas of the other blocks. It is difficult to determine the reason for this temperature difference. This area also corresponds to the high height area ratio (HAR) values seen in Figure 8. The unhealthy blocks clearly have more areas that are hotter. When comparing to the RGB data, these hotter areas tend to lie where the trees are sparser and thus more bare ground is exposed. There are also areas in the healthy blocks that have higher temperatures—this is because there is a high variability in all blocks and even those classified as healthy have unhealthy looking regions.

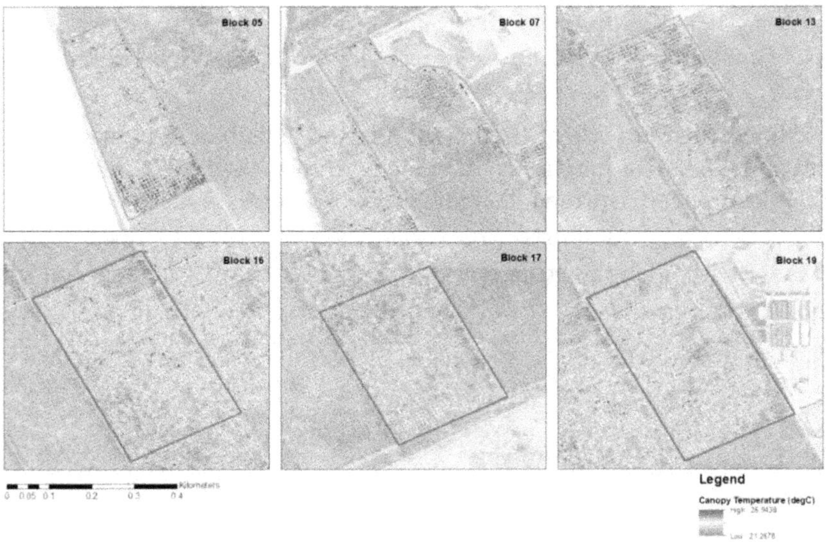

Figure 7. Temperature of the extracted canopy pixels representing individual trees for the test blocks.

Figure 8 shows the values for HAR in each block, where height is divided by canopy area. The healthy test blocks (5, 7 and 13) appear to have mostly values between 0.15 and 0.2 and Blocks 16 and 17 have more tree polygons with a larger height area ratio value, although one area of Block 5 contains trees with larger HAR values. This is to be expected as it indicates that there are more trees with abnormally small canopies, which one would expect in an unhealthier block. However, Block 19 has low HAR values, and in fact looks healthier than some of the "healthy" blocks. The trees in this block are shorter than in the other blocks, and the canopy area is similar, and thus it has the lowest mean height area ratio of all the blocks.

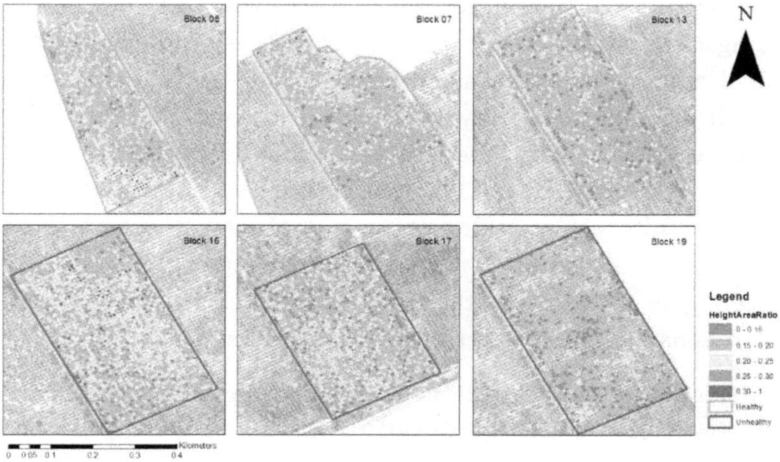

Figure 8. The height area ratios of the derived tree polygons representing individual trees for the test blocks.

Based on Figures 7 and 8, it is evident that within the blocks there is a lot of variation. This is especially true for Block 5, where there is a clear distinction in temperature and height between the north and south of the block.

ANOVA analysis was conducted on 10 randomly selected (within a height threshold; see Section 2.2.3) 'healthy' trees and 10 'unhealthy' trees per block, in order to evaluate whether the two factors of assumed health and the block containing the tree have an effect on tree temperature, and also whether there is interaction between the two (Table 4). There is a main effect of both health (probability (p) value = 0.000611) and block number (p value = 1.06E-14), and there is no interaction effect (p value = 0.834). The statistically significant outcome for block number indicates that the differences in the mean temperature of all the trees in a block are large enough to be considered as different. When comparing the mean temperature of all healthy trees and all unhealthy trees there is a statistically significant difference.

Table 4. Results of ANOVA analysis for temperature in relation to block number and the assumed block health.

	Df	Sum Sq	Mean Sq	F Value	Pr (>F)	Significance 1
Block No.	5	20.515	4.103	21.161	1.06E-14	*** .
Health	1	2.417	2.41	12.465	0.000611	***
Block No.: Health	5	0.407	0.081	0.419	0.834311	
Residuals	108	20.94	0.194			

[1] *Significance levels: 0 '***'; 0.001 '**'; 0.01 '*'; 0.05 '.'; 0.1 ' '; 1.*

The results of post-hoc Tukey's HSD test reveal which mean differences contribute most to the statistical significance of the ANOVA results. A subset of results in Table 5 show six comparisons between healthy and unhealthy trees for each test block. When looking at the p value, we see that they are all greater than a significance level α = 0.05, indicating no statistical significance. This suggests that when comparing the means of healthy and unhealthy trees within each block there are no statistically significant differences between these means. The differences in means of healthy and unhealthy trees range from 0.1 ◦C to 0.45 ◦C. This leads to the conclusion that the significant results seen in Table 4 are a result of comparing trees from different blocks. However, in all cases, we do see that the unhealthy trees have a mean temperature higher than the healthy trees, which does match the hypothesis that unhealthy trees have higher temperatures due to stressors [6,17].

Table 5. Selection of Tukey's HSD comparing mean temperatures of selected healthy (n = 10) and unhealthy trees (n = 10) per block.

	Mean Temperature (°C)			p Value
	Unhealthy	**Healthy**	**Difference**	
Block 5	24.37052	24.10113	0.26939	0.967108
Block 7	23.30163	23.04954	0.25209	0.979962
Block 13	23.02733	22.77498	0.25235	0.979803
Block 16	23.80391	23.4071	0.39681	0.682276
Block 17	23.40484	23.31995	0.08489	0.999999
Block 19	23.68493	23.23745	0.44748	0.50232

3.3 Individual Tree-Level Analysis

Individual tree analysis has been carried out in order to compare the temperature and height values per tree to the results of the vegetation analysis. Block 11 has been selected for further analysis as there are a wider variety of trees, such as those with wider and smaller canopies. The 100 randomly selected test trees are shown in Figure 9, overlaid on the RGB image, the Chlorophyll Index Green (CIG-Table A1) vegetation index and thermal imagery. The hyperspectral image has geometrical distortion across the whole image, and it is especially notable in the northern region, and along the edges. However, when comparing with the visual (RGB) image, definite patterns are visible when comparing the vegetation index values and the tree canopy density that is visible in Figure 9a. In the south of the block, for example, there is more dense vegetation, and on the right side on the bottom there are patches of sparser vegetation. The regions of sparse vegetation in the top right and dense vegetation in the central region are also visible in the thermal imagery. The randomly selected test points cover a good range of these areas, with points in both the dense and sparse regions of canopy.

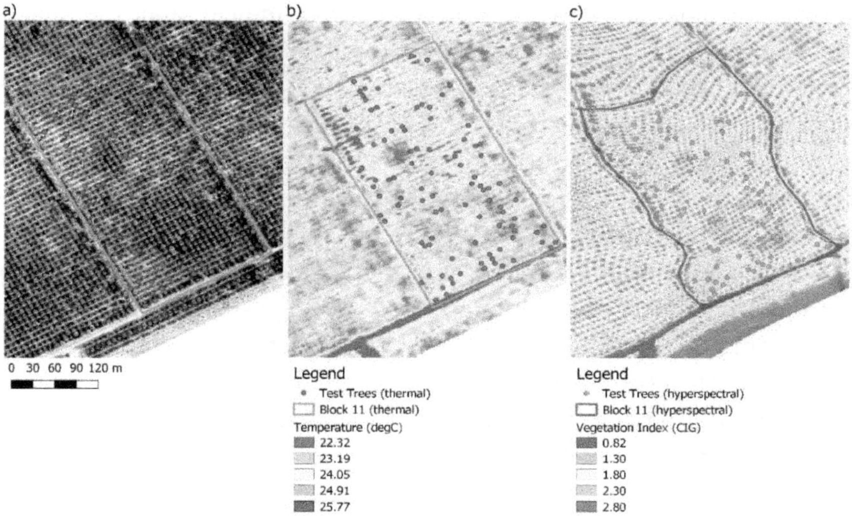

Figure 9. Illustrations of the selected test trees in Block 11 for (a) the visual RGB imagery, (b) the thermal imagery, and (c) the hyperspectral imagery represented as calculated Chlorophyll Index Green (CIG) (Table A1).

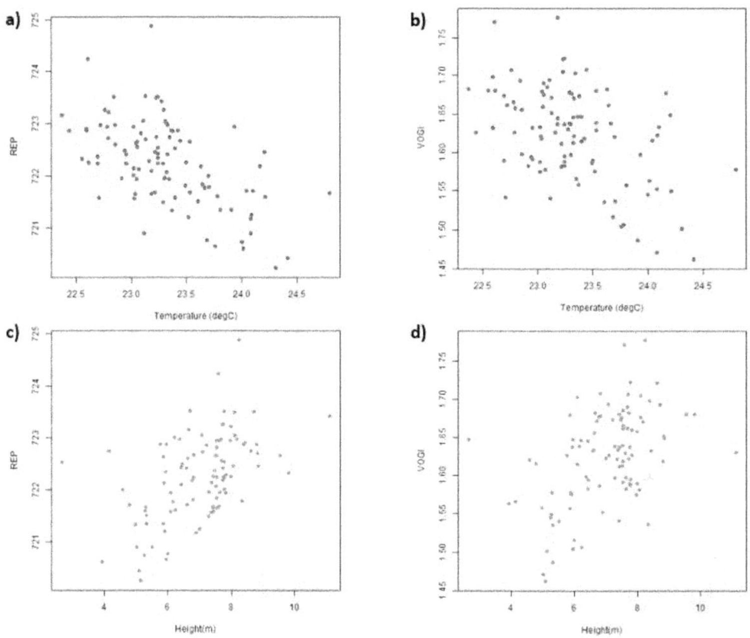

Figure 10. Correlation analysis for 100 test trees in Block 11: (a) and (b) show the correlation between temperature and two selected VIs, red-edge position (REP) and Vogelmann red edge index (VOGI), and (c) and (d) show the relationship between these indices and height.

The correlation analysis between the vegetation indices and the height and temperature data show some interesting results. The R-squared values reach a maximum of 0.313 between the vegetation index red-edge position (REP) and temperature, and 0.253 between REP and height (Figure 10). The vegetation index Vogelmann red edge index (VOGI) also has a relatively strong correlation value with both temperature (R^2 = 0.227) and height (R^2 = 0.213) (Figure 10) and these indices both use bands in the red edge positions. Green NDVI (gNDVI) and Chlorophyll Index Green (CIG) also show relatively high correlation values (Table A2). These R-squared values are quite low in general, but as Figure 10 shows there is a trend between the two most highly correlated VIs (REP and VOGI) plotted against temperature and height for the 100 test trees. It is evident that temperature has a negative correlation with the VIs, indicating that at higher temperatures, these VI values tend to be lower. Height has the opposite correlation—taller trees have higher VI values.

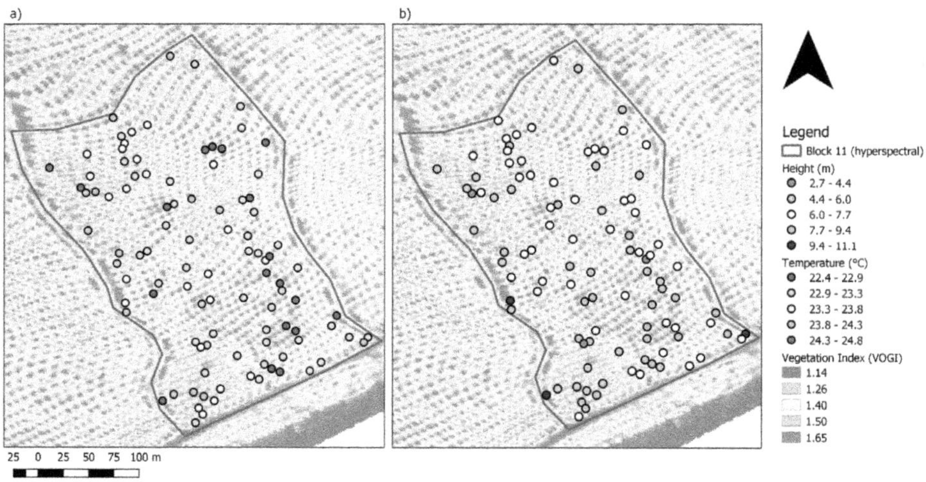

Figure 11. Illustration of the spatial patterns between tree-based point observations of (a) temperature and (b) height, with the vegetation index VOGI visualized as background map for comparison.

In Figure 11 the temperature and height of the test trees are plotted overlaying the VOGI vegetation index image. It is evident that the regions with lower VI values tend to have higher temperatures and shorter trees; the opposite is generally true in regions with high VI values.

To get a clearer impression of the relation between the temperature and height values and VIs, the values have been visually displayed in relation to the VOGI vegetation index. It can be noted that the regions with lower VI values tend to have higher temperatures and shorter trees; the opposite is generally true in regions with high VI values.

4. DISCUSSION

This study explored various options for assessing vegetation health of date palm plantations using a combination of remote sensing data sources, specifically LiDAR, thermal, hyperspectral and visual RGB imagery. In order to differentiate between more general stressors and those that only affect individual palms, the plantations have been assessed on both a block and individual tree level. Although no direct field data was available for validation at tree level, exchange with the plantation manager provided support for the identified spatial patterns. The main contribution that this research can make is as a suggestion for methods to be used in further research. That being said, the preliminary results do show an ability to successfully differentiate between groups of trees and can be compared to earlier studies [6,16,17,19].

4.1 Exploring the Main Findings of the Study

On a block level, it is evident that the remote sensing images do highlight differences between the blocks; the results of the thermal box plots, for example, show up to a maximum of 1 °C difference between the blocks (Figure 5). These box plots show variation across the data—the spatial maps in Figures 7 and 8 illustrate how this variation is spatially distributed. Block 5 is a clear example, as it shows two very distinct areas in the north and the south of the block, where the trees are both much warmer and shorter. These inter-block variations can be seen to a lesser extent within each block. Therefore, it could be useful to add a further level to the current analysis, namely the development of management zones within the blocks that are based on tree characteristics.

ANOVA analysis was carried out to compare the temperature of individual healthy and unhealthy trees across each block, a method that has been applied in previous studies [6,17]. In the current study, the ANOVA results were statistically significant (Table 4) when comparing block number (p value = 1.06E-14) and tree health (p value = 0.000611) independently of one another. Tukey's HSD test (Table 5) shows the most significant differences come from comparing the trees in different blocks. There is no statistically significant result when comparing means of healthy trees in one block with means of unhealthy trees in that same block. The differences in means when comparing the unhealthy and healthy trees of one block range between 0.1 and 0.45 °C, which may not be practically significant to the extent that significant differences between trees can be identified based solely on temperature. When comparing the individual tree differences found in this study with those found by Cohen et al. (2012), who noted that trees under an 80% water deficit were warmer than the control treatment by approximately 1 ◦C following four months of the assigned water treatment, the differences in temperature seen in the current study seem notable [6]. Cohen et al. (2012) used comparable thermal images with a resolution of 1.8 m [6]. Golomb et al. (2015), using thermal images with a resolution of 0.5 m, also found statistically significant results when comparing mean Crop Water Stress Indices (CWSI) values of control trees and infested trees using ANOVA, with a statistical significance of $p = 0.001$ [17]. When looking at the absolute temperature difference, a preliminary study by Soroker et al. (2013) noted a mean temperature difference of 1.5°C between trees infested with *R. ferrugineus* and control trees [16].

Spatial resolution for the thermal data was 1.8 m, which is comparable to the study of Cohen et al. (2012), but lower than for Golomb et al. (2015), where the pixel size was 0.5 by 0.5 m [6,17]. This spatial resolution could have several implications for the results. Based on the RGB images, a large palm tree canopy is approximately 6–8 m in diameter and is covered by approximately 9 to 16 pixels for 1.8 m resolution. However, due to the leaf structure of the date palm canopy there are areas where the background soil will be visible in the pixel, and temperature of the background soil will also be included in the pixel value—it is probable that only one or two pixels include only pure canopy. In studies looking for a large-scale pattern such as the effect of irrigation on an entire plantation, these changes could be visible at a resolution of 1.8 m because every tree in a block is affected by the lowered irrigation [6]. However, for analysis on an individual tree level this resolution may be too coarse to detect whether a tree has a markedly different temperature in comparison to the surrounding trees. Therefore, a maximum resolution of 0.5 m such as in the thermal images acquired by Golomb et al. (2015) might be required to be able to effectively assess individual tree stressors [17].

Considering the results of the hyperspectral data assessment several conclusions can be drawn for individual tree analysis based on the vegetation index analysis. In comparison with other studies, VOGI was one of the most indicative VIs of palm stress in the study by Shafri et al. (2012) on oil palms [19]. gNDVI was also found to identify *R. ferrugineus* infestation of date palms [18]. These were also two of the most indicative VIs when looking at correlations on an individual tree level in this study (Figure 10), indicating that VIs which focus on red-edge and green bands may be most suitable for date palm health analysis.

4.2 Added Value of Combined Datasets and New Indicators

Evaluating the combination of remote sensing data sources recorded at the same timestep makes this study unique in the sphere of date palm management. Several studies have shown that the combination of remote sensing sources can provide multivariate indicators with improved characterisation of the vegetation health status [20,26].

From a methodological perspective, the ability to combine datasets has proven to be very relevant when identifying trees. Using the tree objects derived from the LiDAR data to select the canopy pixels for temperature analysis provided an alternative to the method proposed in Cohen et al. (2012), which utilised the watershed algorithm to extract canopy pixels [6]. The LiDAR data again proved useful for temperature data analysis as the pixels below a certain height could be removed to reduce the chance of soil pixels being included in the analysis. A drawback is that an extra sensor is required, which will add costs to the data acquisition process. The high-resolution RGB data was compared with the outcome of the canopy extraction method to check whether the trees were delineated well (Figure 3). Correlation analysis between thermal and hyperspectral datasets showed promising relationships (Figure 10). The expected trends are visible, with negative correlation between temperature and several of the VIs. A positive correlation was observed for height and the VIs. Whilst the R-squared values between the variables are quite low, there is a clear spatial pattern between the indicators and VIs as seen in Figure 10. This suggests that these data, especially thermal and hyperspectral imagery, could be combined to identify stressed trees. The temperatures of the trees range from 22.5°C to 24.5°C (Figure 10), a difference of 2 ◦C. Cohen et al. (2012) were able to illustrate differences in temperature of stressed and healthy palms of 1°C [6]. This provides an indication that using thermal data could prove useful for diagnosis of individual tree health. Based on the correlation results, the hyperspectral data should be useful for reinforcing this diagnosis.

Based on the results of the tree point extraction using NDVI, it can be concluded that this method has high potential for identifying tree objects. Over 90% of trees were identified based on a comparison of field measured trees and NDVI, assuming accurate tree detection (Section 3.1). Misidentifications occur along the edges of the block where the trees are often close together and individual tree canopies are not distinct. As these results are based on distorted data one can assume that, when using a geometrically corrected dataset, identification accuracy will be higher. Using this identification method could prove to be an alternative to expensive LiDAR techniques. It should be noted that one of the reasons this method works well is because the canopies are distinct in a data palm plantation; however, in denser canopies such as the edges of date palm blocks, identification may be less accurate.

In this study, the height area ratio (HAR) index was used in order to assess the health of the blocks. When conducting analysis on the height of the trees, it became apparent that there were large differences in height of trees in certain blocks. For example, Block 19 had lower trees than the other blocks (Figure 6). From this information, one could conclude that either Block 19 is less healthy than the other blocks, or that the trees are younger in that block. Therefore, adding canopy area as a variable when considering tree health adds another consideration to the analysis. By determining the ratio between height and canopy area, one can determine whether a tree has an abnormally small canopy in relation to height—this could be an indication that the tree is experiencing growth limitations. In a study by Shendryk et al. (2016), a width height ratio was included amongst hyperspectral and LiDAR indicators to determine the most "useful" indicator for assessing tree health using Principal Component Analysis (PCA) [20]. Of the indicators derived from the recorded tree height, the width height ratio was the most important in their study. The conclusion that taller trees have smaller canopies in comparison to shorter trees indicates that using HAR alone does not give an accurate representation of tree health, as the growth of the tree affects this value for reasons other than health. Therefore, using an index that normalises height area ratio based on the height of a tree could prove more meaningful when making conclusions about tree health.

4.3 Data Limitations

As has been discussed in the Section 2.2.1 on preprocessing, there are some issues with data quality, especially for the hyperspectral dataset used in this study. Both the geometric distortions and the unknown scaling factor mean that the values used in the analysis might not be fully representative and certainly cannot be directly compared to other hyperspectral datasets that could be used in the future. For example, one step involved using bilinear interpolation for geo-correction, which resulted in multiple raster points with the same value because the hyperspectral image was effectively stretched to fit the corrected dataset.

Due to the distortions of the hyperspectral analysis, it was not possible to compare the trees on an individual level without manually matching the points. This meant that a relatively small sample size of 100 trees has been used as representative of over 10,000 trees. It is also possible that errors have been made when manually identifying the corresponding trees, as in some cases it was difficult to count rows and columns accurately within the distorted data.

Additionally, there are issues within the thermal dataset; specifically, for some limited areas blurring was noticed. This is due to the ortho-mosaicking technique used to merge the snapshot images. Trees are not identifiable in some regions, meaning that extracting the temperature per tree will result in values that are not an accurate representation of tree temperature. This was considered when selecting a block for individual tree analysis carried out in Section 2.2.4, as it was important that the blurring effect was minimal.

Finally, without ground truth data, it is not possible to determine a measure of actual tree health. This means that all assumptions made about tree health are based on indirect field observations and the relative health indicated by the RGB imagery (Table 2). The remote sensing datasets have been compared in various ways to determine whether the expected relations for healthy and unhealthy trees can be seen, and in some cases the initial assumptions were correct. However, as was seen in Figure 5, Block 5 had a much higher temperature than expected despite being considered as a healthy block. This suggests that there were factors not considered in classifying the blocks, and further field information was required. Following the analysis, further field information was provided, and the farm manager indicated that Block 5 had relatively poor soil quality in the southern region. Validation with field data is an essential step for future studies to explore the methods discussed in this paper further. As a result of the limited field data, especially on an individual tree level, the conclusions that can be made are limited to identifying that there are differences in date palms or date palm groups, and these differences cannot be attributed to a specific stressor.

4.4 Future Developments

Overall, this study reiterates the findings of earlier research indicating that different remote sensing datasets could prove useful for management of date palm plantations [6,16,17,19]. Thermal and hyperspectral data are the most promising, especially since identifying trees using NDVI is comparable to the results of LiDAR analysis. For future studies the opportunities of high-resolution RGB imagery to identify variation in palm canopy geometry could be investigated by using a combination of RGB based vegetation indices [27] and object-based image analysis.

Remote sensing data will prove most useful when provided as part of an integrated plantation management platform, where field data is being collected as the flight is taken, and ideally field data will be available over an extended period of time. The imagery could provide useful insights at all steps of an integrated system, including; field registration, field surveys, data acquisition, and monitoring. For example, the step of tree identification using NDVI or LiDAR would be useful for field registration. A topic that could be interesting to explore would be identifying within-block parcels with differing characteristics to be classified as management zones, which would allow for the adoption of precision agriculture techniques. Regular monitoring using a combination of satellite and aerial, or even drone imagery [28], potentially using the methods as explored in this article, could add value to the current monitoring systems that are in place. Based on a time-series analysis approach, changes in vegetation reflectance properties can be indicative of palm health, production and disease infestation processes over the growing season. For example, Jimenez-

Brenes et al. [29] showed how multi-temporal UAV-based images can be adopted to support pruning management in olive tree orchards. The ground monitoring system should also be well designed and could include ground sensors [30] and standardised methods of recording tree characteristics. Based on these spatial-temporal data, spatial analytics and geostatistical functions can be adopted to evaluate infestation processes of pests and diseases both at farm and regional scale-level [31]. A design based on the DateGIS platform approach could prove useful, as it aims to integrate imagery data with a ground data recording system where farm employees can update information about specific trees in the field [32].

5. CONCLUSIONS

This research explored the options of using multiple types of high-resolution remote sensing imagery in order to assess its potential for detecting various health aspects of date palm plantations. The factors which affect health were considered on two levels; those that impact large areas of the plantation such as a block, and those that affect the individual trees. Based on the spatial variation within blocks, it is suggested to also consider a third level in future studies; namely, areas within blocks with similar characteristics. Despite issues with the hyperspectral and thermal data preprocessing and quality, the methods developed in this study provide new options for future analysis and indicate that remote sensing data could aid plantation management and supplement precision agriculture techniques. Specifically, using a combination of high resolution thermal and hyperspectral imagery can give an indication of individual tree health, and by using these indicators together it could be possible to define a status for each tree. There are many new developments to be made in this area of research—combining remote sensing with detailed field data could provide an early indication of *R. ferrugineus* infestation; and comparing flights at different time steps could provide insight into the rates of change of date palm health. Overall, this study adds to the currently small body of research regarding using remote sensing imagery for date palm plantation management by using a combination of data sources and by providing suggestions for future research in this topic.

Appendix A

Table A1. Formulae for all hyperspectral vegetation indices used for the hyperspectral analysis.

Vegetation Index	Formula	Reference
SRI	$\frac{R_{800}}{R_{670}}$	[19,33]
NDVI 800	$\frac{R_{800}-R_{660}}{R_{800}+R_{660}}$	[19,34]
MSR	$\frac{R_{800}/R_{670}-1}{(R_{800}/R_{670})^{-1}+1}$	[19,35]
NDVI 750	$\frac{R_{750}-R_{660}}{R_{750}+R_{660}}$	[36]
VOGI	$\frac{R_{740}}{R_{720}}$	[19,37]
CIR	$\frac{R_{780}}{R_{710}}-1$	[38,39]
gNDVI	$\frac{R_{801}-R_{550}}{R_{801}+R_{550}}$	[18,40,41]
CIG	$\frac{R_{780}}{R_{550}}-1$	[38]
PRI	$\frac{R_{531}-R_{570}}{R_{531}+R_{570}}$	[42,43]
Optimum NDVI	$\frac{R_{738}-R_{610}}{R_{738}+R_{610}}$	[19]
SIPI	$\frac{R_{800}-R_{445}}{R_{800}+R_{680}}$	[18,44]
MCARI/OSAVI [705,750]	$\frac{[R_{750}-R_{705}-0.2(R_{750}-R_{550})](R_{750}/R_{705})}{(1+0.16)(R_{750}-R_{705})/(R_{750}+R_{705}+0.16)}$	[45–47]
TCARI/OSAVI [705,750]	$\frac{3[R_{750}-R_{705}-0.2(R_{750}-R_{550})(R_{750}/R_{705})]}{(1+0.16)(R_{750}-R_{705})/(R_{750}+R_{705}+0.16)}$	[46,47]
REP	$700+40\frac{R_{670}+R_{780}/2-R_{700}}{R_{740}-R_{700}}$	[48]

Appendix B

Table A2. Results of R-squared analysis of the vegetation indices contained in Table A1 and their relations to temperature and height.

Vegetation Index	Temperature (R-Squared)	Height (R-Squared)
REP	0.313	0.253
VOGI	0.227	0.213
gNDVI	0.206	0.189
CIG	0.196	0.204
CIR	0.196	0.202
TCOS750	0.191	0.103
SIPI	0.137	0.161
SRI	0.119	0.147
NDVI800	9.106	0.146
NDVI750	0.088	0.136
MSR	0.08	0.14
NDVI730	0.071	0.112
MCOS750	0.054	0.134
PRI	0.001	0.014

References

1. Stringer, L.C. Reviewing the links between desertification and food insecurity: From parallel challenges to synergistic solutions. **Food Secur. 2009,** 1, 113–126. [CrossRef]
2. Godfray, H.C.J.; Beddington, J.R.; Crute, I.R.; Haddad, L.; Lawrence, D.; Muir, J.F.; Pretty, J.; Robinson, S.; Thomas, S.M.; Toulmin, C. Food security: The challenge of feeding 9 billion people. **Science 2010**, 327, 812–818. [CrossRef] [PubMed]
3. Arias, E.; Hodder, A.J.; Oihabi, A. FAO support to date palm development around the world: 70 years of activity. **Emir. J. Food Agric. 2016**, 28, 1–11.
4. Al-Dosary, N.M.N.; Al-Dobai, S.; Faleiro, J.R. Review of the management of red palm weevil Rhynchophorus ferrugineus olivier in date palm *Phoenix dactylifera L. Emir. J. Food Agric.* **2016**, 28, 34–44. [CrossRef]
5. Shabani, F.; Kumar, L.; Nojoumian, A.H.; Esmaeili, A.; Togheyani, M. Projected future distribution of date palm and its potential use in alleviating micronutrient deficiency. *J. Sci. Food Agric.* **2015**, 96, 1132–1140. [CrossRef] [PubMed]
6. Cohen, Y.; Alchanatis, V.; Prigojin, A.; Levi, A.; Soroker, V.; Cohen, Y. Use of aerial thermal imaging to estimate water status of palm trees. *Precis. Agric.* **2012**, *13*, 123–140. [CrossRef]
7. Elshibli, S.; Elshibli, E.M.; Korpelainen, H. Growth and photosynthetic CO2 responses of date palm plants to water availability. *Emir. J. Food Agric.* **2016**, 28, 58–65. [CrossRef]
8. Tripler, E.; Ben-Gal, A.; Shani, U. Consequence of salinity and excess boron on growth, evapotranspiration and ion uptake in date palm (*Phoenix dactylifera* L., cv. Medjool). *Plant Soil* **2007**, 297, 147–155. [CrossRef]
9. Askri, B.; Ahmed, A.T.; Abichou, T.; Bouhhlila, R. Effects of shallow water table, salinity and frequency of irrigation water on the date palm water use. **J. Hydrol. 2014**, 513, 81–90. [CrossRef]
10. Yaish, M.W.; Kumar, P.P. Salt tolerance research in date palm tree (Phoenix dactylifera L.), past, present, and future perspectives. *Front. Plant Sci.* **2015**, 6, 348. [CrossRef]
11. Abraham, V.A.; Al Shuaibi, M.A.; Faleiro, J.R.; Abozuhairah, R.A.; Viyasagar, P.S.P.V. An integrated management approach to for red palm weevil *Rhynchophorus ferrugineus* Oliv. A key pest of date palm in the Middle East. *Agric. Sci.* **1998**, 3, 77–84. [CrossRef]

12. Faleiro, J.R. A review of the issues and management of the Red Palm Weevil *Rhynchophorus ferrugineus* (Coleoptera: Rhynchophoridae) in coconut and date palm during the last one hundred years. *Int. J. Trop. Insect Sci.* **2006**, 26, 135–154.
13. Ali, A.-S.A.; Hama, N.N. Integrated management for major date palm pests in Iraq. *Emir. J. Food Agric.* **2016,** 28, 24–33. [CrossRef]
14. Gutiérrez, A.; Ruiz, V.; Moltó, E.; Tapia, G.; del Mar Téllez, M. Development of a bioacoustic sensor for the early detection of Red Palm Weevil (*Rhynchophorus ferrugineus* Olivier). *Crop Prot.* **2010**, 29, 671–676. [CrossRef]
15. Dembilio, O.; Jacques, J.A. *Sustainable Pest Management in Date Palm: Current Status and Emerging Challenges,* 1st ed.; Springer International Publishing: Basel, Switzerland, 2015; pp. 13–36.
16. Soroker, V.; Suma, P.; la Pergola, A.; Cohen, Y.; Cohen, Y.; Alchanatis, V.; Golomb, O.; Goldshtein, E.; Hetzroni, A.; Galazan, L.; et al. Early detection and monitoring of Red Palm Weevil: Approaches and challenges. In Proceedings of the Palm Pest Mediterranean Conference, Nice, France, 19–18 July 2013.
17. Golomb, O.; Alchanatis, V.; Cohen, Y.; Levin, N.; Cohen, Y.; Soroker, V. *Precision Agriculture '15*, 1st ed.; Wageningen Academic Publishers: Wageningen, The Netherlands, 2015; pp. 643–650.
18. Bannari, A.; Khurshid, K.S.; Staenz, K.; Schwarz, J. Potential of Hyperion EO-1 hyperspectral data for wheat crop chlorophyll content estimation. *Can. J. Remote Sens.* **2008**, 34, 139–157. [CrossRef]
19. Shafri, H.Z.M.; Hamdan, N.; Anuar, M.I. Detection of stressed oil palms from an airborne sensor using optimized spectral indices. *Int. J. Remote Sens.* **2012**, 33, 4293–4311. [CrossRef]
20. Shendryk, I.; Broich, M.; Tulbure, M.G.; McGrath, A.; Keith, D.; Alexandrov, S.V. Mapping individual tree health using full-waveform airborne laser scans and imaging spectroscopy: A case study for a floodplain eucalypt forest. *Remote Sens. Environ.* **2016**, 187, 202–217. [CrossRef]
21. Alhammadi, M.S.; Glenn, E.P. Detecting date palm trees health and vegetation greenness change of the eastern coast of the United Arab Emirates using SAVI. Int. *J. Remote Sens.* **2008**, 29, 1745–1765. [CrossRef]
22. Vidyasagar, P.S.P.V.; Al Saihati, A.; Al Mohanna, O.E.; Subbei, A.I.; Mohsin, A.M.A. Management of Red Palm Weevil *Rhynchophorus ferrugineus* oliv., a serious pest of date palm in Al Qatif, Kindgdom of Saudi Arabia. *J. Plant. Crop.* **2000,** 28, 35–43.
23. Aleid, S.M.; Al-Kharyi, J.M.; Al-Bahrany, A.M. *Date Palm Genetic Resources and Utilization: Volume 2: Asia and Europe,* 1st ed.; Springer: Dordrecht, The Netherlands, 2015; pp. 49–95.

24. InfraTec. *VarioCam HD Head: User Manual;* InfraTec: Dresden, Germany, 2016; Available online: https://www. infratec.at/downloads/en/thermography/manuals/infratec-manual-variocam-hd-head.pdf (accessed on 11 January 2019).

25. Westoby, M.J.; Brasington, J.; Glasser, N.F.; Hambrey, M.J.; Reynolds, J.M. 'Stucture-from-Motion' photogrammetry: A low-cost, effective tool for geoscience applications. *Geomorphology* **2012**, 179, 300–314. [CrossRef]

26. Dalponte, M.; Reyes, F.; Kandare, K.; Gianelle, D. Delineation of Individual Tree Crowns from ALS and Hyperspectral data: A comparison among four methods. *Eur. J. Remote Sens.* **2015**, *48*, 365–382. [CrossRef]

27. Hamuda, E.; Glavin, M.; Jones, E. A survey of image processing techniques for plant extraction and segmentation in the field. *Computelectron. Agric.* **2016**, 125, 184–199. [CrossRef]

28. Gevaert, C.M.; Suomalainen, J.; Tang, J.; Kooistra, L. Generation of spectral-temporal response surfaces by combining multispectral satellite and hyperspectral UAV imagery for precision agriculture applications. *IEEE J.-Stars* **2015**, 8, 3140–3146. [CrossRef]

29. Jiménez-Brenes, F.M.; López-Granados, F.; de Castro, A.I.; Torres-Sánchez, J.; Serrano, N.; Peña, J.M. Quantifying pruning impacts on olive tree architecture and annual canopy growth by using UAV-based 3D modelling. *Plant Methods* **2017**, 13, 55. [CrossRef] [PubMed]

30. Massimo, P.; Alberto, R.A.; Roberto, M.; Khalid, A.; Ali, A. Devices to detect red palm weevil infestation on palm species. Precis. Agric. 2018, 19, 1049. [CrossRef]

31. Al-Kindi, K.M.; Kwan, P.; Andrew, N.R.; Welch, M. Modelling spatiotemporal patterns of dubas bug infestations on date palms in northern Oman: A geographical information system case study. *Crop Prot.* **2017**, 93, 113–121. [CrossRef]

32. DateGIS Demo—An Integrated Geo-Information System (GIS) for Precision Agriculture of Date Palm Farming. Available online: https://business.esa.int/projects/dategis-demo (accessed on 9 December 2018).

33. Jordan, C.F. Derivation of Leaf Area Index from quality of light on the forest floor. *Ecology* **1996**, 90, 663–666. [CrossRef]

34. Baret, F.; Guyot, G. Potentials and limits of vegetation indices for LAI and APAR assessment. *Remote Sens. Environ.* **1991**, 35, 161–173. [CrossRef]

35. Chen, J.M. Evaluation of vegetation indices and a modified simple ratio for boreal applications. *Can. J. Remote Sens.* **1996**, 22, 229–242. [CrossRef]

36. Apan, A.; Held, A.; Phinn, S.; Markley, J. Detecting sugar cane 'orange rust' disease using EO-1 Hyperion imagery. *Int. J. Remote Sens.* **2004**, *25*, 489–498. [CrossRef]

37. Vogelmann, J.E.; Rock, B.N.; Moss, D.M. Red edge spectral measurements from sugar maple leaves. *Int. J. Remote Sens.* **1993**, 14, 1563–1575. [CrossRef]

38. Gitelson, A.A.; Gritz, Y.; Merzlyak, M.N. Relationships between leaf chlorophyll content and spectral reflectance and algorithms for non-destructive chlorophyll assessment in higher plant leaves. *J. Plant Physiol.* **2003**, 160, 271–282. [CrossRef] [PubMed]

39. Clevers, J.G.P.W.; Kooistra, L. Using Hyperspectral Remote Sensing Data for Retrieving Canopy Chlorophyll and Nitrogen Content. *IEEE J.-Stars* **2012**, 5, 574–583. [CrossRef]

40. Gitelson, A.A.; Kaufman, Y.J.; Merzlyak, M.N. Use of a green channel in remote sensing of global vegetation from EOS-MODIS. *Remote Sens. Environ.* **1996**, 58, 289–298. [CrossRef]

41. Blackburn, G.A. Relationships between spectral reflectance and pigment concentrations in stacks of deciduous broadleaves. *Remote Sens. Environ.* **1999**, 70, 224–237. [CrossRef]

42. Gamon, J.A.; Peñuelas, J.; Field, C.B. A narrow-waveband spectral index that tracks diurnal changes in photosynthetic efficiency. *Remote Sens. Environ.* **1992**, 14, 35–44. [CrossRef]

43. Barton, C.V.M. Advances in remote sensing of plant stress. *Plant Soil* **2012**, *354*, 41–44. [CrossRef]

44. Peñuelas, J.; Baret, J.; Filella, I. Semi-empirical indices to assess carotenoids/ chlorophyll a ratio from leaf spectral reflectance. *Photosynthetica* **1995**, 31, 221–230.

45. Daughtry, C.S.T.; Walthall, C.L.; Kim, M.S.; Brown de Colstoun, E.; McMurtrey, J.E., III. Estimating corn leaf chlorophyll concentration from leaf and canopy reflectance. *Remote Sens. Environ.* **2000**, 74, 229–239. [CrossRef]

46. Haboudane, D.; Miller, J.R.; Tremblay, N.; Zarco-Tejada, P.J.; Dextraze, L. Integrated narrowband vegetation indices for prediction of crop chlorophyll content for application to precision agriculture. *Remote Sens. Environ.* **2002**, 81, 416–426. [CrossRef]

47. Wu, C.; Niu, Z.; Tang, Q.; Huang, W. Estimating chlorophyll content from hyperspectral vegetation indices: Modelling and validation. *Agric. For. Meteoro*l. *2008*, 148, 1230–1241. [CrossRef]

48. Clevers, J.G.P.W.; de Jong, S.M.; Epema, G.; van der Meer, F.; Bakker, W.H.; Skidmore, A.K.; Addink, E.A. MERIS and the red-edge position. *Int. J. Appl. Earth Obs. Geoinf.* **2001,** 3, 313–320. [CrossRef]

Chapter 13

DETERMINATION OF CULTIVATED AREA, FIELD BOUNDARY AND OVERLAPPING FOR A PLOWING OPERATION USING ISO 11783 COMMUNICATION AND D-GNSS POSITION DATA

1. INTRODUCTION: AN OVERVIEW

Over the last decades, economies of scale have been achieved by using machines with increasing working width and speed. However, the technical optimization of machines for crop production is increasingly exhausted. Nevertheless, management processes in agriculture become more complex by, e.g., implementing Precision Farming (PF) technologies, machinery sharing or contracting [1]. Furthermore, machinery operation has a remarkable effect on the total cost of crop production [2] and affects other cost factors such as time and farm inputs. Thus, there is an increasing demand for ways to detect and minimize inefficiencies within the whole processes. Tsiropoulos et al. [3] emphasized that performance analyses are also very valuable for tractor–implement combinations and commercial telematics solutions for implements are already on the market, e.g., CLAAS telematics on implements (TONI). The efficiency of an agricultural process can be assessed with regard to several parameters, e.g., time, cost and single machine-related parameters [2–10]. Beyond those, area is a very important indicator, because plant production processes are always strongly linked with a spatial component. Area optimization, in terms of minimizing the overlapped areas, combined with time optimization, by minimizing the non-working time [11], could lead to efficient agricultural processes. Easily available and detailed area-related information could help farmers, contractors or machinery rings facilitate documentation and invoicing.

Earlier studies (e.g., [5]) have used machine data to derive basic area-related information. Meanwhile, similar functionalities are implemented in commercial solutions. However, common hectare counters perform a rather indirect calculation without really examining the actually covered area. Commercial PF terminals can set the estimated cultivated area in relation to deposited values for field boundaries and, thus, give a rough estimation of the overlapping. Recent scientific works have already encompassed an automated field recognition [6,9]. However, there is still a lot of manual effort necessary for farmers and contractors to define and then deposit the field boundaries in digital field records. These tasks are usually inconvenient for farmers and contractors because they are time-consuming and error-prone and partly demand ever new familiarization with the operation. First approaches to facilitate these processes are represented by commercial solutions for an automated derivation of field boundaries from GNSS data, e.g., FARMDOK [12].

Past research works have also dealt with area-related analyses within a field. First examples are studies with the aim to detect the effective cutting width of combine harvesters and, thus, enable more accurate yield measurements [13,14]. The efficiency of a harvesting operation by means of coverage analyses was examined by Adamchuk et al. [2]. The presented information was assumed to deliver valuable information for the improvement of traffic patterns through optimization of the harvester route during non-harvest portions of the operation. However, a method that allows differentiated area-related analysis is still lacking. Especially for soil tillage, overlapping is still causing additional wear as well as fuel and time consumption. Meanwhile, lateral overlapping can be minimized with automatic steering systems. However, the accuracy of such systems is usually limited. Since the working width of a moldboard plow is usually low in comparison to other tillage implements, an evaluation of the lateral overlapping is still of interest. Furthermore, the overlapping between the headland and the main cropping area is still highly influenced by the operator and the field geometry. This interface area is of special interest for plowing with a moldboard plow: When it is elevated and lowered in the headland, the plow forms an inconsistent tillage operation and undesirable triangular shapes of unplowed segments. In [15], this problem is addressed by discussing section control technology for plowing operations while a first commercial application has been introduced by KUHN [16]. In general, it seems that there is not yet a way to enable a detailed quantification and differentiated assessment of the overlapping.

The increasing utilization of controller area network (CAN) data on agricultural machinery and the ISO 11783 (commonly designated as ISOBUS) compliant machine communication networks, as well as affordable positioning systems using global navigation satellite systems (GNSS), are promising data sources for information-driven crop production. In recent years, researchers have been using CAN-Bus data for various purposes and it is becoming clear that these data will be used in the future for optimizing agricultural processes. Infield tractor load states are defined in [10] by determining tractor's engine performance from CAN-Bus data. The results indicate that information acquired from a tractor's CAN-Bus is reliable enough for an assessment of different machine-related parameters. CAN-Bus data are acquired from axle housing loads and the driver's operation signals for three different applications, namely plowing, subsoiling and implement transportation, in [17]. Georeferenced CAN-Bus data for different analysis purposes have also been used [3,6–9]. According to Iglesias et al. [18], the functionality of ISOBUS compliant agricultural machines is increasing. Thus, the potential to gain sophisticated information from these data should also increase. However, there is still a need for algorithms and methods to handle the increasing amount of data that are collected [6].

The main aim of this study was the development of algorithms that allow detailed area-related analysis of a plowing operation, which involves the in-work status of the plow as well as the specific step-by-step way the area is covered, by using low-cost and embedded sensor data from ISOBUS. From this information, the cultivated area and the field boundary were determined. Furthermore, a differentiated analysis of the overlapping was performed by developing different indicators to quantify the lateral overlapping between single cultivated passes and the overlapping between the headlands and the main part of a field. To fulfill these aims, ISOBUS messages, as well as D-GNSS position data from a plowing operation, were post-processed. This encompassed the determination of the in-work sequences, a distinction of the different field parts, the modeling of single cultivated tracks, and the calculation of different area-related parameters based on these tracks. The main contribution of this paper is providing methods for an automated derivation of accurate and sophisticated, area-related information from continuous data records.

2. MATERIALS AND METHODS

2.1 Instrumentation

Data were acquired during plowing with a four-furrow reversible VariOpal 7 moldboard plow (Lemken GmbH & Co. KG, Alpen, Germany) mounted on a 6210R tractor (Deere & Company, Moline, Illinois, USA) (Figure 1). The working width of the plow was hydraulically adjustable to a maximum value of 0.54 m per share, resulting in a maximum total working width of 2.16 m. To record ISOBUS communication data, a GL2000 CAN-Bus data logger (Vector Informatik GmbH, Stuttgart, Germany) was configured and connected to the diagnostics interface of the tractor. The acquired data were georeferenced as the logger was also recording absolute positioning from a low-cost D-GNSS receiver with a specified circular error probable (CEP) of 2 m. The receiver's antenna was placed on the tractor cabin, on the longitudinal axis of the tractor. The correction data originated from the European geostationary navigation overlay service (EGNOS). The logged data were stored on a 2 GB storage card and, when triggered by the operator, they were wirelessly transmitted to a cloud-based server for further processing. The raw data analyzed for the present research work originate from the work performed in [7,9], where the data acquisition system is also described in detail.

Figure 1. The mounted four-furrow moldboard plow that was used for the tillage operation.

2.2 Data Acquisition

The GL2000 data logger was configured using its own configuration tool to filter the tractor's ISOBUS communication data and record only messages with a specific parameter group number (PGN). An example of an ISOBUS message as it was retrieved from the data logger is presented in Figure 2. The ISOBUS messages that were used for the present research work are listed in Table 1, with their PGN indicated as a hexadecimal value according to part 7 of the ISO 11783 standard [19]. They were recorded with a frequency of 10 Hz. One of the recorded D-GNSS messages was also relevant for the area-related analysis and it is listed in Table 1. The D-GNSS messages did not originate from the CAN-Bus, but the data logger embedded them into the log files with a frequency of 1 Hz. The structure of these messages was basically the same as shown in Figure 2. However, their ID was defined specifically by the logger's manufacturer. The acquired D-GNSS position data are presented in

Figure 3. and are indicated as green dots. They can be assigned to the field "Schlag 5" of the examined farm, with a size of 2.90 ha as this was indicated in the field catalogue.

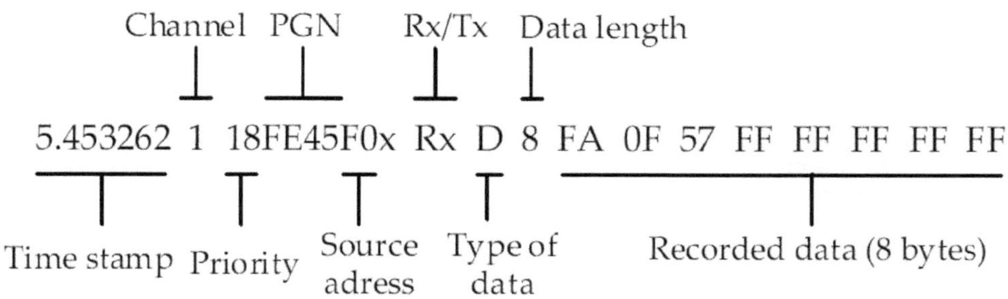

Figure 2. An example of an ISOBUS message as retrieved from the data logger.

Table 1. Messages used for the area-related analysis.

Message	PGN/ID	Content
RHS WBSD Msg2	FE45 FE48 1FFFFFD1	Rear Hitch Status Wheel Based Speed, Direction and Distance D-GNSS Latitude and Longitude

Figure 3. Satellite view with the raw D-GNSS position data (green dots).

2.3 Data Analysis Flow

For data analysis, the MATLAB R2016b (The MathWorks Inc., Natick, Massachusetts, USA) programming environment was used. The main data analysis steps are presented with the main MATLAB functions in Figure 4. From the data editing to the modeling of the plow points, there was a continuous flow. The absolute values for the cultivated area and the area of the field boundary, as well as the overlapping, were directly derived from the modeled plow points. The calculation of indicators for the overlapping, except the plow points, required the identification of the cultivated area and the boundary value.

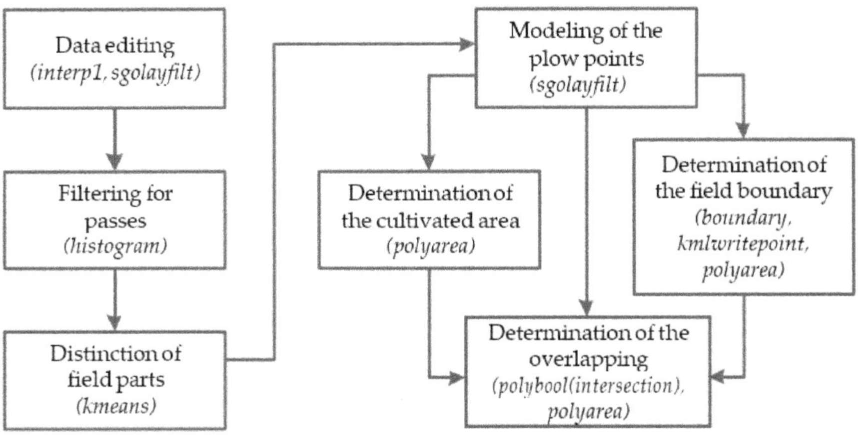

Figure 4. Flowchart of the main data analysis steps. The utilized MATLAB functions are also indicated.

2.4 Data Editing

The first important information from the ISOBUS was the wheel-based machine speed (WBMS), which is the value of the speed of the machine as calculated from the measured wheel or tail shaft speed [19]. It is specified in meters per second and, similar to the corresponding timestamps, it was derived from the WBSD messages using the suspect parameter number (SPN). Expressed in hexadecimal values, the WBMS has the SPN "1862". In the next step, the RHS messages from the ISOBUS were split into their time stamps and the rear hitch position (RHP), which has the SPN "1873". The RHP is the measured position of the rear three-point-hitch, expressed as a percentage of full travel, whereby 0% indicates the full down position, and 100% the full up position [19]. Finally, the D-GNSS latitude and longitude according to the world geodetic system 84 (WGS 84), as well as their timestamps were deduced from Msg2 (see Table 1). The acquired data for the signals RHP and WBMS are presented in Figure 5a,b, respectively.

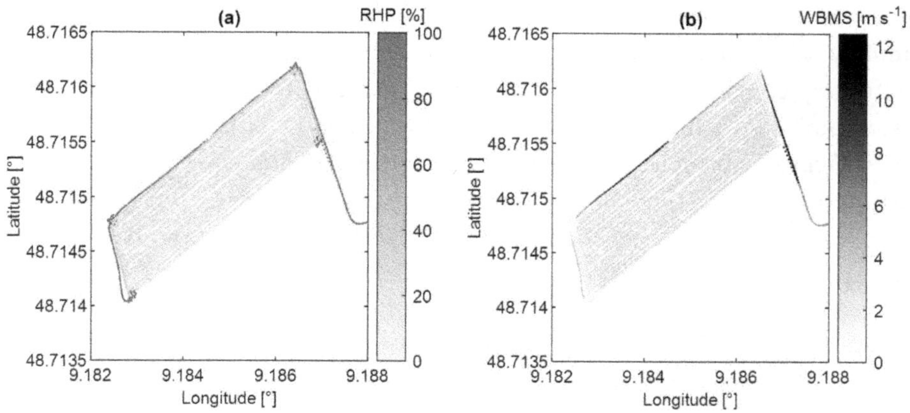

Figure 5. Acquired data for: (a) rear hitch position (RHP) signals; and (b) wheel-based machine speed (WBMS) signals.

To enable a concurrent analysis at a specific point and to increase the resolution of the D-GNSS information, a linear interpolation of the WBMS, as well as the D-GNSS latitude and longitude, was performed using the MATLAB function "interp1". The RHP time stamps were the query points for the interpolation, whereas the sample points were the time stamps of the WBMS and D-GNSS. As the positioning error of the used D-GNSS receiver, as well as the excitation of the tractor cabin, caused a jittering of the recorded track, a Savitzky–Golay filtering was applied using the MATLAB function "sgolayfilt". The polynomial order of the filter was set to one because it best fit the chronologic sequence of the recorded track. A frame length with a value of 25 was considered to give a smoothing that represents the real track of the rover in a suitable way.

2.5 Filtering for Passes

A pass was supposed to contain all the in-work points from a lowering up to a lifting of the plow. The in-work points were defined as the timestamps with their corresponding RHP, WBMS and D-GNSS information, where the plow was assumed to be set in operation. The analysis started with the second value of each dataset, as every RHP value was compared to the preceding one. As soon as it fell below a threshold of 60%, the event was recognized as a lowering point. As an in-work condition for the subsequent points, the WBMS at a query point had to exceed a threshold of 0.134 m s^{-1}. As soon as the RHP exceeded again the threshold of 60%, a lifting event was detected. To facilitate geometric calculations, the D-GNSS information assigned to in-work points was transformed from latitude and longitude to northing and easting using the universal transverse Mercator (UTM) projection.

The above-mentioned in-work thresholds for the RHP and WBMS were determined manually after some trials on the data, without a deeper examination of the real tractor–implement combination or plowing process. Using the MATLAB function "histogram", two histograms were created in the run-up to the filtering for in-work points. Figure 6a,b shows the histograms for the RHP and the WBMS, respectively. The bins indicate the number of elements for each parameter. The range of the RHP values was set from 0% to 90% to exclude the numerous 100% counts originating in a large part from road transport. To enable a focus on the in-work points, the range of the WBMS values was set from 0 to 4 m s-1. The RHP histogram clearly indicates the distribution of the in-work positions. The upper edge is at about 50%. To confirm that all real in-work points were included, the threshold was set to 60%. As it can be observed in Figure 6b, the pattern for the WBMS was not so clear and thus not a useful indicator. The threshold was finally set after the effect of different values was visually analyzed on a plot of the in-work D-GNSS positions.

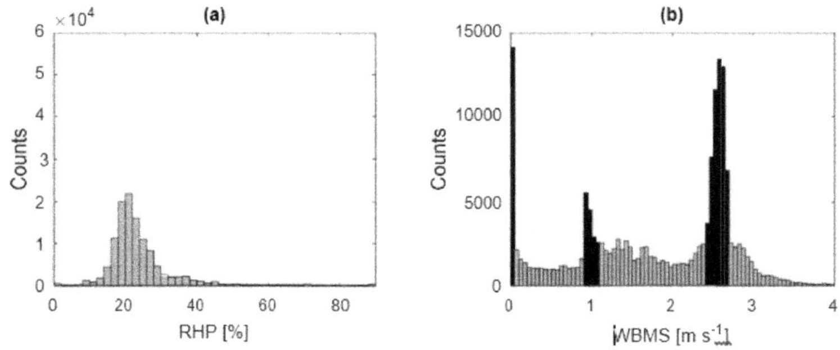

Figure 6. Histograms of: (a) the rear hitch position (RHP); and (b) the wheel-based machine speed (WBMS).

2.6 Distinction of Field Parts

To enable a distinct analysis of the lateral overlapping between the passes and the overlapping between the main part and the headlands, a distinction of the field parts had to be performed. As a preparatory step, the rough direction of every pass was expressed by a pass-vector created from the lowering to the lifting point. First visual assessments of the data resulted that the very first pass was also the first main part pass, which is usual for a plowing operation.

The distinction of the field parts was performed by calculating first the angle between the vector of the very first main part pass and every other pass of a field. To do so, the following formula was applied:

$$\alpha = \cos^{-1}\left(\frac{\vec{a} \circ \vec{b}}{|\vec{a}| \cdot |\vec{b}|}\right) \tag{1}$$

where α is the angle between two vectors →a and →. When the value for the angle was within the range of 45–135°, the arbitrary pass was declared as a main part pass. Otherwise, it was declared as a headland pass. Figure 7 schematically illustrates the operating principle of the field part distinction algorithm. The long passes represent the main part, whereas the two headlands are represented by the short passes running perpendicular.

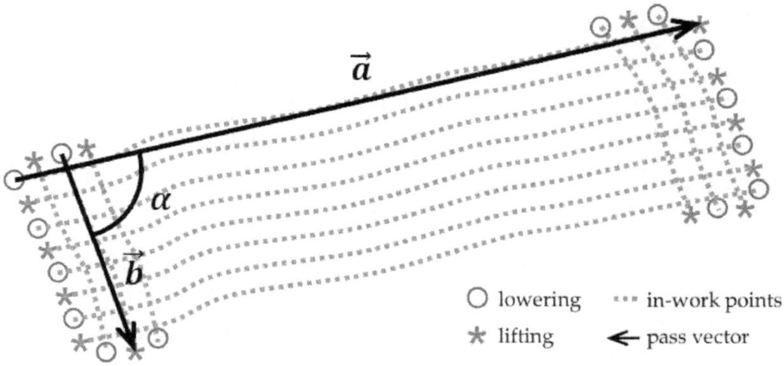

Figure 7. Schematic operating principle of the algorithm for the distinction between main part passes and headland passes.

To distinguish the two headlands of every field, k-means clustering was performed on their D-GNSS data. The MATLAB function "kmeans" was using a k-means++ algorithm according to Arthur and Vassilvitskii [20]. The number of clusters was manually preset to a value of two according to the number of headlands.

2.7 Modeling of the Representative Plow Points

As a preparatory step, relevant sizes were taken from the real tractor–implement combination, whereby the working width was set to its maximum. In Figure 8, the modeling of the representative plow points is schematically illustrated and the relevant dimensions are indicated. Figure 8a shows the plow rotated on the left rotating position, while Figure 8b shows the plow rotated right. The figure is kept abstract because it is supposed to serve only as a schematic illustration. To enable a better overview, only the shares that are operating in each case are presented. A distinction of the two rotating positions is important because they represent two mirror-inverted arrangements of the working tools, which has an important effect on the reconstruction of the cultivated track.

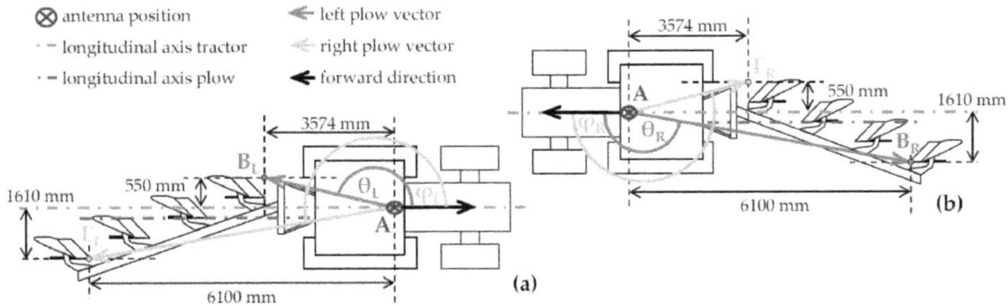

Figure 8. Relevant dimensions and schematic illustration for the modeling of the representative plow points for the two rotating positions of the plow: (a) rotated left; and (b) rotated right.

For every rotating position separately, two points were defined that represented the lateral edges of the working width (points B and Γ). As can be seen in Figure 8a,b, the representative plow points are formed by first drawing a line parallel to the outer edge of the first and last share and crossing it with a perpendicular line, which is touching the front top of the shares. Then, this intersection point is projected vertically to the height of the antenna (Point A) to get the representative plow point. For the subsequent calculations, the left and right plow vectors pointing from the antenna to the representative points of the plow were created. Then, the φ angle (φL and φR for both rotating positions) and the θ angle (θ_L and θ_R for both rotating positions) were determined. They represent the angle between the forward direction and the left and right plow vectors. Furthermore, the length of the left and right plow vectors for both rotating positions (ABL and ABR, and AΓL and AΓR, respectively) was determined. The values for these dimensions, which indicate the position of the representative plow points relative to the antenna position, are presented in Table 2.

Table 2. Relevant dimensions for the modeling of the representative plow points.

Rotated Left		Rotated Right		
Parameter	**Value**	**Parameter**	**Value**	**Unit**
θL	171	θR	165	°
ABL	3616	ABR	6309	mm
φL	195	φR	189	°
AΓL	6309	AΓR	3616	mm

Several adjustment options of the plow have a substantial effect on its effective working width. As the logged data could not deliver exact information about the work state, assumptions were made concerning the main parameters. The working width was assumed to always be set to its maximum during the data acquisition. In fact, this was plausible, as the tractor had a relatively high power compared to the plow's size. Concerning the rotation status, a simplified scheme was developed and applied to simulate this special characteristic. First, a distinction of passes having the same direction as the very first pass of a field part and passes running in the opposite direction was performed. For this, the angle between the vector of the very first and all other passes of a field part was determined using Equation (1). If the angle was between 0° and 45°, the pass was declared as running in the same direction as the very first pass. Then, the plow was assumed to be rotated right according to the definition in Figure 8. If it was between 135° and 180°, it was assumed to belong to a pass running in the opposite direction and, thus, rotated left.

2.8 Determination of the Cultivated Area

Figure 9 schematically illustrates the modeling of the cultivated area for three exemplary passes. To determine the driving direction, for every point of a pass, a vector was created, which originated in the D-GNSS position and pointed to the position of the chronologically subsequent position point. Thus, there was a vector indicating the direction at every point of a pass except for the lifting point, which was the very last point of a pass. For every pass, the left and right plow points marked the outer edges of the track that the plow had left during its operation. Using the previously outlined measures for the points representing the edges of the plow and setting them in relation to a static unit vector, which was parallel to the UTM easting and pointing eastwards, the absolute position of the representative points could be determined for every antenna position within a pass. To model the kinematics of the plow during work and thus smooth the track, a Savitzky–Golay filter was applied on the D-GNSS positions of the tracks using the MATLAB function "sgolayfilt".

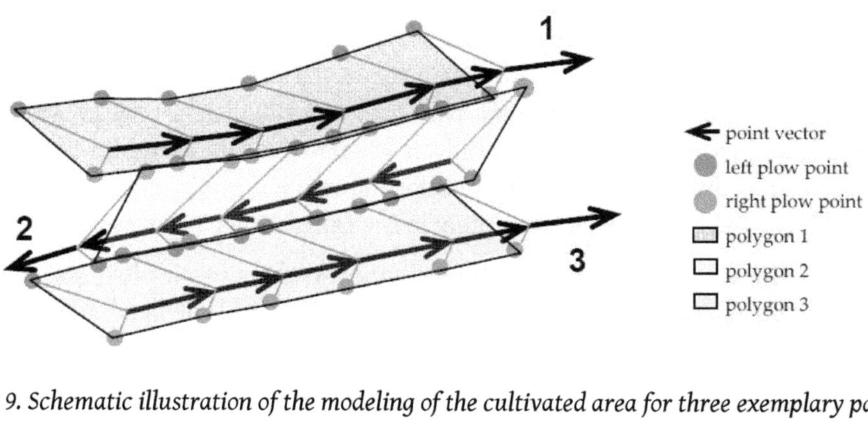

Figure 9. Schematic illustration of the modeling of the cultivated area for three exemplary passes.

In this case, a polynomial order of one and a frame length of 49 were applied. Very short passes with several in-work points below this frame length had to be removed in advance. All the points of a track could finally be connected to a polygon that was supposed to represent the cultivated area of a specific pass. The area of the main part and headland passes was calculated and added up to the total cultivated area of the field (Atotal).

2.9 Determination of the Field Boundary

The automatic determination of the boundary was configured to give the D-GNSS coordinates of the outermost representative plow points encompassing all the cultivated tracks within the field. The maximum shrink factor of the corresponding MATLAB function "boundary" was applied to get the most compact boundary possible. A polygon was formed using the boundary points and its area (Aboundary) was calculated using the MATLAB function "polyarea". To enable an examination of the boundary within a geographic information system (GIS), a keyhole markup language (KML) file was created, which contained all the WGS 84 coordinates of the boundary points. For this, the MATLAB function "kmlwritepoint" was used.

2.10 Determination of the Overlapping

Starting with the first main part pass, they were sequentially analyzed in terms of the lateral overlapping. The overlaps between the cultivated tracks of an arbitrary pass and the five chronologically subsequent passes were identified to ensure that no overlap was ignored. The overlap identification was done by determining for each comparison the intersection polygons and calculating their area. For this, the MATLAB functions "polybool (intersection)" and "polyarea" were used. The same process was performed on the headland passes. However, the overlaps between each cultivated track of a headland pass and all its chronologically subsequent ones were identified.

The overlaps between the main part passes and the headland passes were determined separately. Technically, the analysis procedure was the same as for the lateral overlaps. In this case, however, for every main part pass of a field, the overlaps with all the headland passes were identified. Naturally, it was possible that there was concurrently a lateral overlap and a headland overlap at locations, where main passes reach into the headland. They were respected separately in the calculations. Figure 10

illustrates the conception of the boundary and overlaps in an idealized way. The main part passes are running in an east–west direction and the headland passes are running in a north–south direction.

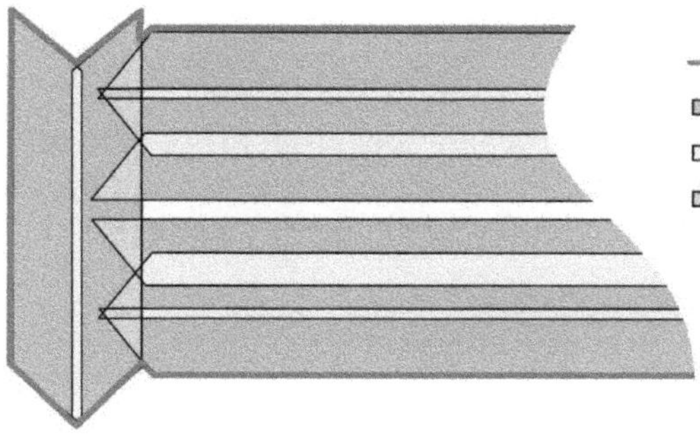

Figure 10. Schematic illustration of the concept of the boundary and overlaps.

To enable a differentiated analysis of the overlapping within the field, different intermediate figures were determined. For the Aoverlap,lat, the area of all lateral overlaps of the field was added up. The total headland overlapping for every field is expressed by $A_{overlap}$,head, which is the summed area of all overlapping polygons resulting from the intersections between the headlands and the main part. The first visual assessments of the modeled cultivated tracks uncovered noticeable lateral overlaps alternating with uncultivated lateral gaps due to the receiver error. Thus, the overlapping was corrected with the area of the gaps (A_{gaps}), which was calculated as follows:

$$A_{gaps} = A_{boundary} - \left(A_{total} - \left(A_{overlap,lat} + A_{overlap,head}\right)\right) \quad (2)$$

A dedicated analysis of the lateral overlapping was enabled by calculating the $A_{overlap,lat,perc}$, which indicates the relation between the area of the corrected lateral overlapping and the total cultivated area within a field:

$$A_{overlap,lat,perc} = \left(A_{overlap,lat} - A_{gaps}\right) / A_{total} \cdot 100\% \quad (3)$$

To enable a distinct assessment of the headland overlapping, the $A_{overlap,head,perc}$ was determined as an indicator for the relation between the summed area of headland overlaps and the total cultivated area of a field:

$$A_{overlap,head,perc} = A_{overlap,head} / A_{total} \cdot 100\% \quad (4)$$

$A_{overlap,perc}$ is considered the adequate indicator for the total overlapping within a field. It was calculated as follows:

$$A_{overlap,perc} = \left(\left(A_{overlap,lat} - A_{gaps}\right) + A_{overlap,head}\right) / A_{total} \cdot 100\% \quad (5)$$

Thus, it is the sum of $A_{overlap,lat,perc}$ and $A_{overlap,head,perc}$.

To compare the $A_{overlap,perc}$ with a more common indicator, which indicates the overlapping percentage within a field, the $A_{overlap,perc,ref}$ was calculated additionally:

$$A_{overlap,perc,ref} = \left(A_{total} - A_{boundary}\right) / A_{boundary} \cdot 100\% \quad (6)$$

The $A_{overlap,perc,ref}$ was inspired by how Demmel et al. [5] set the total cultivated area in relation to the area of the field boundary. This is a common way to express the overlapping by means of commercially available PF terminals.

3. RESULTS AND DISCUSSION

3.1 Distinction of Passes

Fifty-eight passes could be detected by applying the algorithm for the distinction of passes. It matched with the number of lifting and lowering points, which is an important indication for the algorithm's functionality. Figure 11a presents the D-GNSS position data of the lowering, lifting and in-work points with the corresponding RHP values, while Figure 11b highlights a zoomed area for a more detailed representation. Some single lowering and lifting points comparatively far away from the headlands can be observed in Figure 11a. This could indicate, e.g., corrections by the driver where he lifted the plow, moved back some distance and lowered it again to correct a driving error. Another explanation would be wedges at the field boundary where it is not necessary or rather possible to complete the whole pass.

In Figure 11b, the main part running in the southwest-northeast direction can be distinguished from the shorter headland running perpendicular. Furthermore, a varying distance between all passes is obvious. This was well expected by using a low-cost D-GNSS receiver. The work in [21] considers the pass-to-pass average error, which is also defined as the relative error, as the most important error for dynamic positioning. This relative error varies with the same receiver due to the testing date and time as well as vehicle speed and it highly depends on the used receiver.

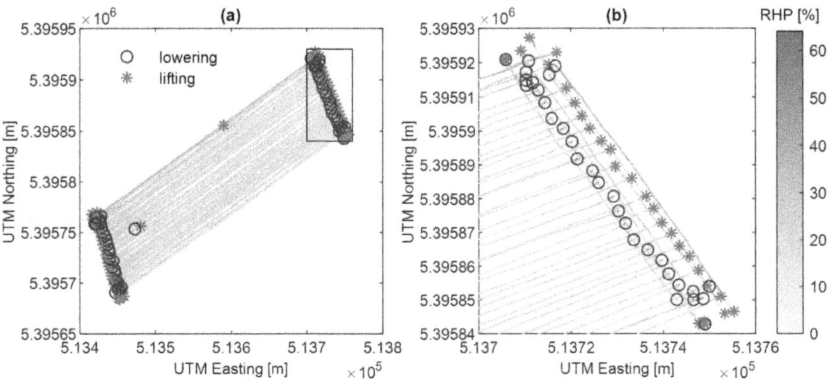

Figure 11. (a) D-GNSS positions of the lowering, lifting and in-work points with deposited RHP values; and (b) a zoomed area.

As the RHP values were slightly below the in-work threshold of 60%, it was easier to visually distinguish the variations. It is interesting that the RHP had comparatively low values for passes at the edges of the main parts and headlands. This was the first indication for a specific driving scheme in these areas. These schemes can be varied and depend on the local conditions as well as the driver's preferences. In the case of a reversible moldboard plow, this could indicate, e.g., a reduced working width by only setting the last few shares in operation or even a deviation from the usual plow rotation scheme.

3.2 Distinction of Field Parts

In Figure 12, the main part as well as the two headlands can be distinguished for the examined field. After the extraction of passes with fewer than 49 in-work points, which was the Savitzky–Golay filtering frame length, 55 passes remained. Of these, 41 corresponded to the main part and seven to each of Headland 1 and Headland 2. It must be noted that the algorithm for the distinction contains simplifying assumptions and thus can only be applied to well-structured fields such as the examined ones. There is an enormous variety of fields with more complex geometries and constellations of main parts and headlands. The thresholds for the angle between the pass vectors could be adapted but the algorithm hardly contains any scope for further adaptions.

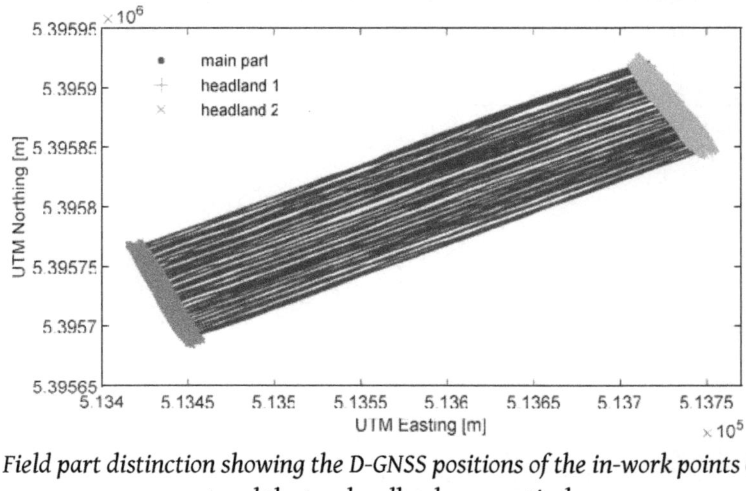

Figure 12. Field part distinction showing the D-GNSS positions of the in-work points of the main part and the two headlands, respectively.

The distinction between the two headlands for every field using k-means clustering was a prerequisite for a differentiated analysis of the overlapping and it enabled making assumptions for the rotation scheme of the plow separately for every headland. Here again, simplifying assumptions were made. For more complex field geometries and fieldworks, it is possible to have some more areas that can be defined as headlands or to even have no headland at all. Even though the number of headlands could be adjusted for every field separately, the developed logic contains hardly any scope for further adaptions. Another aspect is the inability of the algorithm to clearly identify an individual headland. Due to the used k-means clustering, the determination, if a headland is the "Headland 1" or "Headland 2" of a specific field, has a random character.

3.3 Cultivated Tracks

For every point of a pass, except for the last one, two points, which represent the lateral edges of the plow, were created depending on the current direction of movement. Under simplifying assumptions, a distinction was made between the two possible rotating directions of the plow. To get an idea of how a pattern of multiple cultivated tracks within the field looks, Figure 13a shows a zoomed area at the northeast corner of the field. The D-GNSS positions of the antenna are plotted together with lowering and lifting points indicating the passes. Furthermore, the representative plow points are illustrated. "Plow left" and "plow right" represent the left and the right edges, respectively, of the plow relative to the current driving direction (see definition of points B and Γ in Section 2.7). For a better representation, Figure 13b focuses on three exemplary passes (Passes 18–20) from the main part. They are representative of the regular operation scheme of the inner main part passes. The rotating position is not indicated separately, but can be derived from the arrangement of the left and right plow points at the beginning or the end of a pass.

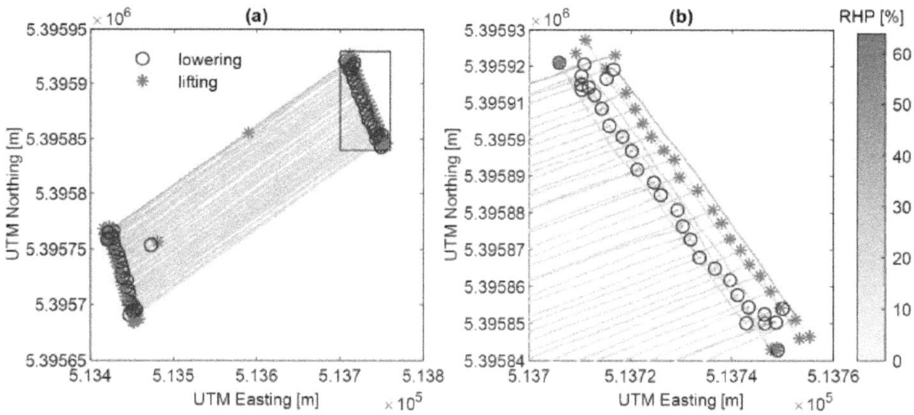

Figure 13. (a) Passes with points representing the plow's lateral edges; and (b) a zoom area with three exemplary tracks.

Regarding the results discussed here, it could be argued that the filtering of the D-GNSS data was done arbitrarily to obtain a subjectively optimal smoothing. However, after several trials, it became clear that the effect of shifting the focus of the filtering to the interpolated raw D-GNSS data or to the cultivated tracks was negligible. Furthermore, the degree of filtering had in general only a minimal effect on the values of the area-related parameters, which are presented in the subsequent sections.

3.4 Cultivated Area

For the examined field, a total cultivated area Atotal of 3.13 ha was determined. Figure 14 presents a part of the cultivated area that corresponds to the same exemplary passes that are presented in Figure 13b. A lateral overlapping can be distinguished between Passes 19 and 20. Furthermore, a gap area is noticeable between Passes 18 and 19. This does not mean that this area was left uncultivated, but it originates from the D-GNSS position error and justifies the choice to subtract from the lateral overlapping area Aoverlap,lat, the total gap area Agaps (see Equation (3)). Furthermore, a remaining jittering of the tracks became obvious. After some visual assessments, this was considered as some kind of noise that had to be accepted to not cause a strong idealization by over-filtering.

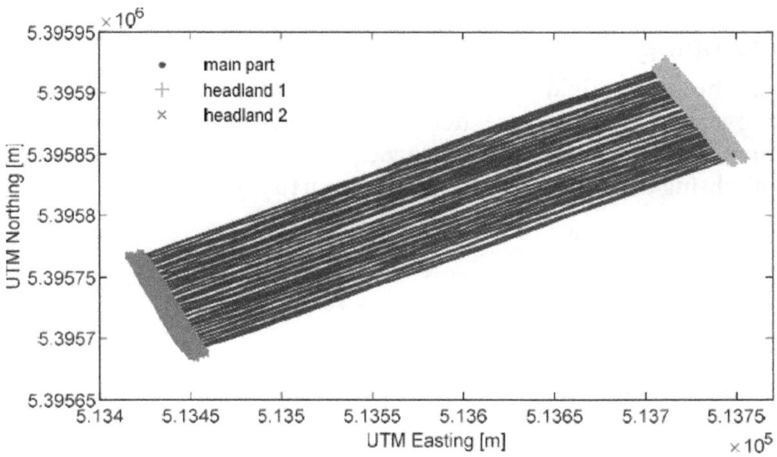

Figure 14. Zoom of the polygons formed from three exemplary cultivated tracks.

3.5 Field Boundary

The area of a boundary should represent the specific field size and location, as documented in the field catalogue (2.90 ha). It was calculated by the developed software with $A_{boundary}$ = 2.94 ha, which results in an overestimation of 1.4%. Figure 15 shows a satellite view of the boundary (indicated in green) for the examined field. It fits the actual field location. Several error sources influenced the determination of the boundary. The erroneous D-GNSS positioning had an effect, especially at the field edges. A further influence negatively affecting the determination of the boundary was that the applied shrinking of the MATLAB function "boundary" could not deliver the closest fit possible. Furthermore, what is assumed to cause error is the deviation from the plow rotation scheme or a variation in working width, especially at the beginning and the end of a field part. Another possible error source is the rolling of the tractor while driving on a slope along the field edges. All these errors could not be monitored and corrected. Considering these errors, the deviation of 1.4% from the actual field size is a strong indication of the efficiency of the developed algorithm.

Figure 15. Satellite view of the examined field and the detected boundary (green line).

3.6 Overlapping

Figure 16a shows the field boundary, the cultivated tracks, and the lateral and headland overlaps. For a more detailed representation, Figure 16b presents a zoomed area at the lower northeast corner of the field. It must be stated that the color highlighting of the headland overlaps is virtually dominant.

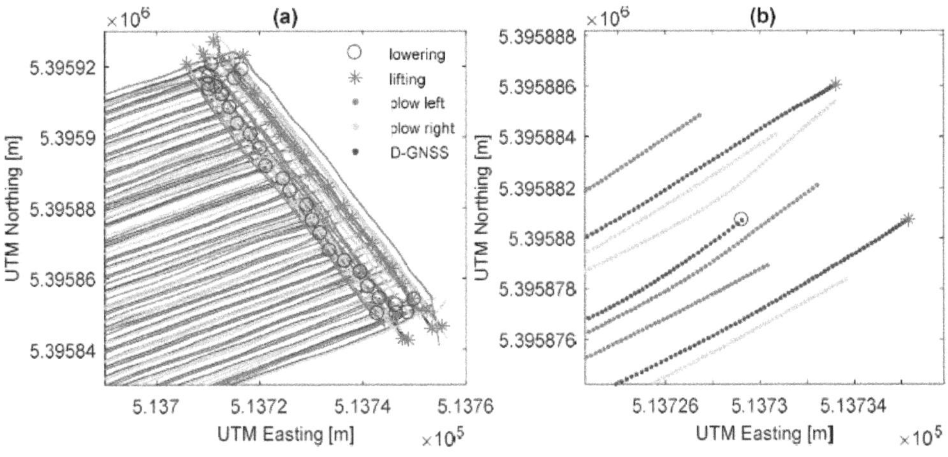

Figure 16. (a) Field boundary, cultivated area and lateral and headland overlaps; and (b) a zoomed area at the lower northeast corner of the field.

In Figure 16b, the fact that the used function for the determination of the boundary could not deliver the closest fit possible is illustrated. Beyond an overestimation of the field size, this also caused an overestimation of the gaps' area A_{gaps} and, thus, an underestimation of the summed area of lateral overlaps $A_{overlap,lat}$. Furthermore, there are noticeable distortions at the tops of several headland tracks, which was a further source of error concerning the boundary. They could be traced back to the error behavior of the used receiver. The discrepancy between different errors depending on the rover's movement has been examined previously (e.g., [22,23]). A further possible source of error was the overall additional excitation of the tractor cabin during headland overlapping.

In Figure 16b, passes can be distinguished where there is an excessive lateral overlap alternating with gaps in the magnitude of the plow's working width. This was mainly due to the erroneous D-GNSS positioning, as discussed in previous sections. The pass-to-pass error of low-cost receivers has been considered previously (e.g., [21]). Furthermore, it can be recognized that two of the first main part passes in the south are almost entirely overlapping, whereas there seems to be no corresponding gap. For the rest of the field, this was also a common pattern for some of the first and last few passes of the main part or a headland, indicating again a specific driving scheme at these sites. This pattern fits the observations discussed in Section 3.1 and led to an overestimation of $A_{overlap,lat}$.

A typical problem with moldboard plows are the undesirable triangular shapes of unplowed segments at the headlands [15]. Headland overlapping is generally an important indicator of the performance of the driver, having a remarkable effect on the field efficiency and causing additional cost. The alignment of the beginning and the end of the main part tracks, which can be observed at the top of the polygons, is highly variable in Figure 16b. Possible reasons for this observation were a varying track-error of the D-GNSS positioning and improper timing by the driver. In general, the headland overlaps seem to be quite long. An obvious reason for that was the remaining error for the setting of the in-work threshold of the RHP, as described in Section 2.5. However, a minimization of the threshold would have had little effect on the length of the pass and probably would have led to an exclusion of points that were only just tilled. The main results in terms of the overlapping indicators are listed in Table 3.

Table 3. Overlapping indicators.

Indicator	Value [%]
Aoverlap,lat,perc	2.05
Aoverlap,head,perc	3.96
Aoverlap,perc	6.01
Aoverlap,perc,re f	6.39

$A_{overlap,lat,perc}$ contains the same information as the efficiency of width use, which is documented in [24]. When transferring the values given in this reference to the $A_{overlap,lat,perc}$, it should not exceed a value of 5%. Considering the numerous previously mentioned error sources and simplifying assumptions, the value of 2.05% seemed usable for a first estimation.

$A_{overlap,head,perc}$ refers only to the overlapping between the main part and the headland passes and was set in relation to the total cultivated area. When comparing the value of 3.96% to the $A_{overlap,lat,perc}$, it became obvious that the headland overlapping had an even bigger influence on the total overlapping within the field than the lateral overlapping between neighboring passes.

$A_{overlap,perc}$ is the sum of $A_{overlap,lat,perc}$ and $A_{overlap,head,perc}$ and indicates the total overlapping within the field boundary. To compare it with a more common way to express the overlapping within a field, the values for $A_{overlap,perc,ref}$ are also indicated in the table. Similar to Demmel et al. [5], the total cultivated area of a field was set in relation to the area of the field boundary. The value was slightly higher than $A_{overlap,perc}$. This was natural because the area of the boundary as a reference was lower than the total cultivated area. $A_{overlap,perc,ref}$ was almost unaffected by the relative error of the positioning; only a small error could be expected for the actual boundary size. This implies that the imperfect information about the actual work state of the plow had the most negative effect on the results concerning the overlapping indicators.

4. CONCLUSIONS

A methodology was developed that enabled a distinction of work sequences from non-work sequences using a few parameters retrieved from a tractor's ISOBUS. A field part distinction was applied as a prerequisite for a differentiated area-related analysis. A modeling of the cultivated tracks, based on the way the plow covered the area step-by-step, was implemented. Based on that, the cultivated area was quantified and the field boundary was detected in terms of location and size. Finally, different indicators were developed and calculated for the overlapping. An assessment with a more common indicator proved the plausibility of the results. Incomplete information about the work state of the implement, erroneous D-GNSS positioning and a limited shrinking of the boundary were identified as the main sources of error.

Automated, detailed detection of the cultivated area and the field boundary can be very valuable, e.g., for documenting and invoicing of agricultural tasks within networked software infrastructures. A differentiated analysis of overlapping can serve as an indicator of efficiency. In future work, the presented parameters for the overlapping analysis should be validated on fields of different sizes and geometries.

If there were more detailed information about the actual work state of the plow, more reliable results could be expected. The new generation of ISOBUS-enabled plows is very promising in delivering this information. Furthermore, the use of more accurate positioning and data acquisition instrumentation is recommended when a higher level of accuracy is required. One possibility would be using real time kinematic GNSS and robotic total stations [25]. Another option would be fusing information from GNSS receivers and inertial sensors, a technology which is already implemented in commercial automated steering systems. Considering these possible improvements, the slight overestimation of 1.4% in terms of the boundary's size is promising for the potential use of more sophisticated applications, e.g., automated steering. With a few adaptions, the used methodology for an area-related analysis could be transferred to a wide range of mounted soil tillage implements and plant production processes.

References

1. Steinberger, G.; Rothmund, M.; Auernhammer, H. Mobile farm equipment as a data source in an agricultural service architecture. *Comput. Electron. Agric.* **2009**, 65, 238–246. [CrossRef]
2. Adamchuk, V.I.; Grisso, R.D.; Kocher, M.F. Machinery Performance Assessment Based on Records of Geographic Position. In Proceedings of the 2004 ASAE/CSAE Annual International Meeting, Ottawa, ON, Canada, 1–4 August 2004.
3. Tsiropoulos, Z.; Fountas, S.; Gemtos, T.; Gravalos, I.; Paraforos, D. Management information system for spatial analysis of tractor- implement draft forces. In *Precision Agriculture 2013, the 9th European Conference on Precision Agriculture,* ECPA 2013; Wageningen Academic Publishers: Wageningen, The Netherlands, 2013.
4. Biller, R.H. Aufbau und Einsatz eines Datenerfassungssystems für Ackerschlepper. *Grundl. Landtechn.* **1985,** 35, 97–104.
5. Demmel, M.; Rothmund, M.; Spangler, A.; Auernhammer, H. Algorithms for data analysis and first results of automatic data acquisition with GPS and LBS on tractor-implement combinations. In Proceedings of the 3rd European Conference on Precision Farming in Agriculture, Montpellier, France, 18–20 June 2001; pp. 13–18.
6. Kortenbruck, D.; Griepentrog, H.W.; Paraforos, D.S. Machine operation profiles generated from ISO 11783 communication data. *Comput. Electron. Agric.* **2017**, 140, 227–236. [CrossRef]
7. Paraforos, D.S.; Vassiliadis, V.; Kortenbruck, D.; Stamkopoulos, K.; Ziogas, V.; Sapounas, A.A.; Griepentrog, H.W. Automating the process of importing data into an FMIS using information from tractor's CAN-Bus communication. *Adv. Anim. Biosci.* **2017**, 8, 650–655. [CrossRef]
8. Brünnhäußer, J.; Knorr, T.; Meyer, H.J. Manufacturer independent system for process and machine data analysis. *Landtechnik* **2014**, 69, 196–200. [CrossRef]
9. Paraforos, D.S.; Vassiliadis, V.; Kortenbruck, D.; Stamkopoulos, K.; Ziogas, V.; Sapounas, A.A.; Griepentrog, H.W. Multi-level automation of farm management

information systems. *Comput. Electron. Agric.* **2017**, 142, 504–514. [CrossRef]

10. Pitla, S.K.; Luck, J.D.; Werner, J.; Lin, N.; Shearer, S.A. In-field fuel use and load states of agricultural field machinery. *Comput. Electron. Agric.* **2016**, 121, 290–300. [CrossRef]
11. Paraforos, D.S.; Hübner, R.; Griepentrog, H.W. Automatic determination of headland turning from auto-steering position data for minimising the infield non-working time. *Comput. Electron. Agric.* **2018**, 152, 393–400. [CrossRef]
12. Anon. Farmdok Was Awarded with the Silver "Innovation Award Agritechnica 2017". Available online: https://www.farmdok.com/en/2017/09/14/dlg-innovation-award-agritechnica-2017/ (accessed on 11 February 2019).
13. Han, S.; Schneider, S.M.; Rawlins, S.L.; Evans, R.G. A bitmap method for determining effective combine cut width in yield mapping. *Trans. ASAE* **1996**, 40, 485–490. [CrossRef]
14. Drummond, S.T.; Fraisse, C.W.; Sudduth, K.A. Combine harvest area determination by vector processing of GPS position data. *Trans. ASAE* **1999**, 42, 1221–1227. [CrossRef]
15. Nielsen, S.K.; Munkholm, L.J.; Aarestrup, M.H.; Kristensen, M.H.; Green, O. Plough section control for optimised uniformity in primary tillage. *Adv. Anim. Biosci.* **2017**, 8, 444–449. [CrossRef]
16. Anon. Silbermedaille AGRITECHNICA 2017 für KUHN-Neuheit "Smart Ploughing"—KUHN.de. Available online: http://www.kuhn.de/de/aus-der-fachpresse-silbermedaille-agritechnica-2017-fr-kuhn-neuheit- smart-ploughing.html (accessed on 4 January 2019).
17. Mattetti, M.; Molari, G.; Sereni, E. Damage evaluation of driving events for agricultural tractors. *Comput. Electron. Agric.* **2017**, 135, 328–337. [CrossRef]
18. Iglesias, N.; Bulacio, P.; Tapia, E. Enabling powerful GUIs in ISOBUS networks by transparent data compression. *Comput. Stand. Interfaces* **2014**, *36*, 801–807. [CrossRef]
19. ISO. *Tractors and Machinery for Agriculture and—Serial Control and Communications Data Network—Part 7: Implement Messages Application Layer;* ISO 11783-7:2015; ISO: Geneva, Switzerland, 2015.
20. Arthur, D.; Vassilvitskii, S. K-means++: The Advantages of Careful Seeding. In Proceedings of the Eighteenth Annual ACM-SIAM Symposium on Discrete Algorithms (SODA '07), New Orleans, LA, USA, 7–9 January 2007; Society for Industrial and Applied Mathematics: Philadelphia, PA, USA, 2007; pp. 1027–1035.

21. Han, S.; Zhang, Q.; Noh, H.; Shin, B. A dynamic performance evaluation method for DGPS receivers under linear parallel-tracking applications. *Trans. ASAE* **2004**, 47, 321–329. [CrossRef]

22. Stombaugh, G.; Cole, J.; Shearer, S.; Koostra, B. A test facility for evaluating dynamic GPS accuracy. In Proceedings of the 5th European Conference on Precision Agriculture, Uppsala, Sweden, 9–12 June 2005; pp. 605–612.

23. Chosa, T.; Omine, M.; Itani, K.; Ehsani, R. Evaluation of the Dynamic Accuracy of a GPS Receiver: Is dynamic accuracy the same as static accuracy? *Eng. Agric. Environ. Food* **2011**, 4, 54–61. [CrossRef]

24. Kollar, L. Gründe und Möglichkeiten für die automatische Lenkung mobiler landwirtschaftlicher Aggregate. *Dtsch. Agrartech.* **1973**, 30, 95–98.

25. Paraforos, D.S.; Reutemann, M.; Sharipov, G.; Werner, R.; Griepentrog, H.W. Total station data assessment using an industrial robotic arm for dynamic 3D in-field positioning with sub-centimetre accuracy. *Comput. Electron. Agric.* **2017**, 136, 166–175. [CrossRef]

Chapter 13

FOOD SYSTEM DIGITALIZATION AS A MEANS TO PROMOTE FOOD AND NUTRITION SECURITY IN THE BARENTS REGION

1. INTRODUCTION: AN OVERVIEW

The food system encompasses all different stages of food production, including harvesting, processing, distribution and storage. Pre-production and post-harvest processes within the food and agriculture sector have witnessed tremendous changes with the application of science and technology. The first industrial revolution of the early nineteenth century had a remarkable impact on agriculture and food processing. It has been labor intensive, and multinational companies have gained profits through value addition to local foods, which are often packaged, stored, and distributed over long distances [1]. The second industrial revolution witnessed mass production from manufacturing plants [2]. Science and technology have also brought about innovation with food products that respond to consumers' demands, leading to the third industrial revolution with machine intelligence, computers, and digitalization [3]. Currently, we live in an increasingly connected society that revolves around the internet, signaling the fourth industrial revolution [4]. The current fourth industrial revolution combines digital, physical and biological systems, and it will encourage local processing with less need for long distances, which can reduce the miles covered by food. Such a revolution can shift the balance towards local and distributed manufacturing in rural communities and sparsely populated regions, such as the Barents region.

The Barents region comprises the northern parts of Finland, Sweden, Norway, and Northwest Russia. The Barents region within the Russian Federation extends geographically into the European Arctic, but its socio-economic characteristics differ from those of the Nordic countries [5]. The region is not as homogeneous as the European High North (EHN), since it includes Northwest Russia.

The traditional foods that are available in the Barents region include fish, potatoes, meat, and berries. Meat mainly consists of reindeer meat but it can also include cow and moose. In many communities of the Barents region, fish is deemed the second most important local food, for the women in some villages it can also be the most important [6]. Wild berries e.g., cloudberry (*Rubus chamaemorus*), bilberry (*Vaccinium myrtillus*), lingonberry (*Vaccinium vitis-idaea*), raspberry (*rubusidaeus*), cranberry (*Oxycoccus*) are widely collected and eaten, and could be available fresh or conserved all year during good seasons [7] (p. 40). In the family gardens of many communities in the region they grow potatoes, strawberries (Fragari avesca) and blackcurrants (Ribes nigrum), as well as some onions and root crops. Mushrooms and herbs are often picked from the forest and incorporated into meals. Animals such as moose (*Alces alces*), rabbit (*Oryctolagus cuniculus*), fowl (*Galloanserae*) and waterfowl (*Aix galericulata*) are not as important because of the low number of wild animals available [8] (p. 347). Dairy farming also exists to produce a wide range of dairy products including milk, cheese, yoghurt, butter and other fermented dairy products [6]. The processing of these foods, in terms of adding value to them, is mainly done by small and medium enterprises in the Barents region. There are some processed products from these traditional foods and herbs in the region. In Finnish Lapland, these include fresh and cold pressed juices made from crowberries, lingonberries and bilberries. Extracts from herbs such as angelica, golden root (*rhodiola rosea*) and nettle are available in the Lapland market. There is also small-scale processing of snack bars from berries and chocolate, spruce sprout syrup and sauces, as well as smoked reindeer chips. When these foods are processed, they are better preserved and can be stored or distributed for longer periods of time. Currently, the value addition to traditional foods in the region can improve with digitalization that targets increased production, better harvesting methods, less waste, as well as better storage and distribution. The introduction of sensors that generates data from food system digitalization will help make accurate predictions in the processes within the food value chains of the countries in the Barents region. Access to food and its utilization in different parts of the Barents region are largely influenced by food markets and the choice of food that is available to the inhabitants of this region. Due to the increasing connection in our society, the last few decades have witnessed new imported foods making their way into the Barents region. Some of these new imported foods, such as tortillas, kebabs, pizza, and sushi, are increasingly popular in the region. This has resulted in changes within the food system of the Barents region, with consequential effects on traditional foods that people have relied upon in the past [6]. The popularity of these foods, at the expense of traditional foods has been shown to increase the risks of cardiovascular diseases, obesity and diabetes in the Barents region [6,9]. It will be important to address the challenges faced by small and medium enterprises in the region for them to expand their businesses and create a niche for their high-quality nutritional food products. This can be achieved by encouraging a better means of harvesting berries and herbs, and highlighting the nutritional benefits of these bioactive ingredients through mobile applications that can appeal to potential consumers, especially the younger generation.

Food security and human security are prevention oriented; food security helps to develop the overall human well-being, thereby strengthening human security [6]. The concept of human security should endorse a people-centered approach, built on the capacity of individuals, and provide key tools for building resilience around food security and nutrition [10]. Food and nutrition are integral to human security, and the link between diet and health is well established [11,12]. The food we consume plays an important role in our health and well-being. Food availability, accessibility, utilization and stability within the food system are essential pillars of food security. In the Barents region, ensuring sustainable food and nutrition security will help to build resilience, and promote culture and ecology that will offer greater human and societal security [13]. Healthy living is associated with clean water, food and air, but also includes the safety and security of life for the individuals, groups and communities. Changes that occur in the climate, environment and land use, together with socioeconomic factors, have been reported to impact on food and water security in the Arctic region, including the Barents region [14]. The trends associated with food consumption are constantly evolving with technological inputs. How food processing methods and the traditional culture related to food have changed over the years with technology are important topics for consideration. The issues of food security, climate change, and population growth are major concerns discussed in many international agreements as in [15–17]. There are only a few studies on food security in the Barents region, and to date, there is no work on food system digitalization in the region. We therefore emphasize the need for small and medium-based enterprises (SMEs) in the food sector to embrace digital solutions in their operations. The purpose of this article is to identify gaps and challenges related to food and nutrition security in the Barents region, improve the sustainability of the food system through digitalization and promote the development of cross-border infrastructures to support digitalization. The authors have drawn largely on their participation in related project workshops and social networking services in this region.

This review article addresses the relevance of digitalization as a disruptive innovation within the food system that will help to promote food and nutrition security. We highlight in Section 2, the impacts of human activities and climate change in the Barents region; Section 3 is on food system digitalization in the Barents region and how other enablers for digitalization, such as internet penetration, will be crucial in achieving the desired objectives of improving food and nutrition security in the region. Section 4 discusses the role of different digital technologies that can transform the food system in the region and the overlapping relationship that exists amongst the consumers, food business, and authorities; Section 5 discusses how information and communications technology can contribute to a sustainable food system; Section 6 looks at future implications for the Barents region. Finally, Section 7 concludes the article.

2. IMPACTS OF HUMAN ACTIVITIES AND CLIMATE CHANGE IN THE BARENTS REGION

The Barents region is characterized by its remoteness, high latitude, harsh climate, and varied ecosystems with the Scandinavian mountain chains in the west, the Arctic tundra in the Kola Peninsula, and the Nenets Area and Novaja Zemlja in the east [18]. The midnight sun may persist up to 24 h a day from May to July. Despite the harsh landscape, it is characterized by boreal forest, thousands of lakes, mountains and fells; this provides the region with an abundance of natural food, made available through hunting, gathering, harvesting and fishing.

As a sparsely populated region with pristine environmental characteristics but rich in natural resources such as oil and gas, it attracts human activities that are related to marine areas or inland mining and mineral activities [19,20]. Human activities have significant impacts on food security in the Barents region; they were reported to be detrimental to the environment of the region [20]. Many towns in Northwest Russia often depend on a single industry that causes pollution that can be a major source of contaminant in their food supply. At the Norwegian–Russian border in Nickel and Zapolyamy, nickel smelters release harmful emissions of sulfur dioxide into the environment [21]. Such emissions can end up in the local food system if not accurately monitored. Dudarev et al. [22] reported on the toxic metal levels in local foods such as fish, berries, mushrooms, and game in the Pechenga district. In Finnish and Swedish Lapland, mining activities are also a cause for concern to the environment as potential risks to food safety [21]. With growing international tourism in the region, there will be more waste generated and it will be important that such waste does not result in environmental pollution. Encouraging industrial symbiosis (where small and medium enterprises find ways to use the waste from one enterprise as raw materials for another at local levels) in the region will help to avoid unnecessary waste and encourage a circular economy. The Finnish Innovation Fund (SITRA) in its current roadmap to a circular economy highlighted industrial symbiosis as a means to tackle the challenges of climate change, depletion of natural resources and to minimize the waste of food resources [23]. Shifting to a circular model for food can have economic, health and environmental benefits in the Barents region.

The Barents region is an area where the impact of climate change is most pronounced, with an average rise in temperature twice that of the global average temperature [24,25]. Digitalization of processes within the food system will help to mitigate the impacts of climate, with generated data. As a result of the warming climate, there will be changes in ocean and atmospheric contaminant transport, which may increase the movement of organohalogens and mercury from lower latitudes to the higher latitudes of the Arctic Barents region [26]. As the water in freshwater lakes gets warmer, tundra ponds and streams are likely to be affected by greater bacterial methylation of mercury and mercury released from thawing permafrost [27]. This will present some challenges for the terrestrial and aquatic food resources in the region. The environmental impacts of climate warming, anthropogenic contaminants and zoonotic disease in subsistence food species and rural drinking water can lead to infections and toxic effects in rural communities [28].

Plant, insect and animal species will be able to expand further north as the climate gets warmer, thus bringing new zoonotic diseases with them. "Higher winter temperatures in the Arctic was noted to likely increase the level of microbes, and consequently, the winter survival of infected animals, raising the risk of hunter and consumer exposure" [29] (p. 3). The warming climate may drive changes in the forage resources of subsistence species, their health, and abundance. Similarly, the expansion of new species northwards has already brought new water-borne diseases such as tularaemia [30], and warmer waters in tundra ponds, estuaries, and nearshore ocean waters now support toxin-producing cyanobacteria and algal blooms [31]. In addition, longer Arctic summers and warmer winters with less sea ice cover will increase the use of northern sea routes by commercial shipping from northern Europe and western Russian ports [25], raising the possibility of new rat-borne infections, such as tick-borne encephalitis [32]. The use of sensors and other digital solutions will lead to a greener means of transport that can prevent such disease outcomes.

Some of the gaps in knowledge for adaptation strategies to better cope with food and water security in the Arctic-Barents region include the availability of better metrics on contaminants and pathogens in the region [29]. It will be important to share data on the microbial and chemical pollutants in the agri-food sector across the region. Therefore, there is a need to ensure food safety and sustainability by utilizing the benefits of digitalization within the food system to accurately measure and predict outcomes from these activities. Efficient and smart manufacturing in the future will need to accurately gather data, with which to predict and inform processing operations. Developing and transforming the food system through digitalization will help in the response to the consequences of these impacts in the Barents region.

3. DEVELOPING THE INFRASTRUCTURE FOR FOOD SYSTEM DIGITALIZATION IN THE BARENTS REGION

Food system digitalization refers to the application of innovative technology to enhance harvesting, processing, distribution and storage operations along the agri-food value chain. The power of mechanization in the early nineteenth century, automation in the 1970s, and the growth of information and the internet in recent decades have had significant impacts on our day-to-day activities, including the food system. Since the current adoption of connected intelligence into the business and social fabrics is happening at an advanced speed, it will change the way we conduct business. The food business operations in the Barents region and its system will not be an exception as it develops novel food products from traditional foods.

The food system has been distorted globally, mainly due to capitalism's drive for increasing profit. Some major worries for the global food system highlighted by United States climate researchers include absurd transnational lengths for supply chains, genetically modified organisms inserted into plant and animal deoxyribonucleic acid (DNA), overuse of pesticides on mono-culture crops, common resources such as water privatized through legislation and patent restrictions on seeds [33]. All components of the food system are challenged to quickly adapt and keep pace with the evolution of technology, trying to evade the disruptive effects of digital Darwinism (a phenomenon where technology and society are evolving faster than businesses can naturally adapt). In doing so, it was anticipated by the World Food System Centre that "the dynamics of the whole food-value chain change, and unforeseen consequences enter the economic landscape through digitalisation" [34]. In responding to these challenges, local food systems need to be revisited to ensure food safety and sustainability in the Barents region. Along the food and beverage supply chain, there are many involved processes, such as workers and touchpoints, which can make it difficult, not only to keep track of food but also to monitor its quality. Quality is of incredible importance in the food industry, as faulty or contaminated foods entering the market can be detrimental to human health [35]. Therefore, food must always remain traceable and safe, and this can be complex to guarantee within the food system with its many working cogs. Digitalization with accurate data on the process inputs and outputs will help to guarantee quality, safety and traceability.

Food system digitalization will consider all the aspects of growing, harvesting, processing, storage and distribution of foods from the region in order to explore their potential in promoting human security. In the Barents region context, food system digitalization can help to transform the local economy by improving the supply of traditional values, increase the yield of natural products and create novel food products with health giving properties. With an improved local economy, the purchasing power of entrepreneurs will increase, leading to an improvement in food and human security in the region. Ultimately, this will result in better food sovereignty across the region as SMEs in the food sector make economic gains.

In an anticipated future scenario, technological advances that are emerging from the Fourth Industrial Revolution (Industry 4.0) will benefit the current food system in becoming greener and sustainable. Industry 4.0 harnesses uncertainties and, in their place, removes assumptions and risky forecasts, enabling a relevant level of actual knowledge and a newfound level of insight. The use of digital technologies, such as precision planting and irrigation techniques, will improve yields organically. "The speed, efficiency, and sustainability of transport will improve radically. Mobile information technologies will improve farmers' understanding of the land they are farming and the markets they are selling to, while also allowing them to communicate with and learn from each other" [36].

The importance of advances in communication, education, and finance to support farmers and their communities cannot be underestimated in the present age. Telecommunication is an important platform in which digitalization can be built and further developed. In order to maximize the benefits of digitalization in the Barents region's food sector, it is important that infrastructure, such as telecommunication networks be developed across the entire Barents region. For instance, in Finland, Norway, and Sweden, the networks are highly developed while in Northwest Russia it needs to improve when digitalization harmonization across the region amongst stakeholders is considered. There is a discrepancy in technological and infrastructural developments in the Barents region. For instance, internet penetration in Finland, Norway, Russia, and Sweden was 92.5%, 98%, 76.4%, and 93.1% respectively [37].

The capacity of the telecommunication network in Northwest Russia remains limited, and the cost of using telecommunications within low-populated northern regions of Russia, as well as that of cross-border services between the Nordic countries and

Northwest Russia, remains extremely high. This is one of the initiatives addressed in the Northern dimension [38]. The Northern dimension is a joint policy initiated in 1999 between four equal partners—the European Union, Russia, Norway, and Iceland—regarding the cross-border and external policies geographically covering Northwest Russia, the Baltic Sea, and the Arctic regions, including the Barents region.

The issues of telecommunication development in the north of Russia are given great attention with the possibility of adopting smart technology with space communication [39]. The task of developing communications appears as one of the strategic documents of the Russian Federation, both at the federal and regional levels [40–42]. The development of telecommunications infrastructure in the Russian Arctic has been an important topic of discussion, both at the state level and in the business environment. Annually, there are many conferences and forums devoted to this issue [43].

In an effort to develop the communication infrastructure in northern Russia, one of the largest projects in the field of telecommunications development is the laying of a fiber-optic backbone along the Northern Sea Route. The project, in different forms, has been discussed since 1999. Currently, the active stakeholders of the project are Russia, Finland, and China [44]. In 2016, the Russian Federation implemented a similar project for the construction of an underwater optical highway linking the largest settlements in the northeast of the Russian Federation (Sakhalin–Magadan–Kamchatka) [45]. Along with the terrestrial telecommunications infrastructure, the satellite communication system is also developing [46,47]. Mobile communications are actively developing as well. Coverage of mobile networks of the third and fourth generations is growing, and already covers the territories inhabited by indigenous peoples [48]. The Nordic industries are also developing a strategy towards the fifth-generation (5G) network. Finland and Sweden are both in the European Union, while Norway is not but aligns strongly with Nordic cooperation for mobile networks. For example, a declaration of intent [Press release, May 2018. https://www.government.se/press-releases/2018/05/new-nordic-cooperation-on-5g/] signed recently states that the Nordic region will be the first interconnected 5G region in the world, and identifies areas in which Nordic cooperation needs to be intensified. The development of communication infrastructures in the Russian North naturally will lead to an increase in internet penetration. In order to reap maximum digital dividends, the economies of a country need to implement the reforms necessary to improve internet access. In addition, they need to focus on the analogue foundations that promote the digital economy, such as skills, institutions, and regulations [49]. It was also suggested that the development of a strategy that will reap digital dividends requires addressing the digital divide by removing the barriers to an internet that is universal, affordable, open, and safe for firms, citizens, and governments [50].

According to Fälström and Jörgensen [51], food system digitalization will need to consider important issues on the Internet of food (IOF). IOF refers to the discussion regarding how digital aspects, technical innovation and new data layers around food will change the global food system. It is related to the complexity of combined food products, their mobility, best practices on quantified food handling, Big Data, hygiene factors in the food sector, environmental control (open reporting sensors), self-configuring sensor nets, edible sensors, and multicast instructions. The authors also suggested the deployment of Internet Protocol Version 6 (IPv6) as a factor for improved food competitiveness, security, better access, identity, and new business models from breaking up structures. IPv6 as a communication tool is able to handle extended space in comparison to IPv4, and will be able to prevent exhaustion that may result from the complexity. In addition, IPv6 provides improved service quality. Automation and modern analytics tools can be deployed to track products and goods from inception to fulfilment. Since the systems in question are designed to track and monitor on their own with little to no input, the user can tap in anywhere along the chain to seek the information he or she needs. The success of the system will largely depend on the exchange of information and knowledge across the region.

One way to encourage knowledge sharing and new partnerships is through an initiative that harmonizes information across the Barents region through data visualization. Data visualization is the modern way of disseminating information that can be used more frequently in the Barents region. The data visualization website contains a broad variety of parameters, such as data on population, unemployment, health, and education that can be visualized on a map (available at https://barents.no/en/focus-areas/european-border-dialogues).

4. TRANSFORMING THE FOOD SYSTEM THROUGH DIGITALIZATION TO PROMOTE FOOD AND NUTRITION SECURITY

Agriculture, rural livelihoods, sustainable management of natural resources and food security are connected within the development and climate change challenges of the twenty-first century [24]. The impact of food processing on the climate is well established and it will be necessary that value addition to traditional foods in the region be well managed throughout the value chain. Food processing contributes to climate change, eutrophication (the process by which there is a gradual increase in the concentration of phosphorus, nitrogen, and other plant nutrients in an aging aquatic ecosystem such as a lake), acid rain, and the depletion of biodiversity. Each of these parameters can be monitored with digital accuracy by sensors. The efficient processing of traditional and local foods incorporating digitalization in the value chain will be important for keeping the global temperature at 1.5 to 2°C above the pre-industrial level. Several researchers have shown that food processing, which often involves water and energy expenditures, needs to be climate smart [52–54]. Since the Barents region and other parts of the world at higher latitudes are witnessing the effects of climate change to a greater degree than elsewhere, climate-smart agriculture and digitalization of the food system can also be part of the solution.

Industry 4.0 is transforming products, the operations of companies, and how their production is managed. This is a giant leap for manufacturing innovation that employs smart devices that take control of machines on the shop floor, communicating autonomously "device-to-device" to manage manufacturing operations and their distribution. "The entire manufacturing value chain is transformed by industry 4.0 and they can be monitored from concept to completion and beyond" [55]. Small and medium-based food enterprises can benefit from Industry 4.0 especially when industrial symbiosis is adopted as a way of promoting sustainability in the region. The management of a business enterprise is now able to access important information in real-time as they are able to monitor the productivity and efficiency of both employees and machinery. Another critical benefit this new technology delivers is a consistent feedback flow between companies and their customers, in which products could be improved or highly influenced by the end-user, transforming how products are designed and produced [56].

"The fusion of 'Big data', the 'Internet of Things', and advanced analytics is providing manufacturers with unprecedented insights into manufacturing performance, customer behaviour, and new product development" [57]. Other enabling tools, such as cloud computing and cyber physical systems, have also been introduced as digital enablers in manufacturing. For example, cyber physical systems research can increase food consumption efficiency and the overall food production capability through precision agriculture, intelligent water management, and more efficient food distribution [58]. Automation, intelligence, and collaborations are also relevant with particular reference to smart manufacturing, smart products/services, and smart cities.

Blockchain technology creates a shared, distributed ledger of transactions over a decentralized peer-to-peer network. As an emerging technology, its suitability needs to be examined against the use cases requirements [59,60]. In the food sector, it makes it possible to track the source of various food items starting from the farm to supermarket shelves. By creating a blockchain network, contaminated products can be traced to their sources faster. Some of the uses of blockchain that will improve business include reducing carbon footprints, the security of the Internet of Things, contract fraud reduction, secure real-time payment, supply chain efficiency, clarity in business agreements and transactions, increased traceability, and improved customer experience. The use of blockchain in food processing will ensure safety, since food ingredients can be easily tracked, monitored and reported on, thereby building trust among consumers [61].

Blockchain, as a distributed, shared ledger for recording transactions, will revolutionize how the food industry can optimize the trading and shipping of food, trace contaminated items, and reduce fraud and waste. Blockchain holds incredible promise in delivering the transparency needed to help promote food safety across the whole supply chain. Within the food supply chain, the different actors, i.e., growers, suppliers, processors, distributors, retailers, regulators, and consumers, can access information about the origin and state of food [62]. This will help to resolve issues of authenticity and food fraud that may result from geographic origin, production method, processing technology, ingredients composition, etc., [63].

In practical terms, a framework can be constructed that can be used for structuring the discussion around the Internet and food, thereby furthering the discussion on how to facilitate openness and innovation in the field of food, with the goal to feed the planet in a healthy and sustainable way. The increasing activity of info-communications in virtual space leads to an avalanche-like growth for information stored and circulating on the Internet. This opens up new opportunities for increasing the security of the population of the territories in all aspects, including food security. Possible ways of realizing this potential based on modern information technologies are diverse. They include monitoring the actual diet of the inhabitants of the territories, monitoring and predicting consumer demand, the formation of new chains of production and food supplies.

The overlapping relationship that exists between the consumers, the food business, and the authorities is shown in Figure 1. The figure shows how the main actors contribute to the digitalization of the food system in different ways, and the benefits of digitalization in the food system. General feedback shown in the picture may reflect in different layers, processes and effects.

The benefits of digitalization seem to appear in two general ways. The first, "extensive" one consists of an improvement of almost all traditional processes within the food system by using information and communications technology (ICT). The second, "intensive" way is connected to a restructuring of processes within the system. Along with the revolutionary changes of Industry 4.0, one particular restructuring is a shift from hierarchical organizations to horizontal ones, based on peer-to-peer (P2P) interactions. In this article, the authors use P2P as a common name of architectures implementing horizontal instead of hierarchical structures of interactions, i.e., 'peer-to-peer. This kind of interaction appears in the food system in different layers, from IOF to informal customer and producer networks within social networking services (SNS) and to the network of food businesses. The most valuable advantages of any P2P system are its scalability, flexibility, and reliability [64]. Transformations from hierarchical to horizontal structures will potentially improve the sustainability of the regional food system, making it more adequate to the actual demands of the actors and expanding it to trans-regional and cross-border contexts.

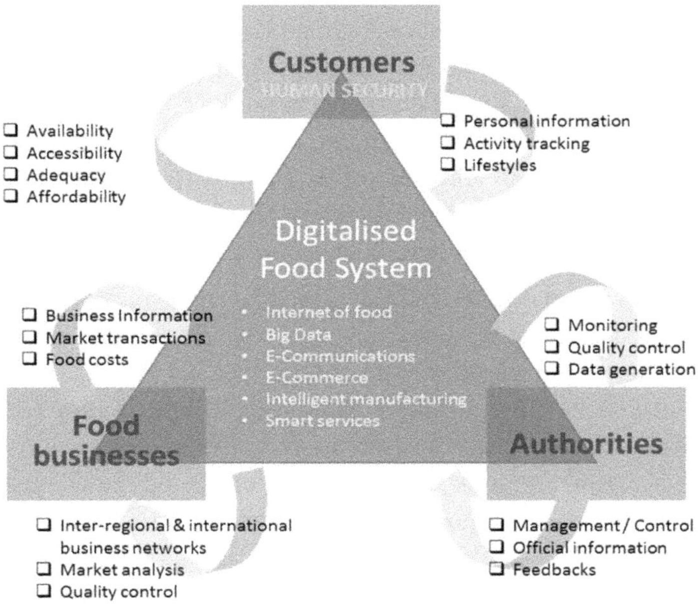

Figure 1. An overlapping relationship within a digitalized food system.

Due to the harsh climate and poor development of transport infrastructure in parts of the Barents region, the development of horizontal relations will be especially important for the region. Regional food markets in the region are specific, characterized by small volume, low density and unevenness of the population as the main consumer of products [65]. The consequence of these features results in vulnerability of the food system in the Barents region, due to dependence on unreliable transportation, especially for the northern delivery dependent territories [66]; the higher cost of products, due to transportation costs, result in an imbalance in the structure of food consumption by the population [67].

Strengthening horizontal interaction within the region will reduce these problems due to more effective use of local opportunities for food supply amongst the population. Digitalization, in a broad sense, covers the sphere of interpersonal communications, which will provide a more efficient exchange of information at all levels—from state to personal. This will create the right conditions and lead to the emergence of alternative channels for providing the population of the region with better access to food, including the crowd principle basis, which uses the resources of the population itself. An example of the realization of this potential at the interpersonal level is of a number of groups in the "VKontakte" social network, which provide the ordering and delivery of products from Scandinavian countries to the cities of northwest Russia [68]. Such means are especially important within the human security perspective, since they provide a more efficient personal satisfaction at individual levels, including specific requirements for food.

The opportunities provided by food system digitalization can be considered in terms of what is currently available and what will be available in the future, such as the Internet of food (IOF) mentioned earlier. These opportunities will create new possibilities to monitor a regional food system to inform more contemporary and adequate decisions for a better secure regional food system.

Digitalization results in the widespread use of ICT that generates and accumulates huge amounts of data that opens the way to obtain distinct raw data, in addition to official statistics, to characterize the peculiar aspects of the food system at the regional level. For example, in Russia, official statistics mostly focus on supporting the tasks of the federal level administration, while the tasks and instruments of the regional administration are essentially different (customs tariffs, compensation fees, excises, sales taxes, quotas, etc.). These statistical records are mainly economic indicators to characterize food security amongst the population as a whole at the federal level. However, specific indicators that focus on local and regional levels are missing. Such an approach makes the Russian food security monitoring system state security focused rather than being human security focused at local and regional levels. For instance, the peculiar requirements for the diet of indigenous people from the northern regions based on cultural preferences are not accounted for in the monitoring system. Therefore, additional data is necessary for effective monitoring of the food system of the Barents region taking into account its peculiarities, since "food security is not only about reliable access to nutritious food. It is also linked to climate change, wildlife management, pollution, economic vulnerability and cultural security" [69]. Digitalization provides access to data that concerns several actors in the food system, including persons and households. Beneficial information can be gathered from this data through text analysis in general and opinion-, knowledge- or data-mining technologies, as well as demand analysis and personalization techniques.

An important task that provides a technological basis for creating information technologies to support food security is the formation and maintenance of a knowledge base (KB) that characterizes the overlapping structure of a digitalized food system. Ideally, such a base should describe all components of the food system and their relations to provide the possibility of automated reasoning aimed at supporting decision making in food security and risk management. The creation of a KB is a key task for the potential use of blockchains, the IOF, and other technologies. The KB provides the necessary common conceptual basis for any intelligent distributed system. An essential feature of the knowledge base, which determines the requirements of the technologies for its implementation, is the high dynamism of its content. In our modern world, new products, ingredients, trademarks, supply chains appear and disappear very quickly over time. In such conditions it is extremely difficult to create and maintain the relevance of the knowledge base directly, relying only on the knowledge of experts. This increases the demand for intelligent information systems with feedback that implement the principle of "user as an expert" [70].

Another opportunity for digitalization in transforming the food system is that widely available e-communications will provide a basis for peer-to-peer networking within food systems that will include customer-to-customer, business-to-business, and business-to-customer (C2C, B2B, and B2C, networking respectively). One modern example of such a P2P network that partly cover the Barents region is the REKO system, which implements a new model of a sustainable marketing channel based on SNS interactions. The system unites small food producers (farmers) and consumers and "offer consumers a way of ordering products directly from the producer, without the need for middlemen" [71]. In addition, the system provides the production of food products in volumes and in assortment, taking into account the actual needs of consumers. Such a model of food producing and marketing makes the food system more diversified and resistant to disruptions from individual components including transportation. The example of REKO demonstrates the potential of digital modern ICT in the creation of sustainable distributed food systems for providing the population of the Barents region with quality nutritious food. Technically, there are no obstacles in using such technologies to create similar cross-border networks, including for larger businesses. However, this potential is not used due to organizational and political problems, such as sanction restrictions and differences within food systems regulations in different countries.

5. SUSTAINABLE FOOD SYSTEM AND ICT

The food system strongly relates to many sustainability challenges such as climate change, biodiversity loss, water scarcity, and food insecurity [72,73]. There is a need to deliver sustainable and healthy diets to global populations by responding to the challenges of climate change, rising populations, and decreasing crop yields [15]. In respect of this global challenge, the Barents region specifically can offer a good example by which digitalization in the food system can help to increase yield, adding value to promoting food and human security. This necessitates digital technology to push beyond traditional agricultural practices and be deeply embedded into the food system [74]. For example, in the Barents region, food business operators in the region can add value to carrots, turnips, cloudberries, bilberries, lingonberries, cranberries and other traditional foods as described under the introductory section. The berries are made into juices, jams, sauces, and liqueurs; natural herbs extracts are also utilized. Finnish Lapland angelica herb (Angelica archangelica) is used as a flavouring for pies and ice cream [6]. These food business operators within their local communities are small and medium sized, they are often faced with challenges on novel food requirements and trade restrictions by the European Union [75]. In the Norwegian aquaculture, the biggest challenges are sea lice, diseases, access to sufficient areas and adequate feed resources, pollution in the form of feed spills and faeces from the fish [76]. The challenges faced by the Norwegian aquaculture industry can benefit from digitalization by providing sensors to monitor biomarkers that are associated with these pollutants.

In the Barents region, many communities have low population density; data can be readily collected for future cross-border collaboration within a digitalized food system. The value addition to traditional foods can be promoted with innovative digital technology that promotes environmentally-friendly food products. Community-based initiatives in the region support food security and help to promote sustainability within the communities. For instance, the Norwegian Institute of Bioeconomy Research (NIBIO), in collaboration with community stakeholders in Troms county, aims to market Arctic foods with the goal of highlighting their unique arctic quality, natural production, and lack of pesticides in the food products [77]. Swede and carrots grown at low temperature result in a fresher and better, sweeter taste. Similarly, the trial marketing of midnight sun-grown, oval-shaped almond potatoes was a success in the Finnish Lapland market. The Finnish Lapland potato has been designated as a brand under the EU name protection "Protected Designation of Origin" [78]. The unique quality of these products can serve as a marketing tool that will benefit from e-marketing and digital technology that can prove their authenticity.

The existing digital and communication infrastructure in the region will be necessary for stimulating the drive towards food system digitalization in the future. It will require cross-border collaborative initiatives to share best practices and access the latest trends to boost innovative entrepreneurship in the food sector. This will require that the digital infrastructure be up to speed in ensuring that information gathered can be easily shared across borders in rural communities. In a study related to innovation and entrepreneurship conducted at the Imperial College, London, the authors emphasized that information and communications technology (ICT) is required to successfully combine secure and sustainable food systems [74]. Digital

ICT was used to help coordinate the activities of multiple suppliers in dynamic strategic networks to coordinate the local delivery of food through the application of innovative digital technologies that take into account the interactions between food and agriculture systems, with broader industrial systems, as illustrated in Figure 2. This was achieved by exploiting the timely delivery of information and expertise using mobile devices in rural areas.

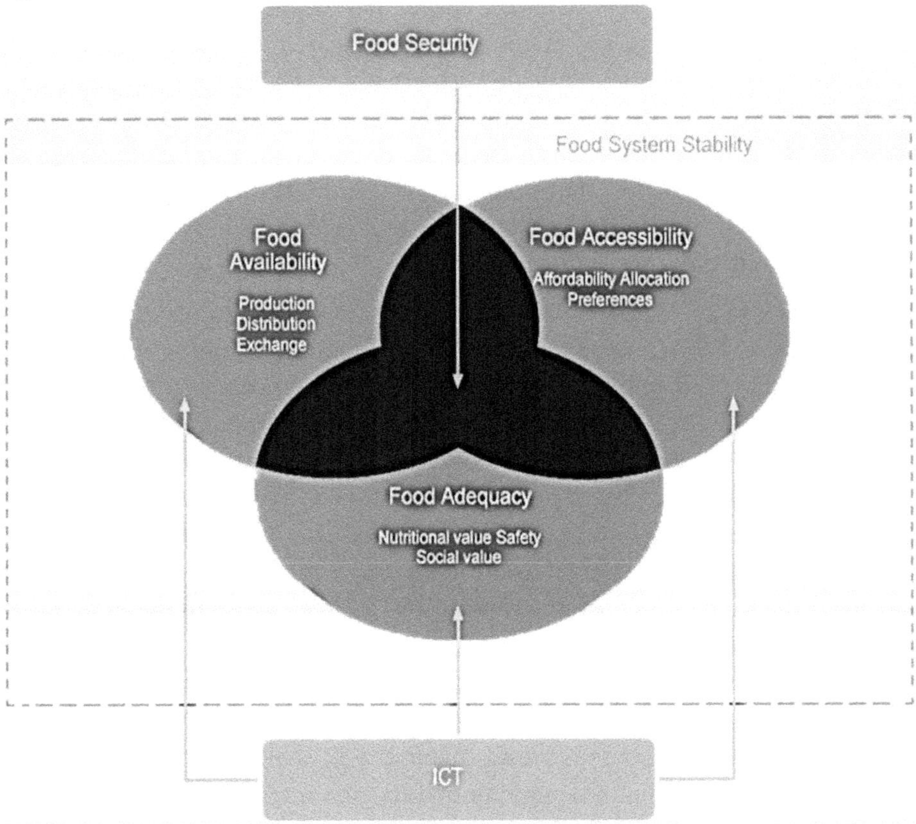

Figure 2. Sustainable food system and information and communications technology (ICT) (Adapted from Berti and Mulligan, 2015).

The food system can be transformed based on the inputs of digitalization on how we produce, process, and integrate information, which will be critical to innovation. Digitalization in food and agriculture will offer new opportunities through the ubiquitous availability of highly interconnected and data-intensive computational technologies as part of the so-called Fourth Industrial Revolution [4]. An interesting aspect of Industry 4.0 is the notion of individualized 'small batch' product offerings. There are interests in data and data analytics as a means to improve production. Small and medium-sized entrepreneurs, as important drivers of economic activities in the Barents region, will benefit from Industry 4.0. It can be applied to all aspects of dairy production, processing of berries, their storage, packaging and distribution in the value chain systems. It also reflects a shift from generalized management of farm resources to highly optimized, individualized, real time, hyper-connected, and data-driven management. The desired outcomes of digital agriculture are more productive, profitable, and sustainable systems. Digital agriculture will leverage the smart use of data and communication to achieve system optimization. The tools that enable digital agriculture are multiple and varied, and include cross-cutting technologies such as computational decision and analytics tools, the cloud, sensors, robots, and digital communication tools. Factories and plants will connect machines and production systems via the internet, allowing information to be exchanged and triggering actions that can enable each entity to control the other independently. Easy data exchange and access to that information can provide 3D printing of foods with more customization possibilities for manufacturers and their customers. As production becomes digitalized, the socio-economic drawbacks can be minimized through new trainings for jobs that are linked to digitalization.

The Finnish Arctic research has a long-term strategy that aims to strengthen Arctic cooperation, as well as new kinds of partnerships, including public and private, in particular by strengthening the business environment and networking of actors in the Finnish Barents region, i.e., Lapland, Northern Ostrobothnia, and Kainuu [79]. The strategy strives to ensure sustainability in line with the Sustainable Development Goals (in particular goals 2, 9 and 12) by 2030. One of the themes at the Natural Resources Institute Finland i.e., *Luonnonvarakeskus* in Finnish (LUKE) is the 'Innovative Food System' that supports a sustainable, profitable, and innovative food chain at all stages. Digitalization will help as a tool to support innovation across the food value chain. The overall aim of the innovative food system is to process wholesome, sustainable food products and support the rotating economy in the food system by utilizing digital and intelligent technology in the Barents region of Finland [76]. The main areas of this theme on innovative primary production systems, circular economy, smart agriculture, healthy and sustainable food and feed, consumers and markets will benefit from digitalization that can be shared through cross-border collaboration in the region.

6. FUTURE IMPLICATIONS FOR THE BARENTS REGION

The previous three sections (Section 3. Developing the infrastructure for food system digitalization, Section 4. Transforming the food system through digitalization, and Section 5. Sustainable food system and ICT) will be relevant for a systemic change in the food system at a regional level; they will play important roles in shaping the future. As shown in Figure 1, a digitalized food system will help to promote both food security and food sovereignty by involving the customers or consumers, food businesses, and the authorities in a digitalized food system. Food sovereignty empowers local and rural communities in value addition to their food crops. It is defined by the World Forum on Food Sovereignty (WFFS) as the ability and the right of people to define their own policies and strategies for the sustainable production, distribution and consumption of food that guarantee the right to food for the entire population [80]. Currently, in Finnish Lapland, there are initiatives such as the "Arctic Bioeconomy, Smart specialisation" that encourage industrial symbiosis with digital solutions. In Loue near Tervola in Finnish Lapland, there are processing facilities for meat, ice-cream, biogas and a cattle barn with robots for milking. These facilities are shared within the same premises with a vocational college with active collaboration with the Natural Resources Institute Finland (LUKE).

There are also 'Facebook', and 'Whatsapp' groups that encourage virtual meetings between local farmers and consumers. Similar practices exist in other communities of the Barents region. In the near future, as the interest in where food comes from grows, and how it has produced becomes more relevant, it will encourage the development of artisan foods and food sovereignty in the Barents region. Food sovereignty puts emphasis on the promotion of small and medium-sized production where respect for the inhabitants' own cultures as well as the diversity of peasant, fishing and indigenous forms of agricultural production, marketing and management of rural areas play a fundamental role [80]. The unique food crops, herbs, natural products and other indigenous foods that are part of the food culture in the Barents region will also contribute to biodiversity when there is value addition with digital solutions. An integrated farming system that employs sensors to increase yield, fertility and help in the management of pests and diseases will be important for future collaboration in the region. As food products move across border in the Barents region, the role of digitalized packages in e-commerce and e-communication will be very important. Digitalized packages with labels with quick response (QR) codes can provide important information on the nutritional quality of the ingredients, ethical standards, and the possibilities to interact with customers. In addition, they can be useful in product tracking, item identification, time tracking, and general marketing. We developed a framework for knowledge base creation as shown in Figure 3 below.

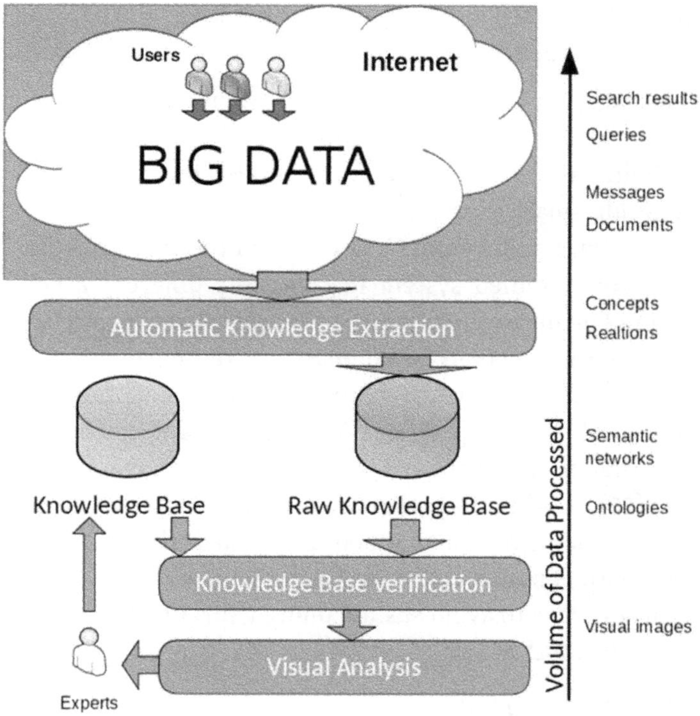

Figure 3. Information technology for knowledge base creation.

We developed a framework for knowledge base creation as shown in Figure 3. This an example of how customers as users will be able to tap into available and documented information of foods and food ingredients. It will collate all information gathered by experts to create the knowledge base within a food system digitalization that can be applied in the Barents region. As there are different languages in the region, there is a need to have a common understanding across the region. In addition, increasing amounts of accumulated information require the use of automated methods for extracting knowledge from raw and inconsistent data, represented in particular by heterogeneous textual sources. Generally, in this case, the task of automated processing should be to summarize the information: a gradual decrease for data to process while simultaneously increasing their information impressiveness. Visual images provide the greatest information impressiveness and at the same time offers the highest speed of information processing by humans. However, the tradeoff for efficiency becomes a reduction in processing accuracy due to the ambiguity of interpreting visual images. A compromise variant of a simultaneously compact, rich information and unambiguously interpreted representation of knowledge is the formalized concept systems presented in the form of an ontology or in the form of another logical system that determines the semantics of a domain. Interactive visual analysis methods can be used in the region which is multi-lingual. This will combine the formal rigor of logical systems with the efficiency of processing visual images.

Digital solutions are important aspects of the Smart specialization strategy in the European Union (EU). The EU recognizes the importance of specific regional knowledge, technical assets and critical mass towards the diversification of regional economies alongside smart specialization [81]. However, there are also challenges in the exchange of data and communication when there is a need to link different systems together in a unified system to cover the different aspects of the food sector in a region. For future collaborations in digitalization of the food system in the Barents region, there should be emphasis on the harmonization of standards through the national and regional governments. The ICT solutions, when developed in isolation from realities and practices of producers and consumers, run the risk of hampering rather than advancing possibilities for sustainability transitions in the food system [82]. Furthermore, there is a need to adopt a holistic approach that takes into consideration links in the food chain; such that ICT solutions consider production practices, communication in food chain, and consumer behavior [83]. This is important in order to avoid sustainability gains in one part of the food chain offsetting changes in another part.

7. CONCLUDING REMARKS

There is an urgent need to have a systemic change in our food system that encourages food security and sovereignty. The basis of the technology that underpins digitalization as described in this paper, when clearly aligned to the food industry, can revolutionize the industry. Digitalization across the food system will positively affect food security and safety, thereby ensuring human security in the Barents region. Such a disruption, coupled with the development of other infrastructures particularly in Northwest Russia, will help to promote cross-border collaboration that will be of benefit to the Barents region. Cross-border collaboration at a regional level that foster sustainability, transparency, and an efficient management of resources will improve with digitalization. This review will be useful in conceptualizing a framework for food system digitalization in the region. The knowledge base can be enhanced through the gathering of metrics that can lead to a more integrated approach for the region in future collaboration on food security and safety. The authors have reviewed the opportunities of food system digitalization, challenges of telecommunication, and the impact of human activities in the Barents region. It is envisaged that policy makers and stakeholders in the region will be better informed to support digitalization in the food sector across the Barents region. The socio-economic challenges associated with digitalization, as with any technological development such as digital inequality and threats caused by the complexity and vulnerability of ICT-based infrastructures, the governance of generated data, cybersecurity, were not fully discussed in this paper. We hope that the food system in the Barents region will embrace digitalization in the future, support and targeted trainings to rural communities will improve the adoption rate of digitalization in the region.

References

1. Allen, R.C. *The British Industrial Revolution in Global Perspective: How Commerce Rather than Science Caused the Industrial Revolution and Modern Economic Growth;* Mimeo: Oxford, UK, 2006.
2. Mokyr, J. *The Second Industrial Revolution, 1870–1914*; Castronovo, V., Ed.; Laterza Publishing: Italy, Rome, 1999; pp. 219–245.
3. Rifkin, J. The Third Industrial Revolution: How the Internet, Green Electricity, and 3-d Printing are Ushering in a Sustainable Era of Distributed Capitalism. *World Financ. Rev.* **2012**, 1, 4052–4057.
4. Schwab, K. *The Fourth Industrial Revolution*; World Economic Forum: Geneva, Switzerland, 2016; 184p.
5. AMAP. Adaptation Actions for a Changing Arctic. Barents Area: Overview Report. Arctic Monitoring and Assessment Programme (AMAP). Available online: https://www.amap.no/documents/download/2885/inline (accessed on 25 July 2019).
6. Hossain, K.; Raheem, D.; Cormier, S. *Food Security Governance in the Arctic-Barents Region*; Springer: Cham, Switzerland, 2018.
7. Nilsson, L.M. *Sami Lifestyle and Health: Epidemiological Studies from Northern Sweden;* Umeå University: Umeå, Sweden, 2012.
8. Muller-Wille, L.; Granberg, L.; Helander, M.; Heikkila, L.; Lansman, A.S.; Tuiku, T.; Berrouard, D. Community viability and well-being in northernmost Europe: Social change and cultural encounters, sustainable development and food security in Finland's North. *Int. J. Bus. Glob.* **2008**, 2, 331–353. [CrossRef]
9. Frenk, J.; Gómez-Dantés, O.; Moon, S. From sovereignty to solidarity: A renewed concept of global health for an era of complex interdependence. *Lancet* **2014**, *383*, 94–97. [CrossRef]
10. FAO. *Food and Agriculture Organization and United Nations Human Security Unit, Human Security & Food Security: Hunger, Food Insecurity, and Malnutrition;* FAO: Rome, Italy, 2016.
11. De Ridder, D.; Kroese, F.; Evers, C.; Adriaanse, M.; Gillebaart, M. Healthy diet:

Health impact, prevalence, correlates, and interventions. *Psychol. Health* **2017**, 32, 907–941. [CrossRef] [PubMed]

12. Kaiser, M.L.; Dionne, J.; Carr, J.K. Predictors of Diet-Related Health Outcomes in Food-Secure and Food-Insecure Communities. *Soc. Work Public Health* **2019**, 34, 1–16. [CrossRef] [PubMed]
13. Raheem, D. Food and Nutrition Security as a Measure of Resilience in the Barents Region. *Urban Sci.* **2018**, 2, 72. [CrossRef]
14. Nilsson, L.M.; Destonuni, G.; Berner, J.; Dudarev, A.; Mulvad, G.; Odland, J.O.; Rautio, A.; Tikhonov, C.; Evengård, B. A call for urgent monitoring of food and water security based on relevant indicators for the Arctic. *AMBIO J. Hum. Environ.* **2013**, 42, 816–822. [CrossRef] [PubMed]
15. Godfray, H.C.J.; Beddington, J.R.; Crute, I.R.; Haddad, L.; Lawrence, D.; Muir, J.F.; Toulmin, C. Food security: the challenge of feeding 9 billion people. *Science* **2010**, 327, 812–818. [CrossRef]
16. Vermeulen, S.J.; Campbell, B.M.; Ingram, J.S. Climate change and food systems. *Annu. Rev. Environ. Resour.* **2012**, 37, 195–222. [CrossRef]
17. Wheeler, T.; Von Braun, J. Climate change impacts on global food security. *Science* **2013**, 341, 508–513. [CrossRef]
18. BEAC, Barents Euro-Arctic Council. History. 2016. Available online: http://www.beac.st/en/About/Barents- region/history (accessed on 10 February 2019).
19. Salminen, M.; Hossain, K. Digitalisation and human security dimensions in cybersecurity: an appraisal for the European High North. *Polar Rec.* **2018**, *54*, 108–118. [CrossRef]
20. Cormier, S.; Raheem, D. Food security in the Barents region. In Society, *Environment and Human Security in the Arctic Barents Region, Routledge Explorations in Environmental Studies;* Hossain, K., Cambou, D., Eds.; Routledge: Abingdon, UK, 2018.
21. Ojala, T. Challenges in sustainable use of natural resources of the North. In *The Finnish Committee for European Security (STETE) Nordic Forum for Security Policy Final Report;* **2014;** pp. 45–47. Available online: http://www.widersecurity.fi/uploads/1/3/3/8/13383775/nordic_forum_2014.pdf (accessed on 18 August 2018).
22. Dudarev, A.A.; Dushkina, E.V.; Sladkova, Y.N.; Chupahin, V.S.; Lukichova, L.A. Evaluating health risk caused by exposure to metals in local foods and drinkable water in Pechenga district of Murmansk region. *Meditsina Truda I Promyshlennaia Ekologiia* **2015**, 11, 25–33.

23. SITRA. Finnish Innovation Fund. Finnish Road Map to a Circular Economy 2.0 (2016–2025). 2019. Available online: https://www.sitra.fi/en/articles/improving-resource-efficiency-industrial-symbiosis- opportunities-smes/ (accessed on 2 May 2019).

24. Tubiello, F. Climate Change Adaptation and Mitigation: Challenges and Opportunities in the Food Sector. In Proceedings of the High-level conference on world food security: the challenges of climate change and bioenergy, Rome, Italy, 3–5 June 2008.

25. IPCC. *Climate Change 2013: The Physical Science Basis. Contribution of Working Group I to the Fifth Assessment Report of the Intergovernmental Panel on Climate Change;* Stocker, T.F.D., Qin, G.K., Plattner, M., Tignor, S.K., Allen, J., Boschung, A., Nauels, Y., Xia, V.B., Midgley, P.M., Eds.; Cambridge University Press: Cambridge, UK; New York, NY, USA, 2013; 1535p. [CrossRef]

26. Kallenborn, R.; Halsall, C.; Dellong, M.; Carlsson, P. The influence of global distribution on fate processes and distribution of anthropogenic persistent organic pollutants. *J. Environ. Monit.* **2012**, 14, 2854–2869. [CrossRef]

27. Stern, G.A.; Robie, W.M.; Outridge, P.M.; Wilson, S.; Chetelat, J.; Cole, A.; Hintelmann, H.; Loseto, L.L.; Steffen, A.; Wang, F.; et al. How does climate change influence arctic mercury? Sci. *Total Environ.* **2012**, *414*, 22–42. [CrossRef]

28. Fisk, A.; DeWit, C.; Wayland, Z.; Kuzyk, N.; Burgess, R.; Letcher, R.; Braune, B.; Norstrom, R.; Blum, S.P.; Sandau, C.; et al. An assessment of the toxicological significance of anthropogenic contaminants in Canadian Arctic wildlife. *Sci. Total Environ.* **2005**, 351, 53–93. [CrossRef]

29. Berner, J.; Brubaker, M.; Revitch, B.; Kreummel, E.; Tcheripanoff, M.; Bell, J. Adaptation in Arctic circumpolar communities: Food and water security in a changing climate. *Int. J. Circumpolar Health* **2016**, 75, 33820. [CrossRef]

30. Rydén, P.; Sjöstedt, A.; Johansson, A. Effects of climate change on tularemia activity in Sweden. *Glob. Health Action* **2009**, 2, 2063. [CrossRef]

31. Carey, C.C.; Ibelings, B.W.; Hoffmann, E.P.; Hamilton, D.P.; Brookes, J.D. Eco-physiological adaptations that favour freshwater cyanobacteria in a changing climate. *Water Res.* **2012**, 46, 1394–1407. [CrossRef]

32. Revich, B.; Tokarevich, N.; Parkinson, A. Climate change and zoonotic disease in the Russian Arctic. *Int. J. Circumpolar Health* **2012**, *71*, 18792. [CrossRef]

33. Holt-Giméénez, E.; Wang, Y. Reform or Transformation? The Pivotal Role of Food Justice in the U.S. Food Movement. Rac*e Ethn. Multidiscip. Glob. Contexts* **2011**, 1, 83–102. [CrossRef]

34. WFS. World Food System Centre ETH Studio AgroFood, Zurich, Switzerland. 2018. Available online: http://www.worldfoodsystem.ethz.ch/research/flagship-projects/eth-studio-agrofood.html (accessed on 12 December 2018).

35. Di Renzo, L.; Colica, C.; Carraro, A.; Goga, B.C.; Marsella, L.T.; Botta, R.; Sarlo, F. Food safety and nutritional quality for the prevention of non-communicable diseases: the Nutrient, hazard Analysis and Critical Control Point process (NACCP). J. Transl. Med. 2015, 13, 128. [CrossRef]

36. Struebi, P. How the Fourth Industrial Revolution Can Radically Improve Our Food Supply Chain. Huffington Post. New York, NY, USA, 2016. Available online: https://www.huffingtonpost.com/patrick-struebi/how-the- fourth-industrial-revolution-can-radically-improve-our-food-supply-chain_b_9011556.html?guccounter= 1 (accessed on 12 November 2018).

37. Internet Live Stats. Number of Internet Users Worldwide. 2016. Available online: http://www. Internetlivestats.com/Internet-users/ (accessed on 28 February 2018).

38. Muilu, T. A Northern dimension for the European Union: Background and proposals. Revival. In *Globalization and Marginality in Geographical Space (2001): Political, Economic and Social Issues of Development at the Dawn of New Millennium;* Taylor & Francis: Abingdon, UK, 2017; pp. 79–89.

39. Lagunov, A.; Terekhin, V. Modelling the Barents territory coverage area of satellite KA-SAT. In *Smart Technologies, Proceedings of the IEEE EUROCON 2017-17th International Conference on Smart Technologies, Ohrid, Macedonia, 6–8 July 2017*; Curran Associates, Inc.: New York, NY, USA, 2017; pp. 588–593.

40. Russian Arctic Policy. Basis of State Policy of the Russian Federation in the Arctic for the Period up to 2020 and the Further Perspective. 2018. Available online: http://static.government.ru/media/files/ A4qP6brLNJ175I40U0K46$\times$4SsKRHGfUO.pdf (accessed on 27 February 2018).

41. Russian Arctic Strategy. Strategy for development of the Arctic Zone of the Russian Federation and Ensuring National Security for the Period up to 2020. 2018. Available online: fromhttp://static.government.ru/media/files/2RpSA3sctElhAGn4RN9dHrtzk0A3wZm8.pdf (accessed on 20 March 2018).

42. Murmansk Strategy. Strategy of Social and Economic Development of the Murmansk Region until 2020 and for the Period up to 2025. 2018. Available online: http://minec.gov-murman.ru/activities/strat_plan/sub02/ (accessed on 17 February 2018).

43. Communications in the North. The 6th Conference "Communication in the Russian North". Official Conference Website. 2018. Available online: http://www.xn--80adblbaj2c5ace3kob.xn--p1ai/ (accessed on 27 February 2018).

44. Commission for Arctic Development. The Russian Federation and Finland Discussed the Project of Laying a Fibre-Optic Communication Line along the Northern Sea Route. 18 October 2017. Available online: https://arctic.gov.ru/News/1959c8dc-31b4-e711-80d7-00155d006312?nodeId=4f828d76- 8f58-e511-8259-e82aea5c46bb&page=1&pageSize=10. (accessed on 27 February 2018).

45. Rostelecom. Underwater Fiber-Optic Line of Kamchatka-Sakhalin–Magadan. 2018. Available online: https://www.rostelecom.ru/projects/FarEast_FOCL/ (accessed on 27 February 2018).

46. D-Russia. RTComm Built a Satellite Station in One of the Northernmost Cities in Russia. 2016. Available online: http://d-russia.ru/rtkomm-postroil-sputnikovuyu-stanciyu-v-odnom-iz-samyx-severnyx- gorodov-rossii.html (accessed on 7 March 2018).

47. Electrosvyaz. Communication in Extreme Conditions. Interview with Evgeny Buydinov, Deputy Director General for Innovative Development of the Federal State Unitary Enterprise "Space Communication" (GP KS) Electrosvyaz. 2016. Available online: http://www.elsv.ru/svyaz-v-ekstremalnyh-usloviyah/ (accessed on 27 February 2018).

48. Rostelecom. Coverage Maps for MTS, Megafon, Yota, Tele2, Rostelecom, SkyLink LTE. 3G, 4G, 2G and Cellular. 2018. Available online: https://4g-faq.ru/karty-pokrytiya/ (accessed on 23 February 2018).

49. World Bank. *World Development Report* **2016**: Digital Dividends; World Bank Group: Washington, DC, USA, 2016. [CrossRef]

50. Kelly, T.; Liaplina, A.; Tan, S.W.; Winkler, H. *Reaping Digital Dividends: Leveraging the Internet for Development in Europe and Central Asia*; World Bank Publications: Washington, DC, USA, 2017.

51. Fälström, P.; Jörgensen, J. Internet of Food. When Food Gets Networked, Life Changes. 2015. Available online: https://stupid.domain.name/stuff/IoF_Framework-201505.pdf (accessed on 16 March 2018).

52. Dinesh, D.; Frid-Nielsen, S.; Norman, J.; Mutamba, M.; Loboguerrero Rodriguez, A.M.; Campbell, B. Is Climate-Smart Agriculture effective? A Review of Selected Cases. CCAFS Working Paper No. 129. Copenhagen. CGIAR Research Program on Climate Change, Agriculture and Food Security (CCAFS): Copenhagen, Denmark, 2015. Available online: www.ccafs.cgiar.org (accessed on 14 April 2019).

53. Neate, P.J. Climate-Smart Agriculture Success Stories from Farming Communities around the World. CGIAR Research Program on Climate Change, Agriculture and Food Security (CCAFS) and the Technical Centre for Agricultural and Rural Cooperation (CTA). 2013. Available online: http://hdl.handle.net/10568/34042 (accessed on 20 May 2018).

54. Whitfield, S.; Challinor, A.J.; Rees, R.M. Frontiers in Climate Smart Food Systems: Outlining the Research Space. Front. Sustain. *Food Syst.* **2018**, 2, 2. [CrossRef]

55. Engineering News. 2017. Available online: http://www.engineeringnews.co.za/print-version/industry-40-and-iot-transforming-the-manufacturing-industry-2017-06-01 (accessed on 10 October 2018).

56. Design and Manufacturing News. Industry 4.0 and IoT Central to Manufacturing Transformation. 2017. Available online: http://www.bizcommunity.com/Article/196/399/162851.html (accessed on 12 October 2018).

57. GAN. Global Africa Network. Industry 4.0 and IoT: Transforming the Manufacturing Industry. 2017. Available online: https://www.globalafricanetwork.com/2017/06/01/company-news/industry-4-0-and-iot- transforming-the-manufacturing-industry/ (accessed on 14 November 2018).

58. Chen, H. Applications of cyber-physical system: A literature review. *J. Ind. Integr. Manag.* **2017**, 2, 1750012. [CrossRef]

59. Lo, S.K.; Xu, X.; Chiam, Y.K.; Lu, Q. Evaluating suitability of applying blockchain. In Proceedings of the IEEE 24th International Conference on Engineering of Complex Computer systems, Fukuoka, Japan, 6–8 November 2017; pp. 158–161.

60. Casino, F.; Dasaklis, T.K.; Patsakis, C. A systemic literature review of blockchain based applications: Current status, classification and open issues. *Telemat. Inform.* **2019**, 36, 55–81. [CrossRef]

61. Lu, S. How Blockchain will Restore Consumer Confidence in Food Safety. 2017. Available online: https://www. fooddive.com/news/how-blockchain-will-restore-consumer-confidence-in-food-safety/503846/ (accessed on 1 August 2018).

62. Manski, S. Building the blockchain: The co-construction of a global commonwealth to move beyond the crises of global capitalism. In Proceedings of the 12th Annual California Graduate Student Conference, University of California, Irvine, CA, USA, 7 May 2016.

63. Galvez, J.F.; Mejuto, J.C.; Simal-Gandara, J. Future challenges on the use of blockchain for food traceability analysis. *TrAC Trends Anal. Chem.* **2018**, 107, 222–232. [CrossRef]

64. Steinmetz, R.; Wehrle, K. What Is This “Peer-to-Peer” About? In *Peer-to-Peer Systems and Applications*; Springer: Berlin/Heidelberg, Germany, 2005; pp. 9–16.

65. Arcticstat-Statistics on Population\Density in All Regions for All Years. 2018. Available online: http://www. arcticstat.org/default.aspx/Indicator/[Population]Density/P/3/default.aspx (accessed on 17 December 2018).

66. Odland, J.Ø.; Donaldson, S.; Dudarev, A.; Carlsen, A. AMAP assessment 2015: human health in the Arctic. *Int. J. Circumpolar Health* **2016**, 75. [CrossRef]

67. Rosstat. Publications Catalog: Federal State Statistics Service. 2018. Available online: http://www.gks.ru/ wps/wcm/connect/rosstat_main/rosstat/ru/statistics/publications/catalog/doc_1286360627828 (accessed on 17 December 2018).

68. V-Kontakte. Russian Version of Facebook for E-Commerce. 2019. Available online: https://www.ecommerce- nation.com/still-dont-know-vkontakte-the-russian-facebook-with-70-million-active-users/ (accessed on 15 January 2019).

69. O’Keeffe, A. Food Security in the Arctic. 2018. Available online: https://griffithreview.com/articles/food- security-in-the-arctic/ (accessed on 17 December 2018).

70. Shishaev, M.; Lomov, P. High Automated Integration of Ontologies on the Basis of Extendable Thesaurus//Information Modelling and Knowledge Bases XXIII. In *Frontiers in Artificial Intelligence and Applications;* IOS Press: Amsterdam, The Netherlands, 2009; pp. 321–330.

71. Szymoniuk, B.; Valtari, H. The REKO System in Finland: a New Model of a Sustainable Marketing Channel. *Probl. Sustain. Dev.* **2018**, 2, 103–111.

72. FAO, Food. *The State of Food and Agriculture: Climate Change, Agriculture and Food Security*; FAO: Rome, Italy, 2016.

73. El Bilali, H.; Mohammad, S.A. Transition towards sustainability in agriculture and food systems: Role of information and communication technologies. *Inf. Process. Agric.* **2018**, 5, 456–464. [CrossRef]

74. Berti, G.; Mulligan, C. ICT & the Future of Food and Agriculture. Industry Transformation – Horizon Scan: ICT & the Future of Food. Telefonaktiebolaget LM Ericsson: Stockholm, Sweden, 2015. Available online: http://gow.epsrc.ac.uk/NGBOViewGrant.aspx?GrantRef=EP/J000604/2 (accessed on 12 February 2018).

75. EU. Novel Food Regulation. European Union (EU) 2015/2283. 2018. Available online: https://eur-lex.europa.eu/legal-content/EN/TXT/?uri=CELEX:32015R2283 (accessed on 15 December 2018).

76. Bellona Report. The Environmental Status of Norwegian Aquaculture. 2003. Available online: https://network. bellona.org/content/uploads/sites/3/The-Environmental-Status-of-Norwegian-Aquaculture.pdf (accessed on 10 May 2019).

77. Johansen, T.J.; Hykkerud, A.L.; Uleberg, E.; Molmann, J. Arctic quality–The effect of high latitude growth conditions on quality of food products. In Proceedings of the Oral Presentation at the 10th Circumpolar Agriculture Conference, Arctic Centre, University of Lapland, Rovaniemi, Finland, 13–15 March 2019.

78. Töyli, P. Introducing Twelve Protected Finnish Products. Aitomakuja Newsletter. 2016. Available online: http://www.aitojamakuja.fi/blogi/?author=3&lang=en (accessed on 10 December 2018).

79. Kurppa, S.; Reinikainen, A. *Tilannekatsaus Luonnonvarakeskuksen (Luke) Arktiseen Biotalouteen Liittyvistä Hankkeista ja Toiminnasta Arktisella Alueella.* Natural Resources Institute Finland Report on Arctic Bioeconomy Research and Development. 2017. Available online: http://jukuri.luke.fi/handle/10024/541454 (accessed on 22 July 2018).

80. Windfuhr, M. *Food Sovereignty and the Right to Adequate Food;* Discussion Paper; FIAN: Heidelburg, Germany, 2003.

81. Stancˇová, K.C.; Cavicchi, A. Smart Specialisation and the Agri-food System. In *Smart Specialisation and the Agri-Food System*; Palgrave Pivot: London, UK, 2019; pp. 43–57.

82. Davies, A.R. Co-creating sustainable eating futures: technology, ICT and citizen–consumer ambivalence. *Futures* **2014**, 62, 181–193. [CrossRef]

83. Svenfelt, Å.; Zapico, J.L. Sustainable food systems with ICT? In Proceedings of the 4th International Conference on ICT for Sustainability (ICT4S 2016), Amsterdam, The Netherlands, 29 August–1 September 2016; pp. 194–201.

Chapter 14

ACQUISITION OF SORPTION AND DRYING DATA WITH EMBEDDED DEVICES: IMPROVING STANDARD MODELS FOR HIGH OLEIC SUNFLOWER SEEDS BY CONTINUOUS MEASUREMENTS IN DYNAMIC SYSTEMS

1. INTRODUCTION: AN OVERVIEW

Sunflower (*Helianthus annuus* L.) is one of the major oilseeds produced in the world. It is cultivated in different climatic zones with varying grain moisture content during the harvesting period. Drying is typically required in order to achieve an optimum final moisture content for safe storage. Excessive moisture levels may lead to a generally increased activity of microorganisms, heating of the product, dry matter losses, and high levels of free fatty acids in the extracted oil [1,2]. Several studies point out that only the non-fat components of the seeds are the critical parts for stability considerations in moisture-dependent storage [2–4]. Water activity a_w holds information on the availability of water for the growth of microorganisms and thus allows inference on threshold levels, above which spoilage is unlikely to occur [5]. It is defined as the partial vapor pressure of water in the measured food, divided by the partial vapor pressure of pure water [6]. This is equal to the equilibrium relative humidity, at which the measured food is in equilibrium with the surrounding atmosphere and does not adsorb nor desorb water. Sorption isotherms describe the relationship between the equilibrium moisture content (MC_e), formed at a given temperature and at the relative humidity, if the food is in equilibrium with the atmospheric surroundings. In general, water activity increases at higher moisture content and, consequently, microorganisms, such as molds, yeasts, and bacteria increasingly grow at $a_w > 0.70$, while enzymatic activity is also promoted by high values of aw [5]. The commonly applied threshold value for safe farm level storage of agricultural products is found at water activities between $0.6 \leq a_w \leq 0.7$ [3,4,7].

In practice, vapor pressure manometers, capacitance hygrometers, and chilled mirror dew point hygrometers represent fast and robust techniques for the indirect measurement of water activity from a common set of partially dried samples [8]. However, these sorption techniques can not generate kinetic data. For the gravimetric measurement of moisture sorption, the static gravimetric method is considered a standard technique. Climatic test chambers have been used before. However, the sorption experiments were realized with discontinuous weight measurements on external balances or with balances inside the test chamber [9,10]. To minimize the negative effects associated with the discontinuous weight measurements, instruments using controlled atmosphere microbalances such as a Dynamic Vapor Sorption apparatus (DVS) have been employed for the automated moisture sorption analysis of food ingredients and other homogenous materials [11–17]. The DVS method is used to measure the equilibrium moisture content of a material at any desired relative humidity and selected temperatures in a short period of time. However, as DVS is designed for extremely small sample mass, bias in sorption measurements may occur when dealing with agricultural products and only a small part of a heterogeneous organ is being measured [7]. Based on the dynamic vapor sorption principle, an innovative experimental system for determining moisture sorption properties of heterogeneous agricultural products is needed. In addition, the system should enable monitoring of moisture sorption by measuring the weight gain or loss at regular time intervals and the automatic acquisition of mass data in more than one high precision ultra-microbalance.

The experimental determination of sorption characteristics allows for the description of moisture diffusivity to be a function of moisture content, partial vapor pressure, and temperature without the requirement to pre-define the mechanisms controlling diffusion [6]. Thus, these isotherms constitute a suitable tool in describing and modeling drying processes for agricultural products. Remarkably, the occurrence of hysteresis between adsorption and desorption is directly affected by the oil content while no significant difference can be observed [4] at values above 48.6%. Most studies on the sorption and drying of sunflower seeds have in common data from different varieties and harvest years, which are combined to provide an empirical basis for model development [3,18,19]. Additionally, seeds are usually remoistened to a desired moisture content, which potentially leads to experimental errors and an alteration of the drying behavior [19]. It is commonly agreed that moisture movement at the surface is negligible, compared to internal resistance, and, thus, the influence of air velocity becomes insignificant after a threshold of approximately 0.1 $m{\cdot}s^{-1}$ [19–21]. Sunflower seeds are comparable to multi-domain composite foods, consisting of a fibrous outer shell, and an oily kernel. Both hulls and kernels show significantly different sorption behavior [2]. In addition, whole seeds and kernels are significantly different in most physical properties, such as volume and equivalent diameter [19,22]. Sunflower kernels show significantly slower moisture diffusivity compared to hulls and, thus, are the limiting factor in drying [3,19].

However, a literature research revealed remarkable differences between the seed/hull-ratio of examined traditional oil-seed varieties on which most sorption and drying studies are based. Given these distinct variations in physical properties, it is obvious that sunflower sorption and drying models should be updated and validated for modern high oleic varieties.

The aim of the present work is, therefore, to develop a robust semi-empirical drying model using a coherent set of experimental sorption and drying data for high oleic sunflower seeds (*Helianthus annuus* L.). The objectives are: (i) to experimentally determine a broad set of equilibrium moisture content data by an automatic, gravimetric analyzer, (ii) to obtain single-layer drying kinetic data in a high precision laboratory dryer at different temperature and absolute humidity of the drying air, and (iii) to establish a generalized single-layer drying model in which its parameters are a function of air conditions. In addition, these related datasets are used to analytically determine moisture effective diffusion coefficients.

2. MATERIALS AND METHODS

2.1 Plant Material

High oleic sunflower seeds (Helianthus annuus L.), F1 of hybrid cultivar 'PR65H22' were harvested mid October 2014 from a farm 30 km east of Würzburg (Germany). The original bulk of approximately 500 kg was reduced to a representative sample of 110 kg. To evaluate the moisture content at harvest (0.317 ± 0.008 $kg{\cdot}kg^{-1}$), 10 samples of 3 g each were used.

2.2 Moisture Content Determination

The moisture content MC of seeds and hulls was measured from samples of 3 g by a standard thermogravimetric analysis in a convection oven at 103 ± 2°C, according to ISO 665:2000 [23] and expressed in kg water per kg dry matter (kg·kg−1). All analyses were performed in triplicates, which is commonly applied in other studies [18,22].

2.3 Determination of Dynamic Vapor Sorption Isotherms

The adsorption isotherms of seeds and hulls were separately determined by using an automated system designed at the Institute of Agricultural Engineering, University of Hohenheim (Stuttgart, Germany) (Figure 1). The system consists of a climatic test chamber (C + 10/600, CTS GmbH, Hechingen, Germany), which regulates air temperature between 10 °C and 95 °C while maintaining a relative humidity between 10% and 98% and ensuring air circulation. A weight measuring system, consisting of five high precision load cells (WZA1203-N, Sartorius AG, Goettingen, Germany) with an accuracy of 1 mg was mounted on top of the chamber to record the change in mass, caused either by adsorption or desorption of water vapor. Each load cell carries a perforated sample holder, which is suspended through the chamber ceiling. Load cell and climate chamber control is realized remotely. Direct control via an attached computer is also possible. Relative humidity is varied gradually at a pre-set temperature, which is held ceteris paribus during an experiment.

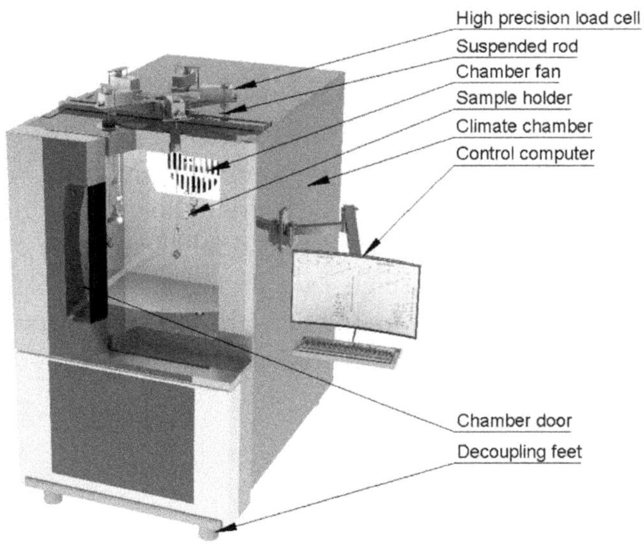

Figure 1. Cutaway view of automated system for dynamic vapor sorption recording.

The test material was dried at 60 °C to a moisture content of 0.006 ± 0.004 $kg{\cdot}k^{g-1}$ and manually cleaned from impurities. The seeds were ground to a particle size of approximately 5 mm. To obtain hulls, seeds were manually hulled. About 12 g of dried seeds or hulls were used per sorption experiment. The samples were loaded into the sample holder and the adsorption isotherms were measured at 25 °C and 50 °C, which increased the relative humidity gradually from 10% to 85% with increments of 10% and a final step of 5%. Mass, temperature, and humidity data were recorded in 20 min intervals. The equilibrium was considered to have been reached when observing a change in weight of less than 5% of the initial sample weight during 10 consecutive measurements. Three repetitions per temperature and material were performed, which resulted in a total of 160 individually determined equilibrium moisture content data points.

2.4 Sorption Isotherm Models

Five commonly applied, three parameter moisture sorption models were tested for their accuracy to describe the experimental sorption data [3,5,6,24]. The models are presented in terms of equilibrium moisture content MC_e (kg·kg^{-1}), water activity a_w ($p_{vs} \cdot p_{sat}^{-1}$), and a, b, c as model constants. The applied model equations and their ranges of validity are shown in Table 1.

Table 1. Models for sorption isotherms. MC_e = equilibrium moisture content. T = temperature in °C. a_w = water activity. a, b, c = model constants [7].

Model	Original Plant Material	Validity (a_W)
Modified Chung-Pfost $MC_e = \frac{-1}{a}\ln\left(-\frac{(T+b)}{c}\ln(a_W)\right)$	Maize and maize components	0.1–0.9
Modified Oswin $MC_e = (a + b \times T)\left(\frac{a_W}{1-a_W}\right)^{\frac{1}{c}}$	Various	0.3–0.5
Modified Halsey $MC_e = \left(\frac{-exp(a+b\times T)}{\ln(a_W)}\right)^{\frac{1}{c}}$	Maize, wheat flour, laurel, nutmeg	0.1–0.8
Modified Henderson $MC_e = \left(-\frac{\ln(1-a_W)}{a(T+b)}\right)^{\frac{1}{c}}$	Maize	-
Modified G.A.B. $MC_e = \frac{ab\left(\frac{c}{T}\right)a_W}{(a-ba_W)\left(1-ba_W+\left(\frac{c}{T}\right)ba_W\right)}$	Various	<0.94

2.5 Thin-Layer Drying Experiments

Thin-layer drying experiments were conducted using a high precision hot-air laboratory dryer designed at the Institute of Agricultural Engineering, University of Hohenheim (Stuttgart, Germany), which allowed the control of the desired drying conditions over a wide range of operating parameters. For the drying experiments, 500 kg of freshly harvested seeds were manually cleaned and 70 subsamples of 1.5 kg each were randomly taken, vacuum-sealed in PEHD bags and stored at 4 °C for no longer than four weeks. The experimental system has been previously described in detail by Argyropoulos et al. [25]. In total, 63 individual drying experiments were conducted at temperatures T between 30 and 90 ± 0.1 ∘C in steps of 10 ∘C and an absolute humidity x of 0.010, 0.015, and 0.020 kg water per kg of dry air. During the individual drying experiments, temperature and absolute humidity were kept constant and a uniform air flow through the sample was maintained at 0.6 ± 0.05 $m{\cdot}s^{-1}$. The initial seed moisture during all drying experiments was 0.317 ± 0.008 $kg{\cdot}kg^{-1}$. An initial mass of 0.400 ± 0.001 kg was evenly spread on a perforated drying tray, which resulted in a layer depth of 15 mm. The tray was supported on PC6 load cells (Flintec GmbH, Meckesheim, Germany). The weight was measured every 10 min. Meanwhile, a bypass valve was opened to prevent floating of the tray. The seeds were dried until a constant weight was achieved, once the total weight change for three consecutive measurements was below 0.5 g. The drying experiments were repeated at least three times for each drying condition. The moisture content of seeds before and after drying was determined as described above (Section 2.2). In total, 63 experiments with 1291 single measurements were conducted.

2.6 Empirical Drying Model

By including equilibrium moisture information from the previous experiments, the normalized moisture ratio (*MR*) was computed. *MR* (*t*) is the average moisture ratio at time *t* (minutes), MC_t the moisture content at time *t*, MC_e the equilibrium moisture content, and MC_0 the initial moisture content ($kg{\cdot}kg^{-1}$). The moisture ratio over time can be well depicted by the semi-empirical page equation (Equation (2)). This approach offers a compromise between inclusion of the physical theory and ease of use, which results in a semi-empirical model [6]. Fickian moisture migration, constant moisture diffusion coefficients isothermal conditions, and negligible shrinkage are the basics of this approach [26]. Two coefficients have to be fitted including k as the rate constant (min-1) and n as the dimensionless coefficient to improve the fit [19].

To provide a generalized, semi-empirical model capable of representing different temperatures and humidity of the drying medium, k and n were transformed to a function of drying conditions at time t. It is important to highlight that the actually recorded conditions from the drying chamber were employed in the model instead of the set-point conditions.

Two different approaches, based on (i) temperature *T* (∘C) and absolute humidity *x* ($kg{\cdot}kg^{-1}$) (Equation (3)) and water vapor pressure deficit ΔP (Pa) (Equation (4)), both described by an Arrhenius-type equation, were followed [27]. ΔP (Pa) was computed as the vapor pressure deficit between the drying air and saturated air under the same temperature conditions with P_{sat} (Pa) derived from the Magnus-equation, T(∘C), and *rh* as the relative humidity (%) (Equation (1)).

$$\Delta P = P_{sat} \times \left(1 - \frac{rh}{100}\right) = 611.2 \times \exp\left(\frac{17.62T}{243.12 + T}\right) \times \left(1 - \frac{rh}{100}\right) \tag{1}$$

$$MR(t) = \frac{MC_t - MC_e}{MC_0 - MC_e} = \exp\left(-kt^n\right) \tag{2}$$

with the following conditions: $k_{T1} < k_{T2}$, $n_{T1} > n_{T2}$ for $T_1 < T_2$ and $x_1 = x_2$, $k_{x1} > k_{x2}$, $n_{x1} < n_{x2}$ for $T_1 = T_2$ and $x_1 < x_2$

$$f(k, n) = d \times \exp\left(\frac{e}{T}\right) + \frac{f}{x} \tag{3}$$

$$MR(t) = \exp\left(-\left(d \times \exp\left(\frac{e}{T}\right) + \frac{f}{x}\right) t^{d \times \exp\left(\frac{e}{T}\right) + \frac{f}{x}}\right) \tag{2a}$$

$$f(k, n) = g \times \log\left(\Delta P\right)^h \tag{4}$$

$$MR(t) = \exp\left(-g \times \log\left(\Delta P\right)^h t^{g \times \log(\Delta P)^h}\right) \tag{2b}$$

Analytical Estimation of Diffusion Coefficients

The high oleic sunflower seeds used in this study are shaped like compressed, oval bodies with a sphericity of 0.45 to 0.55 and an equivalent diameter of 5.48 ± 0.64 mm [22]. Average moisture diffusivity D was assumed to remain constant during the relevant drying period [28]. Under these conditions, the special "short time "solution to the differential diffusion equation proposed by Becker (1959) can be applied. The physical basis of this solution is the restriction of changes in moisture to the vicinity of the surface [29]. The validity is limited to the fast drying region of 0.2 < MR < 1 including the name "short times solution". The mathematical derivation was intensively discussed by Giner and Mascheroni [29], and Becker's short time analytical solution was proven to be accurate, fast, and applicable in the practical drying range for agricultural products. For spheres, the thin layer equation takes the form of Equation (5) where av is the kernel's surface specific area in $m^2 \cdot m^{-3}$.

$$MR = \frac{MC_t - MC_e}{MC_0 - MC_e} = 1 - \frac{2}{\sqrt{\pi}} a_v \sqrt{Dt} + \frac{f''(0)}{2} {a_v}^2 Dt \tag{5}$$

with the following conditions: $MC_t = MC_0$ for $t = 0$, $MC_t = MC_e$ for $t \rightarrow \infty$

The term $f''(0)$ is a shape dependent factor, derived from the slope of Equation (6), expressed as a straight line [29].

$$Y = \frac{1 - MR}{a_v \sqrt{Dt}} = \frac{2}{\sqrt{\pi}} - \frac{f''(0)}{2} a_v \sqrt{Dt} \tag{6}$$

D can be derived from a moisture independent, Arrhenius type relationship, proposed by Sun and Woods [30].

$$D = 1.126 \times R^2 \times exp\left(-\frac{2806.5}{T + 273.15}\right) \tag{7}$$

where R is the seeds' equivalent radius. After correction of the shape dependent factor $f^{||}(0)$, the individual diffusion coefficients D_s can be determined with Equation (5). The limits of validity for spherical bodies, where D_s is no longer essentially constant, are found at $MR = 0.2$ [28]. The same approximation was also adapted by Giner and Mascheroni [31] for wheat as well as by Santalla and Mascheroni [19] for sunflower seeds.

2.7 Statistical Analysis

All statistical analyses were conducted using the R Project for Statistical Computing [32]. The sorption models and individual Page equations were fitted to the experimental data using R's nls2 procedure [33]. The coefficient of determination R^2 and the mean absolute percentage error $MAPE$ with $M_{Ce,exp}$ and M_{Rexp} as the observed and MCe,pre and MRpre as the predicted equilibrium moisture content and moisture ratio were taken as the main criteria for the goodness of fit.

$$R^2 = 1 - \frac{\sum(MR_{exp} - MR_{pre})^2}{\sum(MR_{exp} - \overline{MR}_{exp})^2} \tag{8}$$

$$MAPE = \frac{100}{n}\sum\frac{|MR_{exp} - MR_{pre}|}{MR_{exp}} \tag{9}$$

The effect of independent variables on the overall model constants was determined by regression analysis and analysis of variance (ANOVA). Asterisks mark significance at $p < 0.05$ (*), $p < 0.01$ (**), and $p < 0.001$ (***) level.

3. RESULTS AND DISCUSSION

3.1 Analysis of Moisture Sorption Models

Figure 2a shows the experimentally derived sorption data for hulls and seeds at 25 °C and 50 °C and the curves as predicted by the selected models. At a constant temperature, the equilibrium moisture content increased with increasing water activity, which indicates an asymptotic convergence at $a_w = 1$. In general, at constant water activity, the equilibrium moisture content decreased with increasing temperature. Remarkably, for hulls, this trend was only observed for aw ≤ 0.82 and for seeds for $a_w \leq 0.72$. Above these threshold values, the equilibrium moisture content increased with increasing temperature, which is a phenomenon that usually can be observed in high sugar foods [34,35].

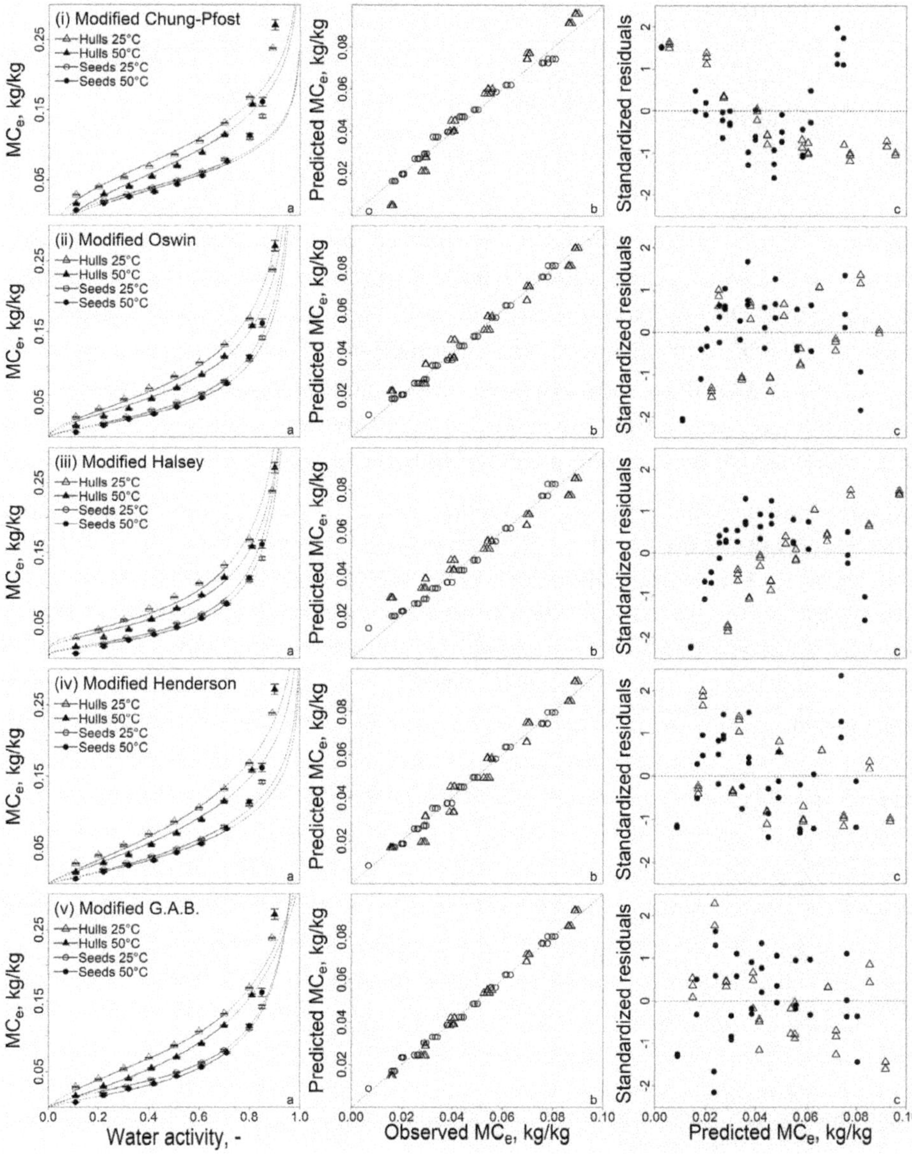

Figure 2. (a) Sorption isotherm for high oleic sunflower seeds (circles) and hulls (triangles) at 25 °C (hollow) and 50 °C (solid) predicted with different models as specified in the plot. Slashed lines are an extrapolation beyond the dataset used for fitting. (b) observed equilibrium moisture content vs. predicted equilibrium moisture content and (c) standardized residuals vs. predicted equilibrium moisture content.

Those sorption models are not designed to depict this inverse phenomenon. Therefore, data at a_w > 0.82 for hulls and a_w > 0.72 for seeds were excluded from the non-linear fitting procedure. A random 60% of the remaining dataset was used for fitting while the entire remaining dataset was used for model performance evaluation in terms of R^2 and *MAPE* value. The fitted curves are shown as solid lines and continued as dashed lines when the above-mentioned limits of a_w are exceeded, which shows what the corresponding models would predict. Figure 2b provides the predicted versus observed plots for employed sorption equations.

In Table 2, the model coefficients determined by the nonlinear least squares procedure and the corresponding performance parameters are summarized. For seeds, the Mod. Oswin, Mod. Henderson and Mod. G.A.B. equation all achieved R^2 > 0.99 and *MAPE* < 10%. However, the residual plots (Figure 2c) reveal a patterned shape for all but the Mod. Henderson equation. For hulls, only the modified Oswin and modified G.A.B. equation showed R2 > 0.99. However, only the modified G.A.B. equation showed random distribution of residuals.

Table 2. Coefficients a, b, and c, coefficient of determination R2 and MAPE for the (i) Mod. Chung-Pfost, (ii) Mod. Oswin, (iii) Mod. Halsey, (iv) Mod. Henderson, and (v) Mod. G.A.B. equation fitted for high oleic sunflower seeds in the range of 10–70% equilibrium relative humidity at 31 degrees of freedom and hulls in the range of 10–80% equilibrium relative humidity and 34 degrees of freedom.

Equation		a		b		c		R2	MAPE, %
(i) Mod. Chung-Pfost		28.181	***	208.987	***	611.811	***	0.988	7.900
(ii) Mod. Oswin		0.048	***	-1.58×10^{-4}	***	1.607	***	0.992	8.131
(iii) Mod. Halsey	Seeds	-3.895	***	-4.60×10^{-3}	***	1.159	***	0.978	13.884
(iv) Mod. Henderson		0.102	***	207.939	***	1.145	***	0.995	5.408
(v) Mod. G.A.B.		2.88×10^{-2}	***	0.919	***	167.831	***	0.994	5.923
(i) Mod. Chung-Pfost		17.761	***	47.204	***	240.673	***	0.985	9.173
(ii) Mod. Oswin		9.94×10^{-2}	***	-0.001	***	1.776	***	0.992	6.618
(iii) Mod. Halsey	Hulls	-3.255	***	-9.87×10^{-3}	***	1.245	***	0.982	9.964
(iv) Mod. Henderson		0.208	***	54.690	***	1.299	***	0.991	6.112
(v) Mod. G.A.B.		0.070	***	0.775	***	122.616	***	0.997	3.207

Significant contribution to the model on the $p \leq 0.05$, $p \leq 0.01$, and $p \leq 0.001$ level, respectively, are indicated by one, two, or three asterisks.

Furthermore, the experimental data for both seeds and hulls indicate the convex shape of a type III sorption isotherm, according to Brunauer's classification [6]. This type is relatively uncommon and not generally found [36]. Physically, the type III isotherm does not show monolayer adsorption [36]. It is well known that plant seed sorption behavior should rather be depicted with type II sorption isotherms, which is generally adopted in many other studies [7,37,38]. Type II isotherms are an extension of Langumir-like monolayer adsorption isotherms to include unrestricted mono and multilayer adsorption [36], which result in the characteristic sigmoidal shape with a nearly straight segment. It is generally assumed that the beginning of this straight line portion represents the most likely point, where a completely saturated monolayer occurs [39]. From Figure 2, however, no transition from monolayer to multilayer adsorption is derivable, which indicates that the surface area was already covered with a monolayer below the lowest measured a_w value. Sunflower seeds are very much comparable to complex multi-domain foods, with the hulls being a fibrous matrix, allowing two to four times faster diffusivity of moisture than the kernel [19]. In addition, it is clear that the oil content in kernels highly affects the equilibrium moisture content. Similar observations have been reported by other studies [2,40]. At a constant temperature, hulls usually reach a higher MC_e than seeds, which again show a higher MC_e than kernels [2,18,19]. The Mod. Henderson was the only model showing both satisfying fit and random residuals for seeds. Opposed to this, hulls alone were best modeled by the Mod. G.A.B. equation. Given the fact that safe moisture levels for oilseeds are found in the region of a_w = 0.64–0.70, the limits of model validity and the observed anomaly at high values of a_w were considered non-critical for the development of a drying model [1,3]. Based on these insights, the fitted Mod. Henderson equation and the coefficients found for seeds were integrated in Equation (2) in order to describe the drying process.

3.2 Modelling of Thin-Layer Drying Behavior

The individual drying data for *T* from 30 ◦C to 90 ◦C and x of 0.010, 0.015, and 0.020 $kg \cdot kg^{-1}$ were described well by the Page equation with high R^2 (>0.99) and low MAPE (<5%) values. The equilibrium moisture content MC_e ($kg \cdot kg^{-1}$) for seeds was calculated with the Mod. Henderson equation with a = 0.102, b = 207.939, and c = 1.145 (Table 2). The kinetic parameter k increased with increasing T at constant x, while n decreased. With increasing absolute humidity at constant *T*, *k* decreased and *n* increased. Equation (10) and Equation (10a) best described model parameters *k* and *n* as a function of *T* and *x*. The resulting course is shown in Figure 3a. Inclusion in Equation (2a) resulted in a temperature and absolute humidity-based generalized model describing the experimental data with an R^2 value of 0.982 and *MAPE* of 9.4%.

$$k = 1.126 \times 10^{-5} \cdot \log(\Delta P)^{4.540};\ R^2 = 0.988,\ MAPE = 4.25\% \qquad (11)$$

$$n = 2.494 \times \log(\Delta P)^{-0.880};\ R^2 = 0.934,\ MAPE = 2.25 \qquad (11a)$$

The description of k and n as a function of the water vapor pressure deficit ΔP was best described by Equation (11) and Equation (11a). The course of k and n vs. ΔP is depicted in Figure 2b. By inclusion in Equation (2b), a ΔP based, generalized, drying model is achieved, which is capable of describing the experimentally derived drying data with an overall R^2 value of 0.988 and *MAPE* of 8.3%. This is slightly better than the model based on T and x.

$$k = 1.126 \times 10^{-5} \cdot \log(\Delta P)^{4.540};\ R^2 = 0.988,\ MAPE = 4.25\% \qquad (11)$$

$$n = 2.494 \times \log(\Delta P)^{-0.880};\ R^2 = 0.934,\ MAPE = 2.25 \qquad (11a)$$

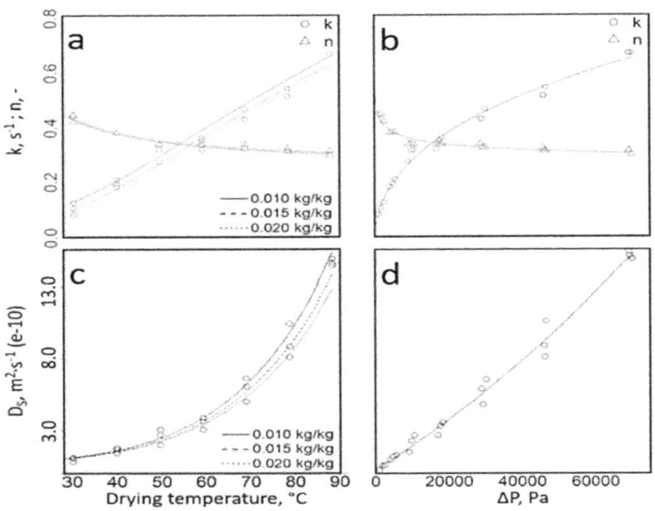

Figure 3. Parameters k, n, and the fitted models from (a) Equations (10) and (10a) for k, n = f (T,x) in ◦C and kg·kg-1 and (b) modeled with Equation (11) and (11a) for k, n = f (ΔP) in Pa. (c) DS = f (T,x) for temperature from 30 ◦C to 90 ◦C and absolute humidity of the drying air of 0.010, 0.015, and 0.020 kg·kg-1 modeled with Equation (12) and (d) DS = f (ΔP) for the same temperature and humidity range, modeled with Equation (13).

Since ΔP is mainly dependent on T and x, it is a promising physical parameter for the description of drying processes. Regarding the slightly superior fit of the generalized model described by ΔP and the requirement of only four model constants instead of six, this relationship was chosen to further describe the drying process. The predicted vs. observed plot in Figure 4a shows a nearly straight line, which indicates a satisfying performance of the generalized model. The residuals plot in Figure 4b shows a random distribution, supporting the validity of the derived model.

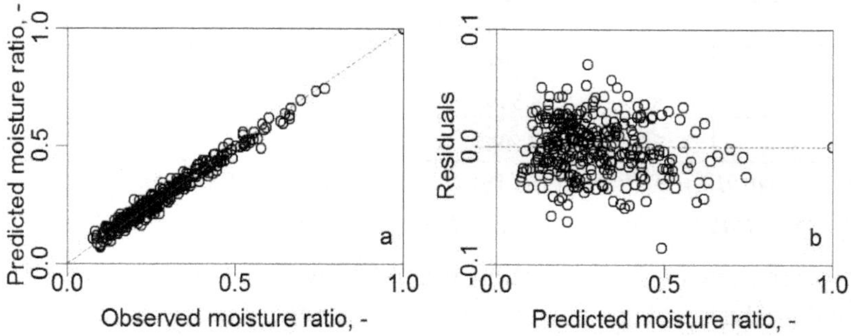

Figure 4. Model performance for MR expressed as a function of ΔP calculated with Equation (2b), (a) predicted MR versus observed MR values are concentrated closely to the perfect fit line of x = y, and (b) the residuals are equally distributed and do not show any trend.

Application of the short times equation indicated an increase of D_S with increasing temperature *T*. With increasing absolute humidity *x*, D_S showed a decreasing trend for temperatures up to 70◦C, from where no obvious trend was derivable. For *T* below 50◦C, the threshold of *MR* < 0.2 usually was not undershot. The shape dependent factor $f^{II}(0)$ found to be 0.428 with a standard error of 1.3%. By inclusion in Equation (6), the diffusion coefficients for short times (0.2 < *MR* < 1) can be calculated (Table 3).

Table 3. Moisture diffusion coefficients D for short times (MR ≥ 0.2), calculated with Equation (5) for different absolute humidity x and temperature T of the drying air, coefficient of determination R2 and MAPE.

x	T,°C	$D\ (m^2 \cdot s^{-1}) \cdot 10^{-10}$		R^2	MAPE, %
	30	0.643	***	0.979	4.088
	40	1.317	***	0.985	6.935
	50	2.620	***	0.991	6.217
$0.010\ kg \cdot kg^{-1}$	60	3.467	***	0.970	10.903
	70	6.277	***	0.989	7.547
	80	10.190	***	0.996	5.059
	90	14.800	***	0.995	7.399
	30	0.593	***	0.984	4.303
	40	1.231	***	0.987	5.591
	50	2.258	***	0.974	11.780
$0.015\ kg \cdot kg^{-1}$	60	3.222	***	0.979	11.238
	70	4.627	***	0.969	11.187
	80	7.796	***	0.990	7.829
	90	14.310	**	0.997	5.611
	30	0.342	***	0.979	3.556
	40	0.978	***	0.980	5.790
	50	1.529	***	0.954	10.944

0.020 kg·kg^{-1}	60	2.634	***	0.988	5.715
	70	5.696	***	0.982	9.844
	80	8.559	***	0.993	5.203
	90	14.490	**	0.999	2.152

Significant contribution to the model on the p ≤ 0.05, p ≤ 0.01, and p ≤ 0.001 level, respectively, are indicated by one, two, or three asterisks.

The mean coefficient of determination was 0.984 with a MAPE of 7.09% for short times. Visual analysis of Figure 3c, which compares the measured data to the fit of Equation (12) revealed an overestimation when the limit of validity is approached. The derived diffusion coefficients, however, are in the same range as the ones reported by Santalla and Mascheroni [19]. In addition to their findings, that initial moisture content did not affect diffusion coefficients. The results from this study indicate an effect of x only in the lower temperature regions up to 70 ◦C. In an analogy with the approach for fitting a generalized empirical drying model, the same procedure was applied for the analytical solution. A generalized analytical model was fitted for $D_S = f(T, x)$, which is well described by Equation (12).

$$D_S = 4.125 \times 10^{-13} \cdot \exp\left(T^{0.4851} \times x^{-0.033}\right);\ R^2 = 0.984,\ MAPE = 12.75\% \quad (12)$$

Integration of Equation (12) in Equation (5) yielded a generalized model with an overall R^2 of 0.966 and $MAPE$ = 7.37%. The description of $DS = f(\Delta P)$ was described well with a similar function:

$$D_S = 1.296 \times 10^{-12} \times \exp\left(\Delta P^{0.1747}\right);\ R^2 = 0.986,\ MAPE = 10.03\% \quad (13)$$

Integration of Equation (13) in Equation (5) resulted in a generalized model based on ΔP, which was fitted with an overall R^2 of 0.976 and $MAPE$ = 6.33%. This is slightly better than the approach based on temperature and absolute humidity. The fit to the experimental data of both Equations (12) and (13) is given in Figure 3c,d.

An applicability to determine the effective diffusion coefficients in the initial, most relevant phase of drying is obviously given and commonly applied [19,29,31]. Especially, the calculation of the shape-depending factor f'(0), based on the physical properties of the investigated product, adds further meaning to the application of Becker's equation, and is expected to increase the applicability of the thin-layer equation for dryer control and simulation.

4. CONCLUSIONS

The embedded systems employed for data recording during the sorption and drying experiments in the current study did allow the acquisition of a large amount of experimental inline-data from one year's harvest. This information was used to fit semi-empirical and analytical sorption and drying models that are simple enough to be run on embedded microcontrollers with limited processing power. The sorption experiments revealed that, for high values of a_w, the equilibrium moisture content increased with increasing temperature. An anomaly, which is usually found in high sugar foods, was not yet reported for sunflower seeds. It is remarkable that most of the commonly applied equations to describe sorption isotherms showed patterned residuals, which restricted their applicability and validity. This study found the Page equation to satisfactorily describe the drying process of high oleic sunflower seeds for a wide range of drying air temperatures and humidity. It also proposes the application of water vapor pressure deficit ΔP to calculate the parameters of the Page equation including both the effect of temperature and the humidity of the drying air. This approach was found to be superior to a prediction based on temperature. Values of diffusivity ascertained with the Becker equation and a correction of the shape dependent factor are comparable to those reported in other studies, which corroborates the applicability and accuracy in thin layer drying of high oleic sunflower seeds.

Abbreviations

a_w	water activity
MC	moisture content
MC_e	equilibrium moisture content
MC_t	moisture content at time t
DVS	dynamic vapour sorption apparatus
kg	kilogram
g	gram
mg	milligram
m	meter
mm	millimeter
rh	relative humidity, %
min	minutes
P_{vs}	water vapor partial pressure, Pa
P_{sat}	saturation vapor pressure, Pa
T	temperature, ◦C
a, b, c, d, e, f, g	model constants
G.A.B.	Guggenheim, Anderson, DeBoer

x	absolute humidity, kg water per kg of dry air
s	second
MR	moisture ratio
t	time
k	rate constant, min^{-1}
n	dimensionless coefficient of page equation
P	pressure, Pa
D	moisture diffusivity, $m^2 \cdot s^{-1}$
a_v	kernel's surface specific area in $m^2 \cdot m^{-3}$
ANOVA	analysis of variance
MAPE	mean absolute perecentage error
R^2	coefficient of determination
p	probability level at which significance is assumed

References

1. Robertson, J.A.; Chapman, G.W.; Wilson, R.L. Effect of moisture content of oil type sunflower seed on fungal growth and seed quality during storage. *J. Am. Oil Chem. Soc.* **1984**, 61, 768–771. [CrossRef]

2. Mazza, G.; Jayas, D.S.S. Equilibrium moisture characteristics of sunflower seeds, hulls, and kernels. *Trans. ASAE* **1991**, 34, 534–538. [CrossRef]

3. Giner, S.A.; Gely, M.C. Sorptional parameters of sunflower seeds of use in drying and storage stability studies. *Biosyst. Eng.* **2005**, 92, 217–227. [CrossRef]

4. Maciel, G.; De la Torre, D.; Izquierdo, N.; Cendoya, G.; Bartosik, R. Effect of oil content of sunflower seeds on the equilibrium moisture relationship and the safe storage condition. *Agric. Eng. Int.* **2015**, 17, 248–258.

5. Mujumdar, A.S.; Beke, J. Grain drying: Basic principles. In *Handbook of Postharvest Technology: Cereals, Fruits, Vegetables, Tea, and Spices;* Chakraverty, A., Mujumdar, A.S., Raghavan, G.S.V., Ramaswamy, H.S., Eds.; Marcel Dekker: New York, NY, USA, 2003; pp. 119–138.

6. Guillard, V.; Bourlieu, C.; Gontard, N. Theoretical background. In *Food Structure and Moisture Transfer;* Springer: New York, NY, USA, 2013; pp. 3–33.

7. Argyropoulos, D.; Alex, R.; Kohler, R.; Müller, J. Moisture sorption isotherms and isosteric heat of sorption of leaves and stems of lemon balm (*Melissa officinalis* L.) established by dynamic vapor sorption. *LWT-Food Sci. Technol.* **2012**, 47, 324–331. [CrossRef]

8. Argyropoulos, D.; Müller, J. Effect of convective-, vacuum- and freeze drying on sorption behaviour and bioactive compounds of lemon balm (*Melissa officinalis* L.). *J. Appl. Res. Med. Aromat. Plants* **2014**, 1, 59–69. [CrossRef]

9. Arslan, N.; Tog˘ rul, H. Moisture sorption isotherms for crushed chillies. *Biosyst. Eng.* **2005**, 90, 47–61. [CrossRef]

10. Stubberud, L.; Arwidsson, H.G.; Graffner, C. Water-solid interactions: I. A technique for studying moisture sorption/desorption. *Int. J. Pharm.* **1995**, 114, 55–64. [CrossRef]

11. Hill, C.A.S.; Norton, A.J.; Newman, G. The water vapour sorption properties of Sitka spruce determined using a dynamic vapour sorption apparatus. *Wood Sci. Technol.* **2010**, 44, 497–514. [CrossRef]

12. Kachrimanis, K.; Noisternig, M.F.; Griesser, U.J.; Malamataris, S. Dynamic moisture sorption and desorption of standard and silicified microcrystalline cellulose. *Eur. J. Pharm. Biopharm.* **2006**, 64, 307–315. [CrossRef] [PubMed]

13. Kohler, R.; Dück, R.; Ausperger, B.; Alex, R. A numeric model for the kinetics of water vapor sorption on cellulosic reinforcement fibers. *Compos. Interfaces* **2003**, 10, 255–276. [CrossRef]

14. Liu, H.; Zhang, L.; Han, Z.; Xie, B.; Wu, S. The effects of leaching methods on the combustion characteristics of rice straw. *Biomass Bioenergy* **2013**, 49, 22–27. [CrossRef]

15. Vollenbroek, J.; Hebbink, G.A.; Ziffels, S.; Steckel, H. Determination of low levels of amorphous content in inhalation grade lactose by moisture sorption isotherms. *Int. J. Pharm.* **2010**, 395, 62–70. [CrossRef] [PubMed]

16. Xie, Y.; Hill, C.A.S.; Jalaludin, Z.; Curling, S.F.; Anandjiwala, R.D.; Norton, A.J.; Newman, G. The dynamic water vapour sorption behaviour of natural fibres and kinetic analysis using the parallel exponential kinetics model. *J. Mater. Sci.* **2011**, 46, 479–489. [CrossRef]

17. Nurhadi, B.; Roos, Y.H. Dynamic water sorption for the study of amorphous content of vacuum-dried honey powder. *Powder Technol.* **2016**, *301*, 981–988. [CrossRef]

18. Santalla, E.M.; Mascheroni, R.H. Equilibrium moisture characteristics of high oleic sunflower seeds and kernels. *Dry. Technol.* **2003**, *21*, 147–163. [CrossRef]

19. Santalla, E.M.; Mascheroni, R.H. Moisture diffusivity in high oleic sunflower seeds and kernels. *Int. J. Food Prop.* **2010**, *13*, 464–474. [CrossRef]

20. Henderson, S.M.; Pabis, S. Grain drying theory IV, The effect of airflow rate on the drying index. *J. Agric. Eng. Res.* **1962**, 7, 85–89.

21. Hutchinson, D.; Otten, L. Thin-layer air drying of soybeans and white beans. Int. *J. Food Sci. Technol.* **1983**, *18*, 507–522. [CrossRef]

22. Munder, S.; Argyropoulos, D.; Müller, J. Class-based physical properties of air-classified sunflower seeds and kernels. *Biosyst. Eng.* **2017**, *164*, 124–134. [CrossRef]

23. *ISO 665:2000 Oilseeds—Determination of Moisture and Volatile Matter Content;* International Organization for Standardization (ISO): Geneva, Switzerland, 2000.

24. Al-Muhtaseb, A.H.; McMinn, W.A.M.; Magee, T.R.A. Moisture sorption isotherm characteristics of food products: A review. *Food Bioprod. Process.* **2002**, 80, 118–128. [CrossRef]

25. Argyropoulos, D.; Heindl, A.; Müller, J. Assessment of convection, hot-air combined with microwave-vacuum and freeze-drying methods for mushrooms with regard to product quality. *Int. J. Food Sci. Technol.* **2011**, *46*, 333–342. [CrossRef]

26. Shahari, N.A. Mathematical Modelling of Drying Food Products: Application to Tropical Fruits. Ph.D. Thesis, University of Nottingham, Nottingham, UK, 2012.

27. Udomkun, P.; Argyropoulos, D.; Nagle, M.; Mahayothee, B.; Janjai, S.; Müller, J. Single layer drying kinetics of papaya amidst vertical and horizontal airflow. *LWT-Food Sci. Technol.* **2015**, *64*, 67–73. [CrossRef]

28. Becker, H.A. A study of diffusion in solids of arbitrary shape, with application to the drying of the wheat kernel. *J. Appl. Polym. Sci.* **1959**, *1*, 212–226. [CrossRef]

29. Giner, S.A.; Mascheroni, R.H. PH—Postharvest technology. *J. Agric. Eng. Res.* **2001**, *80*, 351–364. [CrossRef]

30. Sun, D.W.; Woods, J.L. Low temperature moisture transfer characteristics of wheat in thin layers. *Trans. ASAE* **1994**, 37, 1919–1926. [CrossRef]

31. Giner, S.A.; Mascheroni, R.H. PH—Postharvest technology. *Biosyst. Eng.* **2002**, *81*, 85–97. [CrossRef]

32. R Core Team. R: A Language and Environment for Statistical Computing 2015. Available online: https://www.r-project.org (accessed on 17 September 2015).

33. Grothendieck, G. nls2: Non-Linear Regression with Brute Force 2013. Available online: https://cran.r- project.org/web/packages/nls2/nls2.pdf (accessed on 17 September 2015).

34. Djendoubi Mrad, N.; Bonazzi, C.; Boudhrioua, N.; Kechaou, N.; Courtois, F. Influence of sugar composition on water sorption isotherms and on glass transition in apricots. *J. Food Eng.* **2012**, *111*, 403–411. [CrossRef]

35. Saravacos, G.D.; Tsiourvas, D.A.; Tsami, E. Effect of temperature on the water adsorption isotherms of sultana raisins. *Dry. Technol.* **1986**, *4*, 633–649. [CrossRef]

36. Ling, M. Manometrische Bestimmung der NO2-Sorptionsisothermen von Superberliner Blau—Derivaten und Charakterisierung der inneren Oberflächen mittels der BET—Methode. Ph.D. Thesis, Universität Hamburg, Hamburg, Germany, 2001.

37. Simha, H.V.V.; Pushpadass, H.A.; Franklin, M.E.E.; Kumar, P.A.; Manimala, K. Soft computing modelling of moisture sorption isotherms of milk-foxtail millet powder and determination of thermodynamic properties. *J. Food Sci. Technol.* **2016**, 53, 2705–2714. [CrossRef]

38. Mathlouthi, M.; Rogé, B. Water vapour sorption isotherms and the caking of food powders. *Food Chem.* **2003,** 82, 61–71. [CrossRef]

39. Brunauer, S. *The Adsorption Of Gases And Vapors Vol I*; Oxford University Press: London, UK, 1943.

40. Pixton, S.W.; Warburton, S. Moisture content relative humidity equilibrium, at different temperatures, of some oilseeds of economic importance. *J. Stored Prod. Res.* **1971**, 7, 261–269. [CrossRef]

INDEX

A

Adsorption 285

Airborne 79, 89-92

Algorithm 8, 231

Amounts 86

Analysis 17, 83, 164, 182, 195, 198-199, 202, 204, 208, 212-215, 225, 228, 242-243, 282

Antennas 49

Arctic 251, 269-270, 274

Atmosphere 124

B

Blockchain 256-257, 272

Boxplots 204-205

Business 252, 256

C

Camera 123-125, 142, 145, 147

Canopy 163, 187, 199, 201, 204, 206, 212-214

Cation 118, 213

Climate 250-251, 255, 269

Component 15

Consumers 257, 260

Cosmic 28

Cosmic-ray 27-28, 30

D

Density 125

Detection 10, 111, 118, 171, 194, 219

Detectors 16

Dielectric 52

Diffusion 288

Digital 17, 254, 261, 263

Dimensions 232

Diseases 159, 171

Drainage 3, 10, 45-47, 49-50, 52-60

Droplet 74-75

E

Ecosystems 117

Electric 103, 108

Electron 244

Elements 127

Emission 124, 142

Energy 20, 52, 113-114, 123, 125, 143

Epithermal 23, 25

Experiment 112, 164

Extension 94

F

Farmers 224

Farming 192

Federation 254

Forecasts 83

Forest 133-134

Fungal 6, 160

Fusion 176, 189

G

Gearbox 103

Growth 193

H

Harrowing 3, 62, 65

Health 5, 117, 131-132, 140, 156, 194, 207, 214-216, 268-269

Histograms 230

Horizontal 258
Humidity 33, 75, 276, 280, 289
Hyperbola 52, 54

I

Impression 168
Indicator 235
Indoor 107, 109, 112
Industry 256, 272
Infection 160
Infrared 5-6, 11, 118, 123-124, 135-137, 156, 172
Insects 128
Internet 254
Invasive 121
Irrigation 42, 195
Isotherms 285

K

Kernels 277

L

Landscape 96
Larvae 129
Low-cost 32, 39, 42

M

Management 2, 8, 114, 120, 190, 195, 215-216, 219
Mapping 47, 58
Matlab 228, 234
Mercury 251
Mitigation 82, 96
Mobile 254
Models 279, 282
Modern 260
Monolayer 285
Motors 104
Movement 3, 65-66

N

Neutrons 14-16
Nitrogen 7, 190
Nuclear 29
Nutrition 249

O

Operations 61, 252
Outdoor 111-112

P

Pesticide 93, 96-97
Phoenix 218
Plantation 195, 216
Pneumatic 94
Protection 73

R

Radiation 124, 142-143
Remote 14, 27, 191-192, 195, 215-216, 219-221
Residues 62
Resistance 172
Resolution 203, 212
Revolution 247, 267
Robotic 115
Robots 102

S

Sampling 78-79
Sapwood 121
Sensors v, 1-2, 7, 32, 35, 37, 41-42, 104, 175-176, 179, 186-188, 191, 255
Serpentine 58
Signals 18, 110
Software 105, 116, 180, 197
Sunlight 153
Symposium 29

Symptoms 6, 160, 169
Systems 70, 102, 156, 175, 260, 263

T

Thermal 6, 16, 125-126, 128-129, 135, 138, 140, 143-145, 147, 153, 155, 163, 194, 204, 212-213
Thermogram 129, 147-148, 150-151, 153
Tillage 3, 61-62, 68-70
Tissue 128-129
Topsoil 61
Transects 49, 51, 54, 58
Treatments 86-87, 92
Trials 68, 80, 83

V

Vectors 232
Vegetation 191, 208, 210, 221
Vineyards 4, 103

W

Warmer 251
Wavelength 160
Weather 83, 92
Weevil 219

www.ingramcontent.com/pod-product-compliance
Ingram Content Group UK Ltd.
Pitfield, Milton Keynes, MK11 3LW, UK
UKHW021328190726
13845UKWH00004B/76